Weather and Climate Modeling

Weather and Climate Modeling

Editors

S.V. Singh
Swati Basu
T.N. Krishnamurti

NEW AGE

NEW AGE INTERNATIONAL (P) LIMITED, PUBLISHERS

New Delhi • Bangalore • Chennai • Cochin • Guwahati • Hyderabad
Jalandhar • Kolkata • Lucknow • Mumbai • Ranchi

The cover pictures show the percent departure of Indian southwest monsoon rainfall for the year 2002 (obtained from the website www.tropmet.res.in) alongwith NCMRWF mean analysis at 850 hPa during July 2002. Anomalous northwestlies/westerlies prevailed over the northern plains.

Published by New Age International (P) Ltd., Publishers
First Edition: 2003
Reprint: 2010

Branches:

- 36, Malikarjuna Temple Street, Opp. ICWA, Basavanagudi, **Bangalore**. © (080) 26677815
- 26, Damodaran Street, T. Nagar, **Chennai**. © (044) 24353401
- CC–39/1016, Carrier Station Road, Ernakulam South, **Cochin.** © (0484) 4051303
- Hemsen Complex, Mohd. Shah Road, Paltan Bazar, Near Starline Hotel, **Guwahati**. © (0361) 2543669
- No. 105, 1st Floor, Madhiray Kaveri Tower, 3-2-19, Azam Jahi Road, Nimboliadda, **Hyderabad**. © (040) 24652456
- RDB Chambers (Formerly Lotus Cinema) 106A, Ist Floor, S.N. Banerjee Road, **Kolkata**. © (033) 22275247
- 16-A, Jopling Road, **Lucknow**. © (0522) 2209578
- 142C, Victor House, Ground Floor, N.M. Joshi Marg, Lower Parel, **Mumbai**. © (022) 24927869
- 22, Golden House, Daryaganj, **New Delhi**. © (011) 23262370, 23262368

ISBN(10): 81-224-1456-7
ISBN(13): 978-81-224-1456-1

Rs. 895.00

C-09-11-4155

Printed in India at Glorious Printers, Delhi.

PUBLISHING FOR ONE WORLD
NEW AGE INTERNATIONAL (P) LIMITED, PUBLISHERS
4835/24, Ansari Road, Daryaganj, New Delhi-110002
Visit us at **www.newagepublishers.com**

प्रो० व. सु. राममूर्ति
सचिव
PROFESSOR V. S. RAMAMURTHY
SECRETARY
&
Co-Chairman,
Indo-US S&T Forum

भारत सरकार
विज्ञान और प्रौद्योगिकी मंत्रालय
विज्ञान और प्रौद्योगिकी विभाग
टेक्नोलाजी भवन, नया महरौली मार्ग, नई दिल्ली-110 016

GOVERNMENT OF INDIA
MINISTRY OF SCIENCE & TECHNOLOGY
Department of Science & Technology
Technology Bhavan, New Mehrauli Road, New Delhi-110 016

January 21, 2003

Foreword

The cooperation between India and USA in Science and Technology has been in existence for more than five decades. The scientific communities in the two countries have demonstrated a natural affinity for partnership in pursuit of scientific excellence and achievements. Scientific collaborations got major impetus with the setting up of a Joint Science and Technology Sub-Commission in 1974. Science & Technology Initiative, taken up in 1982, encompassed a few high priority areas. US-India Fund established in January 1987 supported very large collaborative activities in science and technology over a 11-year period. The establishment of the Indo-US S&T Forum is an important milestone in the bilateral relation between India and USA. The 'Weather and Climate Modeling' has been identified as an important thrust area of collaboration under the Forum.

Severe weather events, heavy rain, tropical cyclones and vagaries of monsoon affect the societies significantly world over. However, till date, none of the numerical models in the world are able to simulate the Asian monsoon and the associated rainfall especially over India. For taking stock of the developments in this important area and for planning of future joint collaborative work, a workshop on 'Weather and Climate Modeling' was organised in New Delhi during 7-9 February 2002 under the aegis of Indo-US Science &Technology Forum.

I am pleased that a volume summarising the deliberations made during this workshop has been brought out. This is the first publication of its kind under the aegis of Indo-US Science and Technology Forum. I look forward for enhanced cooperation between the scientists of the two countries on this challenging area of Weather and Climate Modelling.

(V.S. Ramamurthy)

Telegram : SCIENCTECH • Telephone : 6510068/6511439
Fax : 0091-11-6863847, 0091-11-6862418 • E-mail : dstsec@alpha.nic.in

Preface

Understanding and predicting weather and climate is truly a global issue. Severe weather events, heavy rain, tropical cyclones alongwith large scale floods and droughts caused by the vagaries of monsoon affect societies significantly worldwide. Prediction, or even simulation of the Indian monsoon, has been a challenging task for the global meteorological community and only limited success has been achieved so far in this direction. Scientists from India as well as USA have been equally concerned with these issues and have been collaborating under various mechanisms to solve them. In order to take stock of our understanding of weather processes of mutual interest and also to identify the problems for joint collaboration, a workshop on weather and climate modeling was organized in New Delhi under the aegis of the newly established Indo-US S&T Forum. Scientists from both countries presented reflecting the state-of-the-art prevailing in their countries and identified the problems encountered, under four major themes, viz.

(i) Improvement of global forecast system;

(ii) Meso-scale NWP model development, cloud processes and land surface processes;

(iii) Dynamical extended range prediction; and

(iv) Location specific forecast and disaster management.

More than a hundred delegates attended and deliberated in the workshop. Twelve scientists from major establishments in the USA like National Oceanic and Atmospheric Administration, National Centre for Atmospheric Research, Naval Research Laboratory and universities like the Florida State University, University of Maryland and the North Carolina State University participated and presented papers in the workshop. From India, scientists from leading organizations like Indian Meteorological Department, National Centre for Medium Range Weather Forecasting, Indian Institute of Tropical Meteorology, Space Applications Centre, and the Indian Institutes of Technology (Delhi and Roorkee) participated alongwith representatives from user communities like the Indian Air Force, Indian Navy, Indian Council of Agricultural Research, and the Snow and Avalanche Study Establishment.

Considering the quality of presentations and the importance of the subject, it was thought desirable to bring out the proceedings of the workshop summarizing the state-of-the-art on weather forecasting prevalent in both countries. The speakers were requested to submit full papers which were peer reviewed by referees from India as well as USA. These papers have been put together in the form of this publication. While some of the papers reflect the work of the individual author, many cover the work done in the organization of the respective author(s) and in some cases that being done in their country in the field. It is hoped that this publication will be found useful by the weather and climate modeling community in both countries.

The editors thankfully acknowledge the help rendered by Prof. Roger Pielke, Dr. Michael Ek, Dr. Mitch Moncrieff, Dr. W.K. Tao, Dr. Fedor Mesinger all from USA and Prof. U.C. Mohanty, Shri S.R. Kalsi, Dr. G.B. Pant, all from India in reviewing the manuscripts.

S.V. Singh
Swati Basu
T.N. Krishnamurti

DEPARTMENT OF THE NAVY
Naval Research Laboratory
4555 Overlook Ave SW
Washington DC 20375-5320

23 January 2003

Climate and Weather Prediction - An International Challenge

During recent years, we have come to recognize and have begun to understand the global nature of ocean circulation, and the nature of atmospheric forcing on the complex meteorological phenomena, particularly those manifested in the coastal regions of India and indeed the world. The challenges of sea/air predictions to provide accurate nowcasts and forecasts of the mid-ocean as well as the coastal air/sea environment are a unique global issue. Since the atmospheric and the oceanic currents know no national or regional boundaries, it is essential that researchers of all countries closely collaborate in developing realistic and accurate models for weather prediction. It is only fitting that this conference brings to a common forum leading research scientists from the United States and India to discuss current status and future challenges on mathematical modeling of weather and climate with an emphasis on monsoons and tropical cyclones.

With the ever increasing importance of global navigation, the US Naval Research Laboratory (NRL) has played a major role in predicting oceanic and atmospheric behavior, from research and development to providing global weather forecasts. These predictions face the unique need to understand phenomena in the coastal, regional, and global environments including local currents, tides, wave, and surfs.

This conference, supported by the indo-US Forum, is indeed a timely event in recognition of recent climate fluctuations in the Indian subcontinent with their devastating effects on its people and landmass. This workshop is yet another testimony to the long standing research collaborations between the United States and India on critical topics of importance.

I am pleased to be an active participant in this workshop and wish to acknowledge the support of the members of the Indo-US S&T Forum and particularly for the elegant efforts of Dr. S. V. Singh, and his staff.

B. B. Rath
Workshop Co-organizer

Materials Science &
Component Technology Directorate
Naval Research Laboratory
Washington, DC 20375-5341 USA
rath@utopia.nrl.navy.mil

Contents

Foreword v

Preface vii

Recent Development in the NCEP/EMC Global Modeling Branch for 2001 1
—Hua-Lu Pan

Improvements in the NCMRWF Global Atmospheric Modeling System 5
—S.C. Kar, G.R. Iyengar, S. Das, Swati Basu, J.P. George & A.K. Mitra

Data Assimilation Efforts at NCMRWF: Results from Recent Experiments 23
—A.K.Mitra, M. Das Gupta, V.S. Prasad, S.C.Kar & D.Rajan

Almost Anything Can be the Leading Singular Vector 37
—Jon Ahlquist

Challenges of Mesoscale Weather Prediction in North Carolina 50
—Sethu Raman

Mesoscale Modeling Research and Development at NCAR. 55
—Jimy Dudhia

Mesoscale Modeling of Convective Systems over India: Status and Scope 6 3
—U.C. Mohanty, M. Mandal, A.K. Das & A.P. Dimri

Mesoscale Convective Systems during SCSMEX: Simulations with a Regional Climate Model and a Cloud-Resolving Model 77
—W.-K. Tao, Y. Wang, J.-H. Qian, C.-L. Shie, W.K.-M. Lau & R. Kakar

Cloud-Resolving Simulation and Super-parametrization of Tropical Cloud Systems 93
—Mitchell W. Moncrieff

Mesoscale and Cloud-Resolving Scale Simulation of a Heavy Precipitation Episode and Associated Cloud System Using MM5 Model 106
—Someshwar Das

Performance of Mesoscale Eta Model over Indian Region 118
—E.N. Rajagopal & G.R. Iyengar

Challenges of Representing Land-Surface Processes in Weather and Climate Models over the Tropics: Examples over the Indian Subcontinent 132
—Dev Dutta S. Niyogi, Roger Pielke SR., Kiran Alapaty, Joseph Eastman, Teddy Holt, U.C. Mohanty, Sethu Raman, T.K. Roy, Y.K. Xue

A New Approach to the Cumulus Parametrization Issue 146
—T. N. Krishnamurti & J. Sanjay

Dynamical Simulation of Indian Summer Monsoon 162
—K. Krishna Kumar & M. Mujumdar

Operational Models for Long-range Forecasts of Indian Summer Monsoon Rainfall 174
—M. Rajeevan

Monsoon's Feedback on ENSO Evolution 184
—Sumant Nigarm

High Resolution Numerical Modeling on Tropical Cyclone Structure and Track 198
—Chi-Sann Liou

Cyclone Related Disaster Management Efforts in India 209
—S.R. Kalsi

Location Specific Forecast: Interpretation of Outputs from Global and Meso-scale NWP Models 220
—Ashok Kumar, Parvinder Maini, L.S. Rathore & S.V. Singh

Recent Development in the NCEP/EMC Global Modeling Branch for 2001

HUA-LU PAN

Chief, Global Modeling Branch
Environmental Modeling Centre/NCEP
NOAA; 5200 Auth Road; Camp Spring, MD 20746-4304
E mail : hualu.pan@noaa.gov

ABSTRACT

Impacts of two major implementations at the EMC Global modeling Branch during the year 2001 at NCEP are described. Both the implementations resulted in substantial improvements in the operational medium range forecasts.

1. MAJOR IMPLEMENTATIONS

During 2001, the EMC Global Modeling Branch completed two major implementations that have contributed to significant improvements to the operational forecasts using the Medium-Range Forecast (MRF) model. We will describe the changes made and the impact to the skill of the forecasts.

On 13 February 2001, we started to ingest the following satellite data: NOAA-16 AMSU/A and AMSU/B radiances, NOAA-15 AMSU/B radiances and GOES-10 radiances. In addition, we also started to assimilate precipitation estimates from the SSM/I satellite and the HIRS channel 12 over land. The AMSU/B data provide us significant information on the mid and upper troposphere water vapor field. While the impact to the forecast is marginally favorable, we feel that this implementation has improved our ability to assimilate all observations in the tropics and has led to more realistic analysis of the tropical circulation.

On 15 May 2001, we changed the MRF model to include a prognostic equation for cloud water and ice. This change is based on the version implemented in the eta model a few years ago by Zhao and Carr (1997). In addition to the separate condensation and precipitation algorithm, we also made extensive changes to the radiation physics such that the optical path and the emissivity calculations are based on the predicted cloud water and ice. The convection parameterization scheme is also modified to detrain cloud water/ice in addition to water vapor. In order to ensure a more realistic cloud water distribution in the tropics, we also modified the convection scheme to generate randomly selected cloud tops. This set of changes has a dramatic impact on the circulation forecasts of the summer hemisphere. It also led to minor improvement in the winter hemisphere circulation forecasts.

In addition to the cloud water package, we have also included a change to the convection scheme. The reason for the change is a need to reduce the false alarm tropical storm predictions in the MRF forecasts. The inclusion of the cumulus momentum exchange in the convection package led to a dramatic reduction of the false alarm storms while preserving most of the real storms. While the pressure gradient effect of the momentum exchange package is very crudely parameterized as an entrainment effect, this package was included with the cloud water package. We plan to further improve the momentum mixing in the future.

2. PRIMARY IMPACTS

In Figure 1, the frequency distribution of the 5-day anomaly correlation scores for the period July-September 2001 for both hemispheres is shown. It can be seen that the majority of the forecasts verified in the .6 to .9 ranges with few if any forecasts below the .6 mark. The key to medium-range forecasts for all operational centers these days is to be consistent in the day-to-day skills. It is particularly difficult to do this for the summer hemisphere as convection plays a bigger role in the forecasts.

In Figure 2, we show two 132-hour forecasts for the tropics from the 0000 UTC 22 September 2001 initial condition. The first forecast is with the cumulus momentum mixing turned off (Figure 2a) and the second with the mixing scheme turned on (Figure 2b). It can be seen that there are three tropical storms in Figure 2a and only

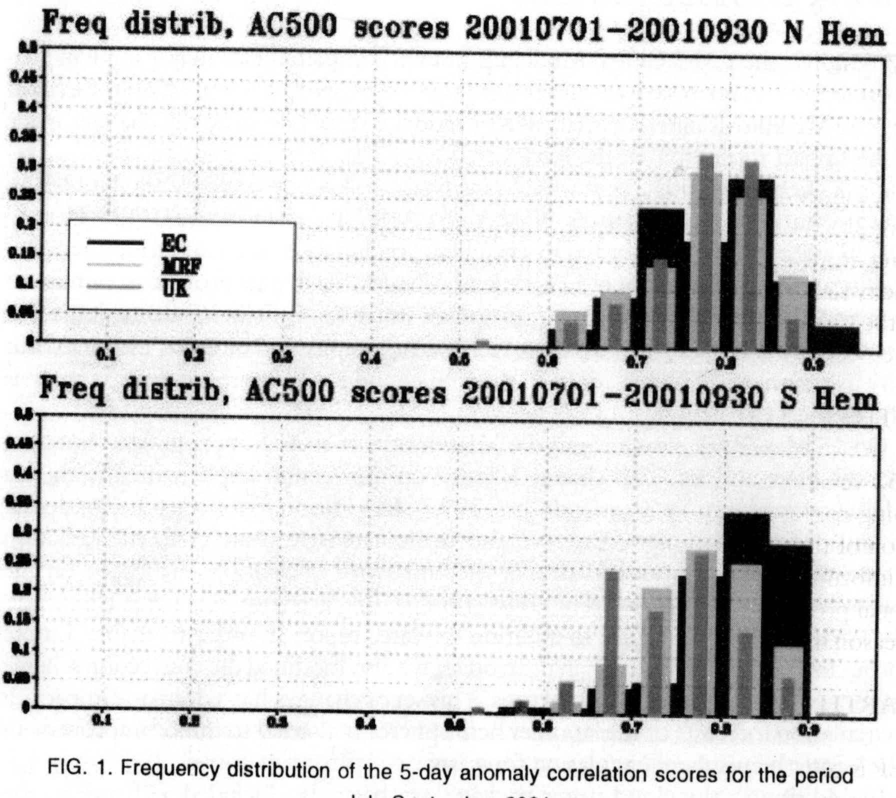

FIG. 1. Frequency distribution of the 5-day anomaly correlation scores for the period July-September 2001

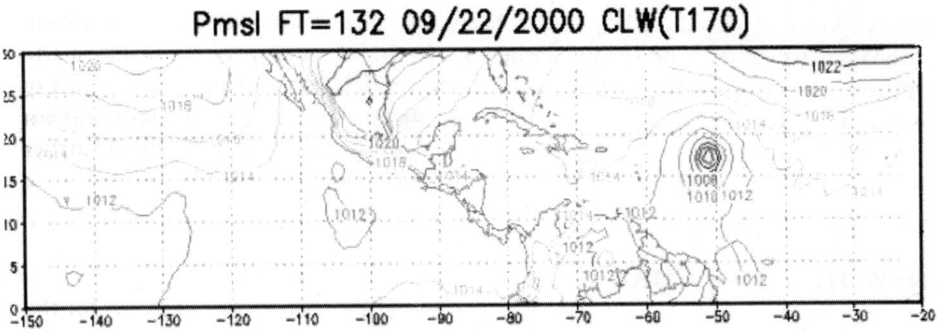

FIG. 2a

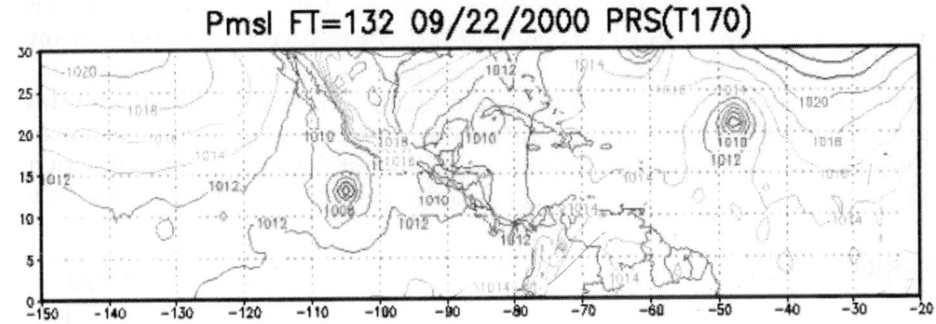

FIG. 2b

one in Figure 2b. It turns out that the only storm developed in reality is the one in Figure 2b. While it is very difficult to determine the reason the cumulus momentum mixing stops certain storms from forming, it is likely that the modification of the vorticity fields near the vicinity of the convection altered the feedback between heating and the circulation. This aspect of the model change deserves much scrutiny in the future.

3. OTHER IMPLEMENTATIONS

A T62 version of the MRF with several different physical parameterizations is now running routinely at NCEP making monthly predictions out to seven months. This version of the model is called the Seasonal Forecast Model (SFM) and is used to provide information to the Climate Prediction Center (CPC) forecasters for monthly and seasonal predictions. A set of 21-year hindcasts is also run to provide calibrations for the seasonal predictions.

4. EARTH SYSTEM MODELING FRAMEWORK

NCEP is collaborating with NASA/DAO, NASA/NSIPP, MIT, GFDL, and NCAR in a project to create an Earth System Modeling Framework (ESMF). This project is currently funded by a NASA Cooperative Agreement Negotiation contract. The idea of the ESMF is to create a modeling framework in which scientists can work on the scientific issues of the complicated earth system models without the worries of the computer platform. The ESMF should provide efficient message passing for different computer configurations for both research and operational use. This should allow for better collaborations amongst all institutions involved and NCEP is committed to cast our global forecast system into this framework.

5. NEW DYNAMIC CORE

In order to take full advantage of the ESMF initiative and to work closely with the Weather Research and Forecast (WRF) effort in the EMC Mesoscale Modeling Branch, we have decided to initiate the development of a new dynamic core. We have also invited scientists from NASA/DAO and from NCAR to participate in the development. We plan to develop a dynamic core with a generalized vertical coordinate. This will allow us to move in steps from the current sigma coordinate to a sigma-pressure hybrid system and possibly to a sigma-theta hybrid system. For horizontal derivatives, we plan to start with the spectral representation but test the semi-Lagrangian system currently been tested in the WRF dynamic core developed in NCEP.

REFERENCES

Zhao, Q, and F. H. Carr, 1997: A prognostic cloud scheme for operational NWP models. *Mon. Wea. Rev.*, **125**, 1931-1953

Improvements in the NCMRWF Global Atmospheric Modeling System

S.C. KAR, G.R. IYENGAR, S. DAS, SWATI BASU, J.P. GEORGE & A.K. MITRA

National Centre for Medium Range Weather Forecasting,
New Delhi – 110 003, India
E mail: sckar@ncmrwf.gov.in

ABSTRACT

Several R&D efforts are going on to improve the performance of the NCMRWF Global atmospheric modeling system. Some of these are described here. A major improvement has been to develop a higher resolution global model at T170L28 resolution. Sensitivity experiments with different planetary boundary layer physics, convection and radiation has led to considerable improvement in the forecasts produced. Ensemble runs using the "breeding method" are being carried out. Substantial efforts have been made in parallelising the model codes to utilize indigenously developed distributed memory parallel processing systems. Seasonal runs using observed SST have been undertaken to understand the intra-seasonal variability of the Indian monsoon. Long-term (multi-year) simulations of the model have been initiated conforming to AMIP-II to examine the diabatic heating errors in the model simulations.

1. INTRODUCTION

Global atmospheric model and global data assimilation are the two most essential components of a medium-range weather forecasting system. Atmospheric processes span a large range of scales. However, due to the computational restraints the numerical models are confined to describe the atmospheric phenomena up to a certain scale. The unresolved processes are parameterized in terms of the resolvable scales of the numerical model. The model errors are very much dependent on the resolution and the type of parameterization schemes used. The weather forecasting system is regularly upgraded through improved physical parameterization, increased vertical and horizontal resolutions and by ingesting more observed data. It has been demonstrated that by properly defining the initial state of the atmosphere and by improving the global model, it is possible to predict weather in medium-range time scale. Rapid improvement in data assimilation and global modeling has taken place during the last decade. With newer types of data being made available, especially through satellites, necessary data assimilation schemes are being developed to include these data. With increasing computing power being made available to modelers, the vertical/horizontal resolutions of the models are also regularly increased. Physical parameterization schemes in the models are also being improved with better understanding of the atmospheric processes through various field campaigns in India and elsewhere. It is therefore essential that along with the improvements in the assimilation forecasting system, evaluation of the same be carried out on a regular basis.

The forecasting skill of Numerical Weather Prediction (NWP) models is limited by the errors in the initial conditions (uncertainties in the initial conditions) and the uncertainties in the representation of the physical and dynamical processes in the numerical forecast model (uncertainties in the forecast model). The true state of the atmosphere cannot be determined precisely due to the observational errors, large data gaps and approximations in the analysis techniques. Due to the chaotic nature of the atmosphere, forecasts started from slightly different initial conditions will have a range of possible solutions rather than a single valid solution. The uncertainties in the initial conditions of the atmosphere give the forecasting problem a stochastic nature. That is, a single deterministic forecast should be replaced by a ensemble (collection) of forecasts starting from slightly different (perturbed) initial conditions. Thus, ensemble forecasting techniques involve generation of multiple forecasts as a function of the uncertainties in the analysis. However, the ensemble forecasting techniques will work only if the initial perturbations represent the uncertainty in the initial conditions and the model errors do not dominate the final error fields .The goals of ensemble forecasting are to increase the average forecast accuracy, to estimate the likelihood of various events, and to estimate the forecast spread which can be used as an indicator of the forecast skill.

Prediction of Asian summer monsoon on different space/time scale has been a challenging task for the scientists. Many modelers are trying to accurately simulate the large-scale summer monsoon rainfall and flow pattern as a benchmark of the model and the modeling capability in general. The same is also true for the medium range weather forecasting capability. The skill of the forecast both in a statistical sense and in case of individual weather systems for the Indian region may not be the same from different NWP models. Even if the principle of the model may be the same, the details of the model in terms of horizontal/vertical resolution, representation of the orography, data usage, choice of different physical parameterization, etc. make an impact on the rainfall forecast skill. It is generally believed that if the model climate is not good, forecasts from the model will not be good. Therefore, the model developers run the model for months/seasons to examine the stability of the model and assess if the climate of the model is reasonable.

It is important to continue research work on the medium-range weather prediction problem for better quality weather prediction. Major thrust areas in this topic are model resolution, air-sea interaction, planetary boundary layer, cloud parameterization, ensemble forecasting, and forecast model evaluation. At the same time, seasonal simulations/annual cycle simulations need to be carried out regularly to understand the issues of weather and climate modeling.

National Centre for Medium Range Weather Forecasting (NCMRWF) is the only organization in India where global data assimilation and medium-range weather forecasting is done on real-time. Research work on data assimilation and global modeling started at NCMRWF in 1989 with the implementation of the ECMWF (cycle-30) global model, (Kar and Basu, 1992) and later R&D efforts were carried out using the COLA R40 model (Basu and Kar, 1992). In 1993, the NCEP weather forecasting system at T80L18 resolution was implemented at NCMRWF. In 1994, NCMRWF started weather forecasting on operational basis using this system. Since then, R&D activities at NCMRWF have been going on using the T80 model based weather forecasting system. Research and development activities at NCMRWF on Global modeling are carried out essentially to improve the weather forecasting skill in the tropics and especially over the Indian region. The global model is also used extensively to understand the physical processes and the Indian summer monsoon system. In this report, we describe, in brief, various R&D efforts carried out recently at NCMRWF on global atmospheric modeling.

2. MODEL RESOLUTION

A major improvement in the NCMWRF global modeling system took place recently, when the vertical and horizontal resolution of the forecasting system was increased from T80L18 to T170L28. The entire upgradation (of the model, analysis, etc.) was planned and executed in-house, (Kar, 2002). This modeling system is based on NCEP T80L18 Model (Kanamitsu, 1989) and subsequent changes made on the T80L18 model at NCMRWF. Vertical domain of the high-resolution model extends from the surface to about 2.0 hPa and is divided into 28 sigma layers. Figure 1 shows the sigma layers of the T170L28 model and corresponding layers in the T80L18 model. It can be seen that in the high-resolution model, more layers have been added in the planetary boundary layer, near the tropopause and in the stratosphere.

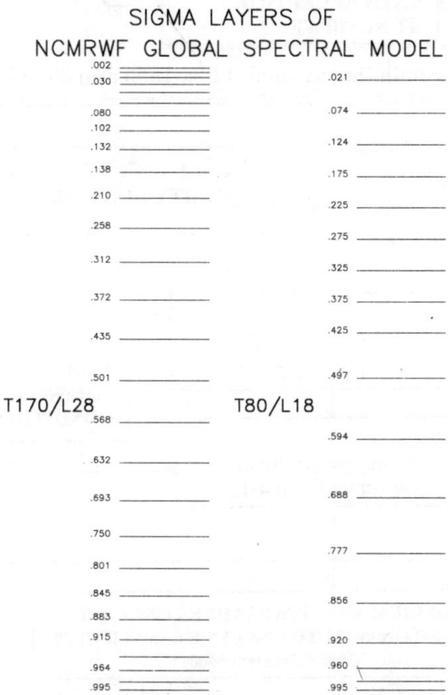

FIG. 1 Sigma Layers in the T170L28 and T80L18 Global models.

The time integration in the model is semi-implicit. The time step is 7.5 minutes for computation of dynamics and physics, except that the full calculation of radiation is done once every 6 hours. Horizontal diffusion is scale-selective, and is of second-order. Both Local and Non-Local (Basu et al., 2002) closure schemes for planetary boundary layer are available in the T170L28 global model. Cumulus convection parameterization scheme in the model is by Kuo (modified by Anthes). Other physical processes in the T170 model are the same as in the T80 model, Kanamitsu (1989). Description of the operational T80 model has been provided in another chapter of this volume. The global data assimilation system has also been upgraded to T170L28 resolution and regular forecasts are being made using the T170L28 based weather forecasting system in parallel to the T80L18 model based system. Figure 2 shows the flow diagram of the newly developed T170L28 global data assimilation forecast system.

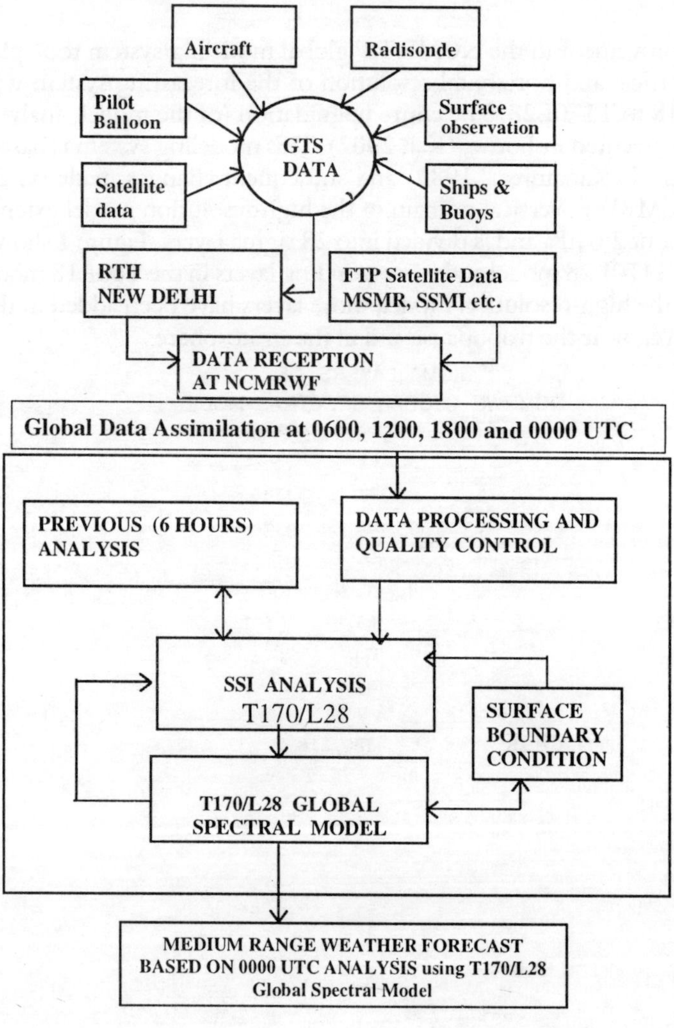

FIG. 2. Flow Diagram of the T170L28 GDAS System.

In order to examine the stability of the T170 model, both the T170 and T80 models were used for one-month integration (for July 2001) with observed SST. The integration started with the T80L18 analysis (interpolated to T170L28) valid for July 1, 2001. Comparison of the monthly mean rainfall over the Indian monsoon region from the T170, T80 models with the observed rain for the month are shown in Figure 3. It can be seen that simulated rainfall from T170 model is more realistic than

(a)

(b)

(c)

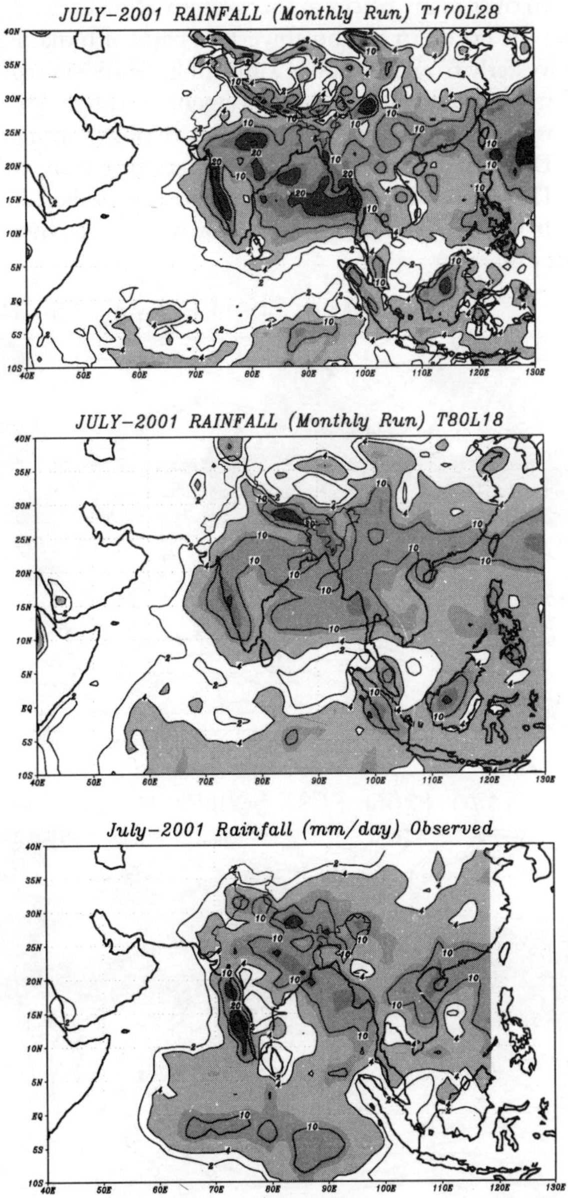

FIG. 3. Monthly mean rainfall for July 2001 (mm day^{-1}) with one moth simulation of the (a) T170 model, (b) T80 model and (c) observed rainfall in the month. Contours levels are 2, 4, 8, 12, 16, 20, 30 mm/day. Shaded area with red color indicates rainfall more than 20mm/day^{-1}.

the rainfall from the T80 model, especially over the Western Ghats, and the Bay of Bengal. It was found that the newly developed high-resolution model is stable and has a reasonably good rainfall distribution, when it is run for one month continuously.

An example of predictions made for the case of a western disturbance using the T170 model is discussed here. A westerly trough affected the northern and north-west parts of India during January 13-16, 2002 after a long dry spell. Under its influence, rain/snow occurred over many places in the western Himalayas. Rainfall activity was also reported over the northern and northwestern parts of India. Figure 4 (a,b) show the mid-latitude westerly trough at 500hPa affecting the Indian region on January 15, 2002. The T170 model could predict the movement of trough 5 to 7 days in advance, and rainfall forecasts from the model agreed well with observation. Seven-day forecast meteograms for Delhi is shown as Figure 4(c) starting from initial conditions of January 8 to 15. This shows that the model consistently predicted rainfall for January 15. The forecasts from the T80 model are not shown, however, those were also reasonably good for this case.

(a) (b)

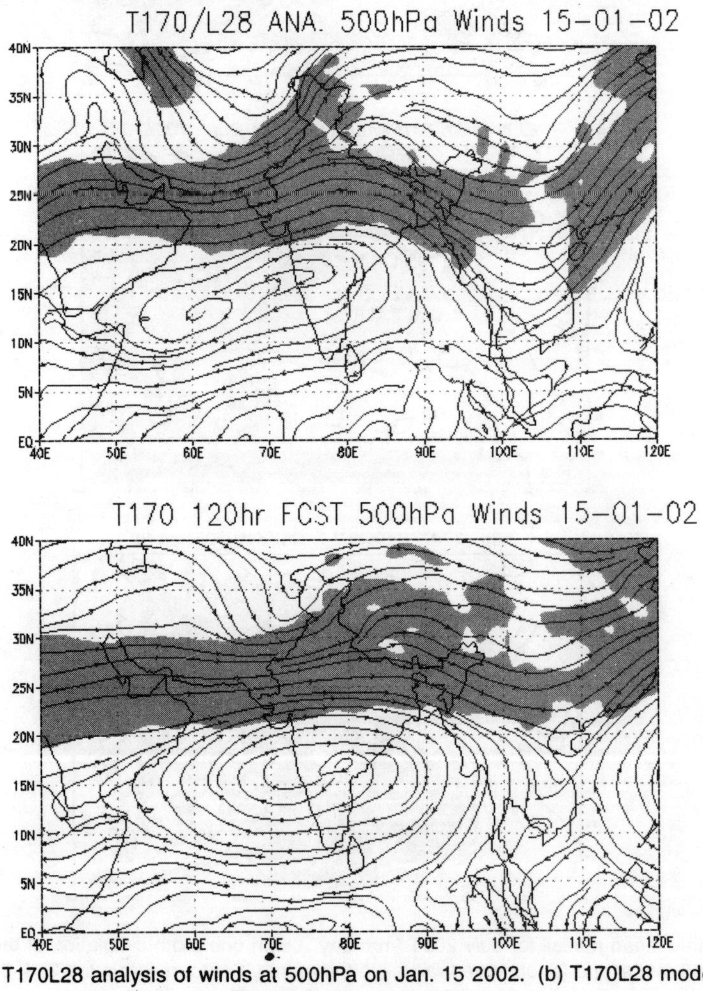

FIG. 4 (a) T170L28 analysis of winds at 500hPa on Jan. 15 2002. (b) T170L28 model forecasts (Day-5) for the same day.

(c)

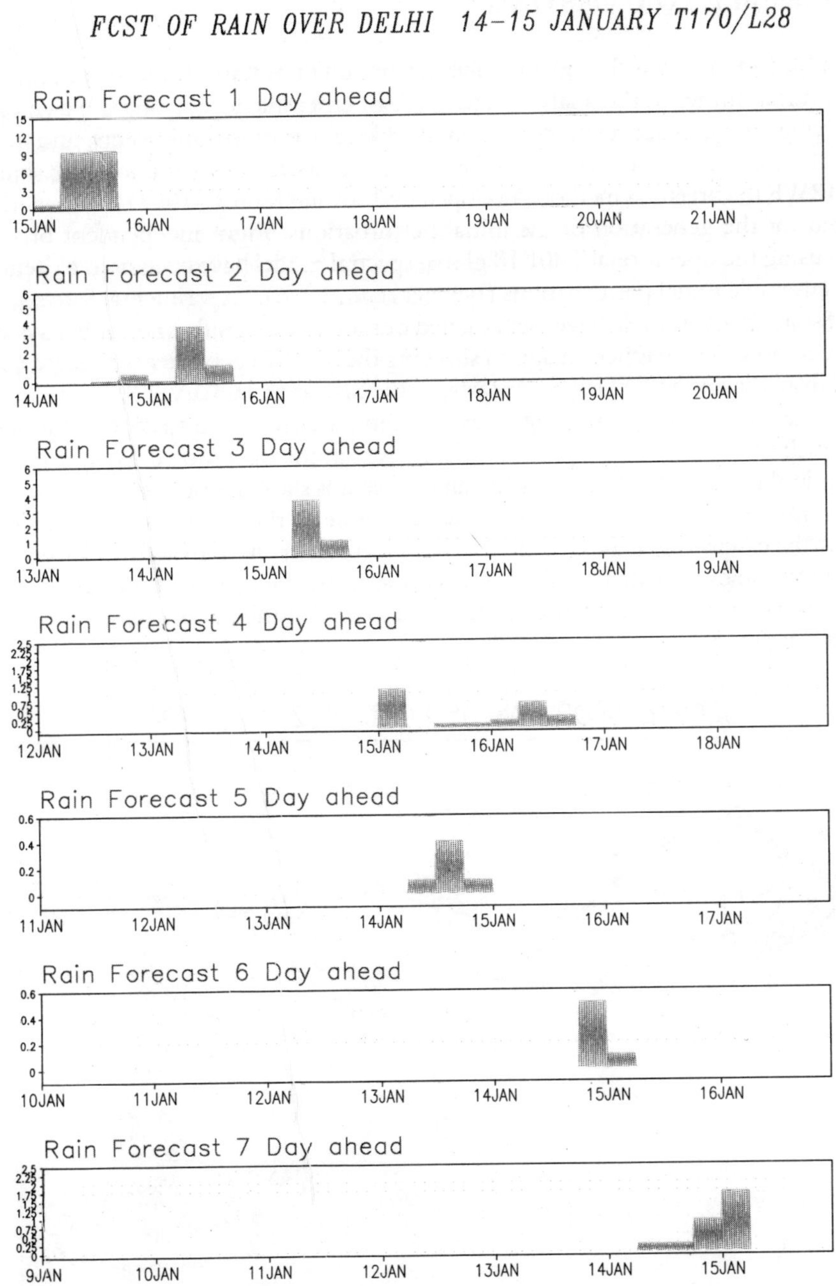

FIG. 4 (c) Meteogram (rainfall forecasts) for Delhi valid for the same day. The lowest panel of the figure shows that the model predicted the rainfall event 7 days in advance, and the model was consistent in the prediction of the event in all the lead time considered (i.e. from 7 days in advance to 1 day in advance). Observed rain at Delhi that day was 1.2 cm.

3. ENSEMBLE FORECASTING

Ensemble forecasting techniques involve generation of multiple forecasts as a function of the uncertainties in the analysis. The goals of ensemble forecasting are to increase the average forecast accuracy, to estimate the likelihood of various events, and to estimate the forecast spread which can be used as an indicator of the forecast skill. At NCMRWF, the breeding method developed by Toth and Kalnay (1993) has been implemented for the generation of the initial perturbations. Eight independent breeding cycles using the operational T80L18 global spectral model have been used to generate eight ensemble initial perturbations (Iyenger *et al.*, 1996). Experimental forecast runs with 8-member ensemble have been carried out for some significant weather systems. Figure 5 shows the spaghetti diagram showing the 500 hPa geopotential height (5600 gpm) from the ensemble Day-5 forecasts for the case of 15 January, 2002. The analysis for this case depicts a trough along 65°E extending southwards upto 25°N. This trough was much weaker and displaced towards the east in the operational day-5 forecast as shown in the figure. The ensemble mean (figure not shown) could show only a marginal improvement over the control forecast. None of the ensemble members could predict the extension of the trough upto 25°N. Another sharp trough was seen along 30°E in the analysis, which could not be predicted by the operational forecast. However some of the ensemble members were able to predict this feature reasonably well as seen in Figure 5.

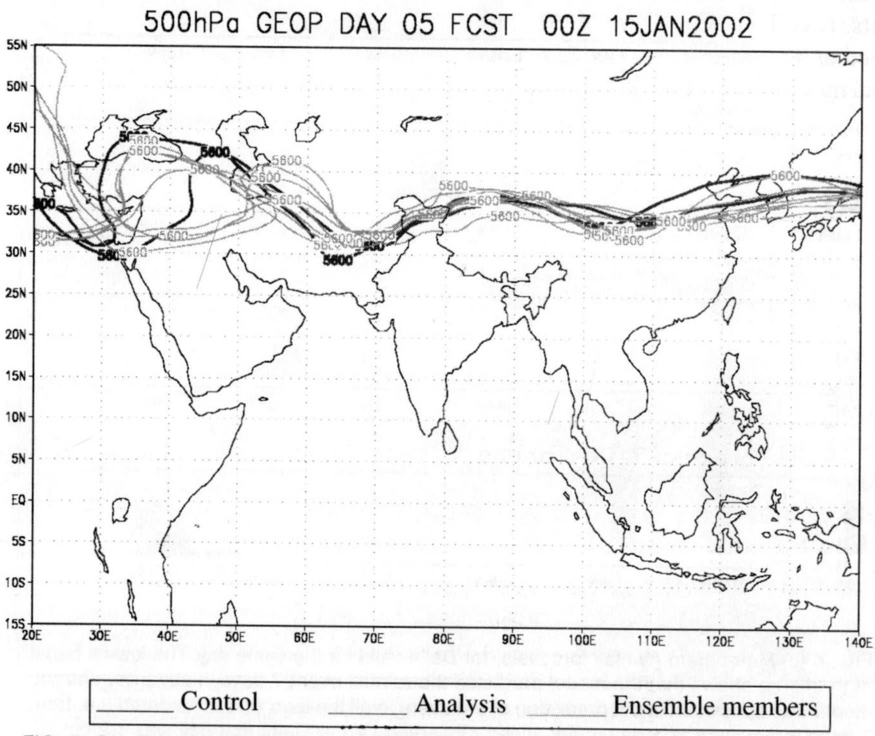

FIG. 5. Spaghetti plot for 500hpa Geopotential height (Day-5 forecasts) for Jan 15, 2002.

4. IMPROVEMENT OF PHYSICAL PROCESSES

4.1 Planetary Boundary Layer Processes

The boundary layer processes in the NCMRWF global model (T80L18) was originally represented by the local-K approach. One weakness of this approach is that it is unable to represent realistically mixing in the convective boundary layer involving the counter-gradient fluxes, since the influence of large eddy transports are not accounted for. This weakness in the scheme affects the profiles of mean quantities where dry convection is of importance in the Planetary Boundary Layer (PBL). Basu et al. (1998, 1999) tested a turbulent kinetic energy (TKE) closure scheme in the model and found that model simulations improved especially for the weather systems associated with summer monsoon, such as monsoon depressions. An alternate approach known as the non-local K closure (NLC) for PBL parameterization has been implemented in the model (Basu et al. 2002). Due to its simplicity and capability to represent the large eddy turbulence within a well-mixed boundary layer, this scheme is widely used in global models. Implementation of this scheme in the NCMRWF global model showed that precipitation distribution over the Indian region is reproduced in a more realistic manner by the NLC and compares better with observations than does the local closure scheme. Two sets of five day integrations were performed using the initial conditions during July 1999 using the local and non-local closure approaches for the boundary layer. The equitable threat score (ETS) obtained for rainfall shown in Figure 6 (a) suggests that the new scheme viz. NLC produces better skill in the heavier rainfall events. RMSE of wind forecasts are reduced and the systematic model errors are improved as shown in Figure 6 (b). This scheme has also been included in the T170 global model at NCMRWF.

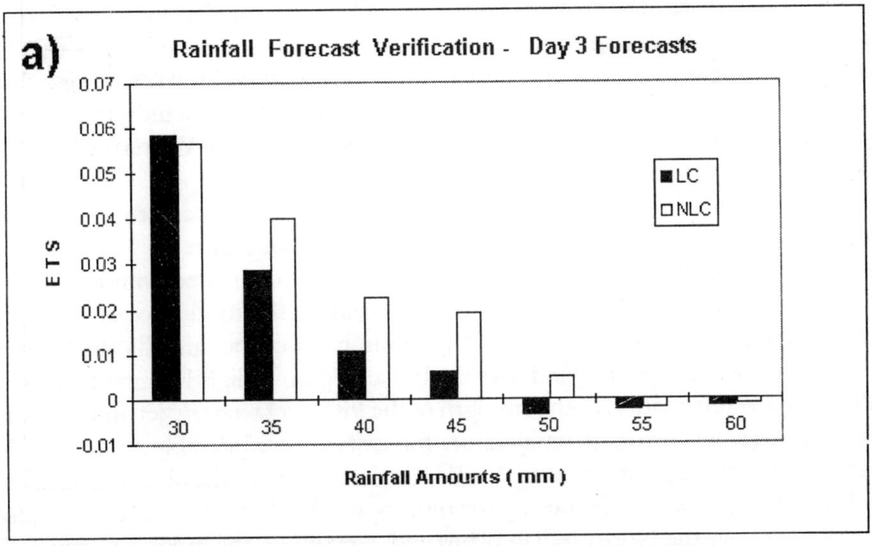

FIG. 6 (a) Equitable threat score for Day-3 forecast rainfall.

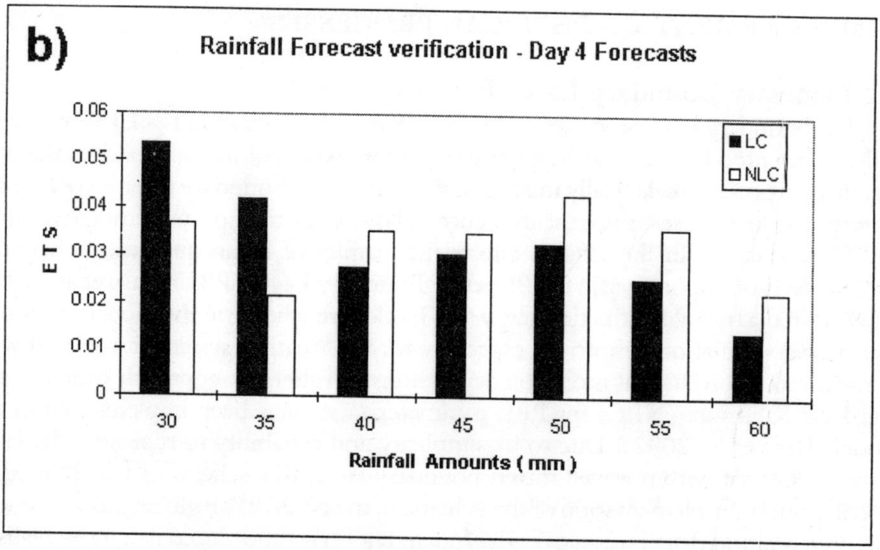

FIG. 6 (b) 850hPa wind rmse with RMSE scheme.

4.2 Parameterization of Cumulus Convection

The NCMRWF global model (T80L18) has been integrated for July (to examine impact on the Indian monsoon simulations) using different cumulus parameterization schemes namely, the Simplified Arakawa-Schubert scheme (SAS), the Relaxed Arakawa-Schubert Scheme (RAS), and the Kuo-type scheme (KUO). Strength of the low Level westerly jet, cross equatorial flow, and tropical easterly jet are better simulated by RAS and SAS as compared to observation than the KUO scheme. RAS and SAS produce strong rising motion owing to strong intensity of convection produced by these two schemes. Cold bias produced by KUO at 200 mb is reduced by RAS and is substantially improved by SAS. Study of observed and simulated rainfall indicated that RAS and SAS produced better distribution of precipitation over the Western Ghat mountains and the Arakan coast, (Das et al., 2001).

Several medium range forecasts (up to 5-days) were made using the global model with these three convection parameterization schemes. In Figure 7, 120 hr forecasts for August 1999 with Kuo, SAS and RAS schemes are compared with the observed rainfall for that month. The RAS and SAS produce fairly good forecasts of rainfall, however, the best forecast is produced by SAS in most of the rainfall categories over the Indian region. RMSE of wind over the Indian region is slightly higher in RAS and SAS due to overestimation of the strength of the low level westerly jet and upper level tropical easterly jet. Errors in temperature forecasts are considerably reduced by RAS and SAS. Systematic errors indicate that KUO tries to weaken the observed southwesterly flow and the low level jet during the monsoon. The RAS and SAS try to intensify the easterlies over the North Indian plains and, strengthen the monsoon trough. Reduction of cold bias and improved simulated temperature by SAS indicate a proper

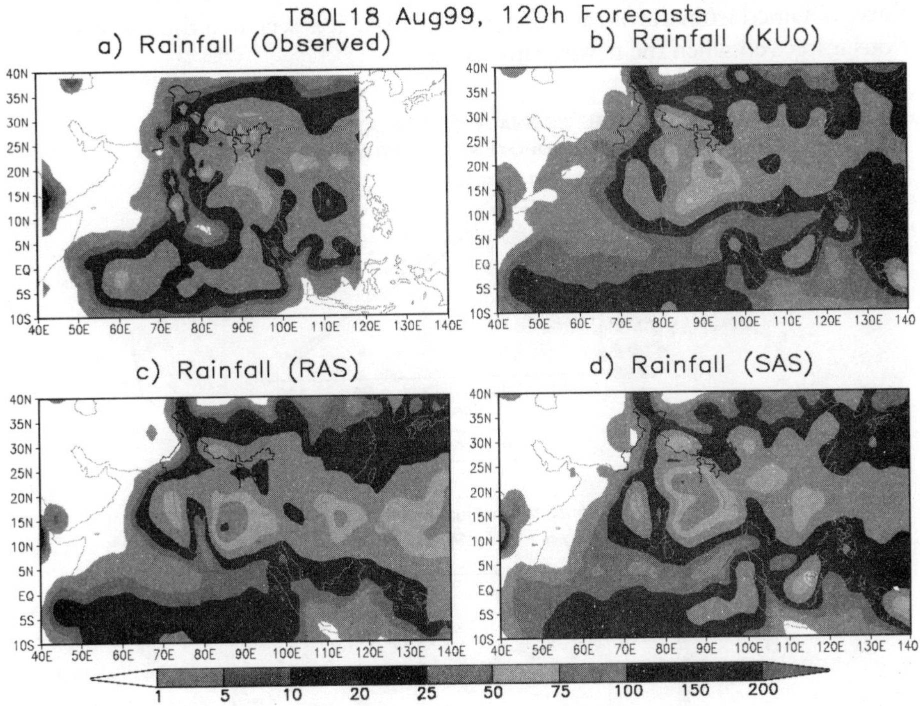

FIG. 7 Observed Rainfall in August 1999 (a) and model forecasts of rainfall at 120 hr. using Kuo (b), RAS (C) and SAS (c) convection schemes.

redistribution of heat by deep convective clouds over the region by this scheme. Evaluation of overall dynamics, thermal structure and rainfall indicates that in general, SAS is able to provide relatively better results compared to other two schemes, (Das et al., 2002), especially over the Indian region.

4.3 Radiation Parameterization

Short wave radiation parameterization scheme of Harshvardhan et. al. (1987) was implemented in the NCMRWF T80L18 global model in December 2000 after extensive performance evaluation studies. In the old short wave radiation scheme, fixed values are assumed for cloud reflectivity and absorptivity. This is a serious shortcoming in the modeling of radiation processes. The new scheme takes care of scattering and absorption processes in the cloudy atmosphere. This scheme uses a two-stream method and delta-Eddignton approximation for the computation of cloudy sky radiation fluxes. Also, in the new scheme, dependence of surface albedo on zenith angle (which is not there in the old scheme) is parameterized. Treatment of clear sky radiation computation by both the schemes is similar. Two sets of five day integrations were performed with initial conditions for a month (August 1999) with old and new short wave scheme. Significant improvement is seen in the forecasts of temperature and geopotential fields in the medium range, (John and Begum, 1997). Figure 8 shows the global anomaly correlation for geopotential height at 500 hPa and zonal fluxes obtained from the old and the new schemes. It is seen that the anomaly correlations improved when the new scheme is used.

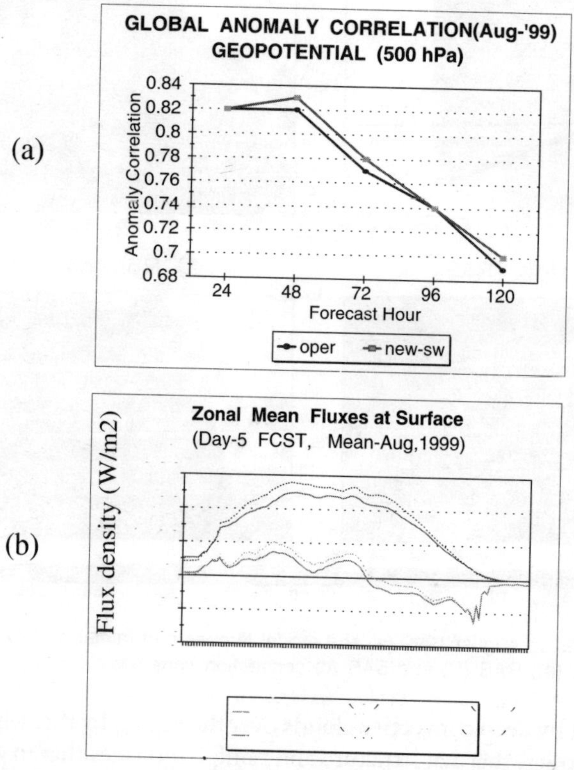

FIG. 8 (a) Global anomaly correlation for Geopotential Height at 500 hPa, (b) Zonal fluxes at surface with Operational SW Radiation scheme and New SW Radiation Scheme.

5. PERFORMANCE EVALUATION

The main goal of a verification process is to constantly check the quality (skill and accuracy) of the NWP products. This includes: (i) establishment of a skill and accuracy reference against which subsequent changes in the analysis-forecast system can be measured, and (ii) identification of the specific strengths and weaknesses of the forecast model and the need for model improvement. Standardised verification scores of NWP products as recommended by the Commission for Basic Systems (CBS) of WMO are computed on a routine basis. The nature and distribution of the systematic error in the operational forecasts are regularly examined at NCMRWF. During the summer monsoon season (June–September) systematic errors of winds at 850 hPa, as shown in Figure 9(a), indicate the presence of anitcyclonic flow over the Bay of Bengal and eastern tropical Indian Ocean and a cyclonic circulation over the eastern/central India. Figure 9(b) shows that at 200 hPa, the forecast wind error over this region is cyclonic. Thus, at both 850 and 200 hPa levels, the systematic errors tend to weaken the climatological monsoon circulation.

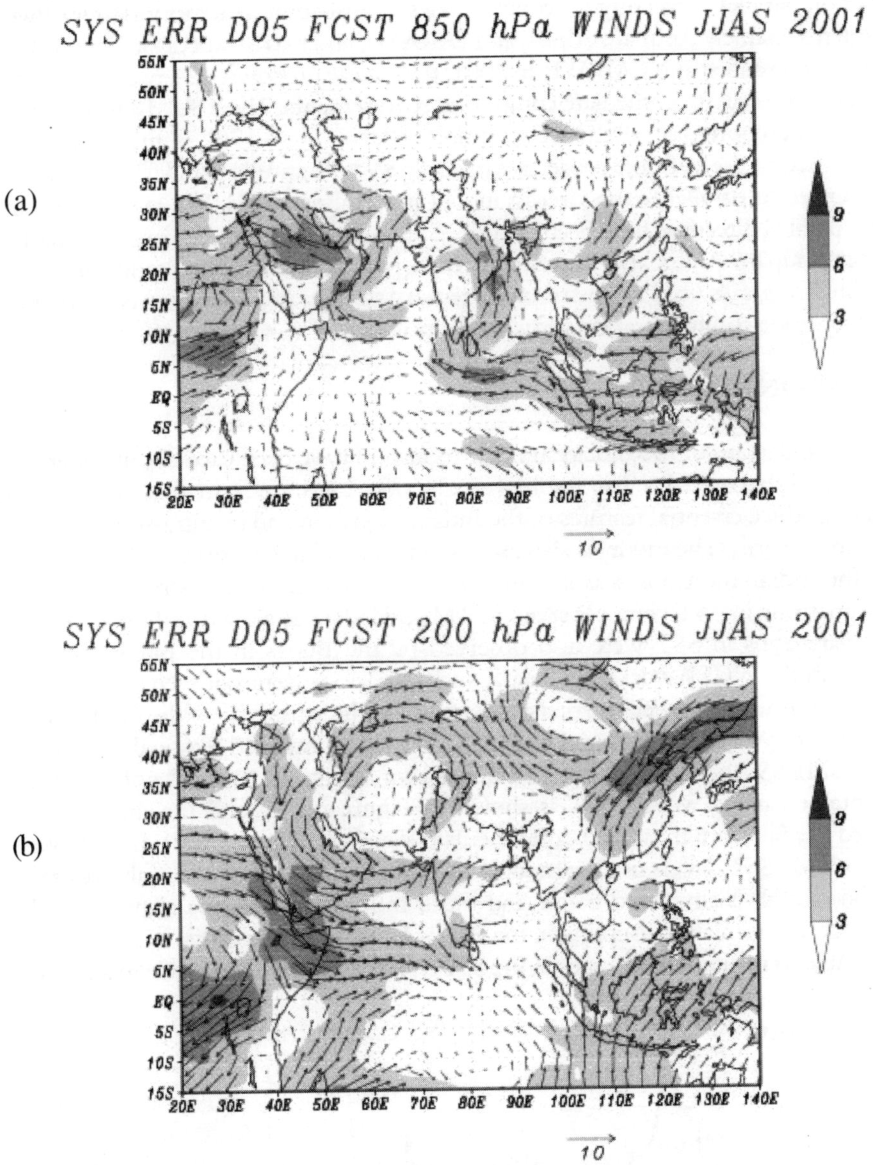

FIG. 9. Systematic Errors of wind in the model forecasts during monsoon season (a) at 850 hPa and (b) at 200 hPa.

6. PARALLEL PROCESSING & OPTIMIZATION

Utilization of parallel processing computer system for NWP based weather forecasting began in India in 1992 when the Cray XMP/14 was upgraded to a dual processor system Cray XMP/216 and NCMRWF weather forecasting packages were ported on to this parallel processing computer. In order to initiate research and development activities on parallel processing of weather codes on distributed memory systems, two such systems (namely Anupam-Alpha, and PARAM), and two 4-processor Origin-200 systems were installed at NCMRWF. In 2001, the Centre procured a 24-processor Cray SV1 system to meet its immediate computing requirements. The T80L18 forecasting system has been ported on all these computing systems. The T170L28 model has also been modified to use multiple processors on Cray SV1. "Latitude parallel" strategy has been followed to parallelize the forecast model and analysis codes. In Cray and Origin systems, parallel directives have been introduced to use multiple processors. The model codes on Anupam-Alpha and Param have been rewritten using the "send" and "receive" directives/calls used for distributed memory systems provided by ANULIB and MPI (inter-node communication software) respectively.

7. MONSOON STUDIES

Seasonal simulations of the Indian summer monsoon have been carried out using the NCMRWF global model at T80 resolution. The simulations indicate that the model is able to reproduce essential features of the Indian monsoon and its intra-seasonal variability satisfactorily. The model is also used to understand the air-sea coupling process during the Indian monsoon season. During 1998 monsoon season, strong intra-seasonal variations in SST were observed in TMI SST data as shown in Figure 10(a). Similar variations in SST were also observed by the Buoys in the Bay of Bengal. However, in the NCEP weekly SST data, such variations were not seen. Several seasonal model simulations were carried out to examine the sensitivity of the model monsoon to monthly mean, weekly and daily SST data sets. Simulated seasonal mean rainfall with TMI SST over the Indian region is shown in Figure 10(b). Rainfall simulations obtained using weekly SST is shown in Figure 10(c) and the difference in simulated rainfall from these two experiments is shown in Figure 10(d). It was found that if the TMI SST data is used in model simulations, intra-seasonal variability of the monsoon rainfall is simulated better and mean monsoon pattern is improved. The study suggests that air-sea coupling in the Bay of Bengal region is quite strong and the model is able to respond to changes in the SST in intra-seasonal time scale in a realistic manner.

(a)

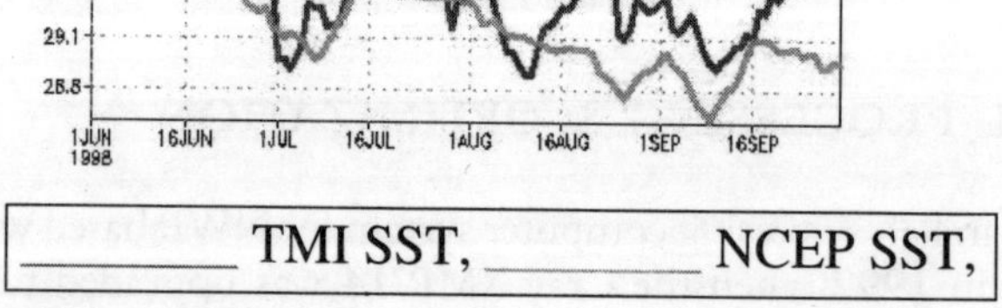

FIG. 10 (a) Intra-Seasonal Variation of SST over Bay of Bengal during Monsoon Season of 1998 as seen in NCEP Weekly SST and TMI daily SST data.

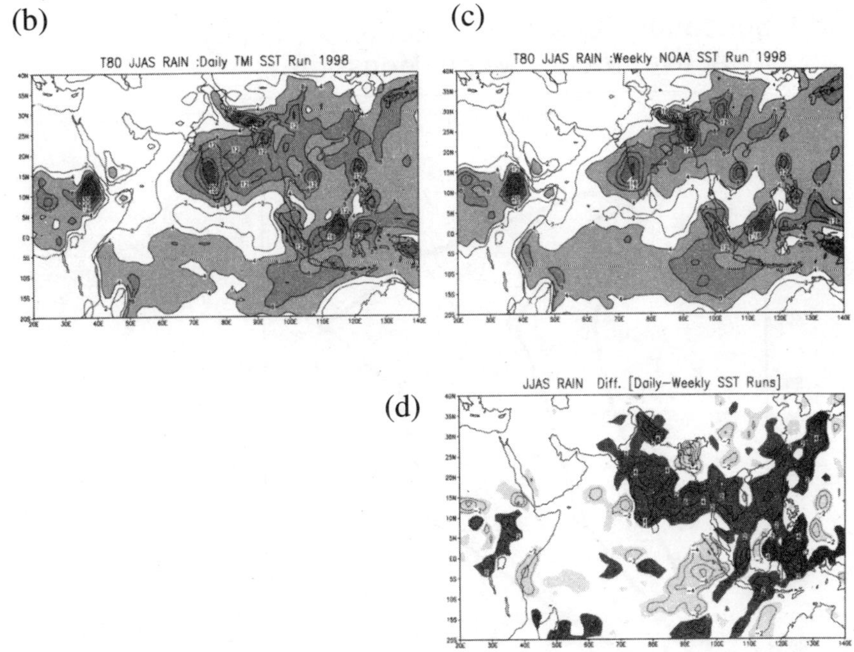

FIG. 10 (b) Seasonal rainfall (mm/day) simulated by the model using daily TMI SST, (c) same as (b) but with weekly NCEP SST and (d) rainfall difference (mm/day) between two experiments. In (b) and (c), contour levels are 2, 4, 8, 12, 16, 20 mm/day. In (c), positive and negative rain areas are shaded with red and yellow colors respectively.

Asian monsoon produces a gigantic heat source in the atmosphere, which impacts the global circulation during both summer and winter seasons. Model simulations are often unable to reproduce key aspects of seasonal and interannual variability of the Indian monsoon because of errors in diabatic forcing. Diagnosis of the diabatic heating errors in the simulations will be helpful in understanding the deficiencies of the model physics. Diabatic forcing as obtained from seasonal model simulations are examined to understand the monsoon dynamics and to check if the model simulates the forcing properly. Figure 11 shows the simulated convective heating during active, weak and normal monsoon period over the Western Ghat region of India. Multi-year simulations of recent climate, using the NCMRWF operational model (T80/L18 resolution) and AMIP surface boundary conditions are being carried out. The simulations conform to the standards of AMIP-II. In Figure 12, we show the total diabatic heating (shortwave, longwave, deep convective, shallow convective, large-scale and heat-transfer due to vertical diffusion) at 400 hPa for the mean June, July and August (JJA) months from a 10-year run (1979-88). This field has been compared with the observed diabatic heating at 400 hPa diagnosed from NCEP reanalysis data (Nigam *et al.*, 2000). It was found that though the structure of the diabatic heating from the model agrees reasonably well with the observed pattern, the model has large heating errors in the monsoon region. Over the north-eastern parts of India and near foot hills of Himalayas,

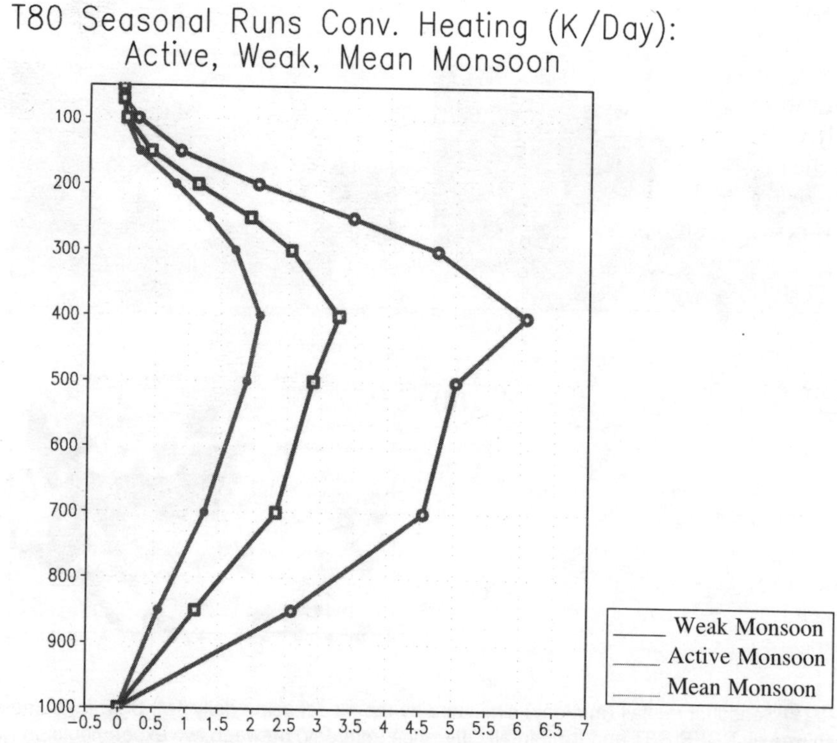

FIG. 11. Convective heating over West Coast of India during active, weak and mean monsoon conditions over India obtained using seasonal monsoon simulation experiments.

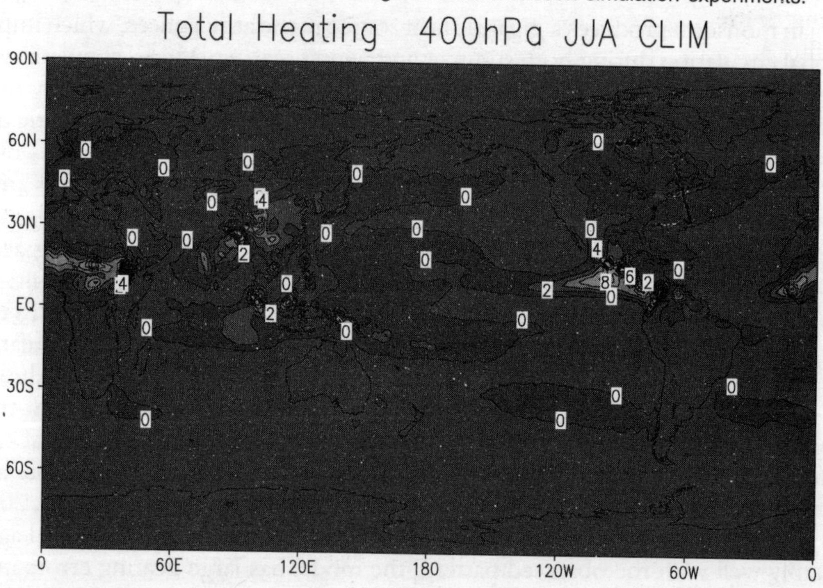

FIG. 12. Total diabatic heating (deg/day) at 400 hPa from model simulations during monsoon (JJA) months (10-year model climatology). Contour intervals are 0, 2, 4, 6, 8 ... deg/day.

the model over-estimates the heating. Since the monsoon process is essentially convective, the total diabatic heating is dominated by deep cumulus convective heating. The work is in progress for comparing the structure and variability of diabatic heating in the simulation with the observational counterparts (diagnosed from NCEP and ECMWF re-analyses). Detailed intercomparisons will help establish the strengths and deficiencies of the model's physical parameterizations. Strategies for model improvement will then be developed from diagnostic modeling, which can reveal the impact of heating errors on circulation.

8. FUTURE PLAN

It is important to understand the causes of systematic forecast errors and eliminate them, since their interaction with physical processes leads to rapid forecast degradation. Therefore, it is essential to continue research work on the medium-range weather prediction problem for better quality weather prediction. NCMRWF proposes to run various global data assimilation forecast systems along with its T170L28 global model to quantitatively evaluate the forecasts. The exercise will demonstrate the skill and shortcomings of this newly developed model. Experience gained from this exercise will help the future development of the model (and increase the model resolution further) at NCMWRF. In order to improve the cloud and convective parameterization schemes of the global model, there are plans to use a high resolution Mesoscale/ Cloud-Resolving model for simulations of mesoscale convective systems over the Indian region and, generate cloud scale parameters to use these to improve the scheme in global model. To improve the parameterization of land surface processes, there are plans to implement NOAH (NCEP, OSU, Air Force and OHD) Land Surface Model (LSM) in NCMRWF global model and to test and validate NOAH LSM over Indian region. For improving the parameterization of boundary layer processes in the model, available surface flux data and flux related data sets through WCRP/SCOR Working group on air sea fluxes and other observing experiments will be acquired and compared against fluxes obtained from NCMRWF model. There are also plans to improve representation of fluxes by improved parameterization for different roughness lengths, calm wind conditions and taking care of interaction of sea-state in determining air-sea fluxes. In order to improve the ensemble forecasting system, its performance over the Indian region will be validated and ensemble technique will be applied for tropical cyclone track prediction. It is also proposed to increase the number of ensemble members to 16 from the present 8-member ensemble.

REFERENCES

Basu, B.K. and S.C. Kar, 1992: A study of 240 hr. integration outputs from a global spectral model for the month of July 1991. *Meteorological Bulletin* No. 1/1992, NCMRWF, 26pp.

Basu, S., Z.N. Begum, and E.N. Rajagopal, 1998: Impact of boundary layer parameterization schemes on the prediction of the Asian summer monsoon, *Bound. -Layer. Meteor.*, **86**, 469-485.

Basu, S., Sethu Raman, U.C. Mohanty and E.N. Rajagopal, 1999: Influence of the planetary boundary layer physics on medium-range prediction of monsoon over India. *Pure Appl. Geophys.*, **155**, 33-55.

Basu, S., G.R. Iyengar and A. K. Mitra, 2002: Impact of non-local closure scheme in simulation of Monsoon system over India, *Mon. Wea. Rev.*,**130**, 161-170.

Das S., A.K. Mitra, G. Iyengar and S. Mohandas, 2001: Comprehensive test of different cumulus parameterization schemes for the simulation of the Indian summer monsoon ", *Meteorol. Atmos. Phys,* **78**, No.3-4, 227-244.

Das S., A.K. Mitra, G. Iyengar and J.V. Singh, 2002: Skill of medium range forecasts over the Indian monsoon region using different parameterizations of deep convection. Under Revision. *Wea. Forecasting,* **17**.1194-1210.

Harshvardhan, R. Davies, D. A. Randall and T. G. Corsetti, 1987: A fast radiation parameterization for general circulation models, *J. Geophys. Res.*, **92,** 1009-1016.

Iyengar G. R., Zoltan Toth, Eugenia Kalnay and John S. Woollen, 1996: Are the bred vectors representative of analysis errors? *Preprints of the 11th AMS Conference on Numerical Weather Prediction,* 19-23 August 1996, Norfolk, Virginia, USA.

John P. George and Z. N. Begum, 1997: Impact of different radiation transfer parameterization schemes in a GCM on the simulation of the onset phase of Indian summer monsoon, *Atmosfera*, **10**, 1-22

Kanamitsu, M., 1989: Description of the NMC global data assimilation and forecast system. *Wea. Forecasting*, **4**, 335-342.

Kar, S.C. and B.K. Basu, 1992: Simulations of 1987 and 1988 summer monsoons and the influence of global SST anomalies.*Proc. of Workshop on Monsoon Variability held at NCAR*, Boulder, Colorado, USA, 1991". ***WMO/TD-No.470**, 2.79-2.85*

Kar, S.C., 2002: Description of a high-resolution Global Model (T170/L28) developed at NCMRWF. Research Report **1/2002**, NCMRWF, 28pp.

Nigam, S., C. Chung, and E. DeWeaver, 2000: ENSO diabatic heating in ECMWF and NCEP Reanalyses, and NCAR CCM3 simulation. *J. Climate,* **13,** 3152-3171.

Toth, Z, and E. Kalnay, 1993: Ensemble Forecasting at the NMC: The generation of perturbations. *Bull. Amer. Meteorol. Soc.*, **74,** 2317-2330.

Data Assimilation Efforts at NCMRWF: Results from Recent Experiments

A.K.MITRA, M. DAS GUPTA, V.S.PRASAD, S.C.KAR & D.RAJAN
National Centre for Medium Range Weather Forecasting
Mausam Bhavan Complex, Lodi Road, New Delhi 110003, India
E mail: akm@ncmrwf.gov.in

ABSTRACT

Data scarcity over the Indian Ocean and adjoining African region poses real problem in defining large-scale features associated with major weather systems of India. To overcome this problem continuous efforts have been made at NCMRWF to utilize various types of satellite observations. Results from the recent experiments at NCMRWF with various types of new satellite data in data assimilation and its impact in medium range weather forecasts are discussed here. Surface wind speed and total moisture content from SSM/I of DMSP was incorporated to study their impact during monsoon conditions. Experiments with same set of derived meteorological parameters from Indian satellite IRS-P4/MSMR were also conducted. For an intense tropical weather system data from TMI/TRMM were also used and compared with DMSP-SSM/I assimilation. Results of inclusion of NOAA/NESDIS derived temperature and moisture information from ATOVS is seen to be encouraging. High density satellite winds from IR, VIS. and WV channels of various geo-stationary satellites (GOES, METEOSAT, GMS, INSAT) became available recently through GTS and from CIMSS, Wisconsin on real time basis. Observation system experiments with different combinations of the products from the above have been performed. Various types of statistical scores for the medium range forecasts have been computed. Impacts of these observations on analysis and forecast system, with special emphasis on different features related to summer monsoon have been studied. Inclusion of these high density satellite derived winds show positive impact both in analysis and forecast. Specifically, the southern hemispheric features, cross-equatorial flow and the lower level somali jet get better defined. Associated rainfall forecasts over Indian subcontinent also show improvement.

1. INTRODUCTION

Real time weather information in medium range time-scale is important for agriculture-based economy of India. National Centre for Medium Range Weather Forecasting (NCMRWF), New Delhi has been providing experimental Agro-meteorological advisory services for different agro-climatic zones to the farming community of India, based on location specific medium range weather forecasts produced from a global numerical weather forecast model since 1994. Accurate forecasts from a numerical weather prediction (NWP) system require an accurate description of the initial state of all the important geophysical variables within the domain of the model and its boundaries. Preparation of the initial state of atmosphere by global analysis (data assimilation) becomes the key issue of NWP for different time scales. Assimilation of satellite observations plays a major role in defining the initial state, and the same is responsible for substantial improvement in forecast accuracy in recent years. Results from the recent observing system experiments carried out at NCMRWF with various kinds of satellite data sets have been discussed in the workshop. Present global data assimilation system at NCMRWF along with brief history about its development is discussed in the following section. Section 3 describes briefly some of the observing system experiments with satellite data carried out at NCMRWF. Results from recent observing system experiments are presented in section 4. Finally, the immediate future plans of NCMRWF associated with data assimilation are listed in section 5.

2. GLOBAL DATA ASSIMILATION SYSTEM — DEVELOPMENTS AT NCMRWF

Operational global data assimilation system of NCMRWF was based on multivariate optimum interpolation (OI) analysis scheme in 1991. Later a new assimilation scheme based on Spectral Statistical Interpolation (SSI) technique of NCEP (Derber *et al.*, 1991; Parrish and Derber, 1992) was adopted and made operational at NCMRWF during 1994, for preparing initial conditions for the T80 global spectral model with 18 vertical levels. Subsequently a modified version of SSI scheme at T80/L18 resolution was implemented at NCMRWF in June 1996 (Parrish *et al.*,1997). Present analysis scheme (SSI) being a 3-D variational type in formulation is capable of including lot more type of satellite information.

2.1 NCMRWF Data Assimilation System

NCMRWF's 6-hourly intermittent (0000UTC, 0600UTC, 1200UTC and 1800UTC) global data assimilation system(GDAS) consists of mainly three components. (i) data processing and quality control (ii) analysis based on Spectral Statistical Interpolation (SSI) and (iii) NWP model for preparation of short range forecast used as background field . Figure 1 depicts the schematic diagram of data assimilation system operational at NCMRWF. Though most of the data used in GDAS are received through GTS, via RTH New Delhi, the amount of data (SSM/I, ATOVS, QuickScat, etc.) received directly through NCMRWF ftp server has also increased significantly in recent years.

Analysis scheme operational at NCMRWF is an adapted version of NCEP's early version of global 3-D variational analysis scheme, known as SSI (Derber *et al.*, 1991; Parrish and Derber 1992, Parrish *et al.*, 1997) In SSI, observation residuals are analysed in spectral space on sigma surfaces. The objective function minimized in SSI is defined in terms of the deviations of the desired analysis from the first guess field (which is taken as six hour forecast of the operational T80-Global spectral model) and the observations, weighted by the inverse of the forecast and the observational error variances respectively. The objective function is defined as follows,

$$J(X)=1/2[(X-X_b)^T B^{-1}(X-X_b) + (Y_{obs}-R(X))^T O^{-1} (Y_{obs}-R(X))]+J_c \quad ....(1)$$

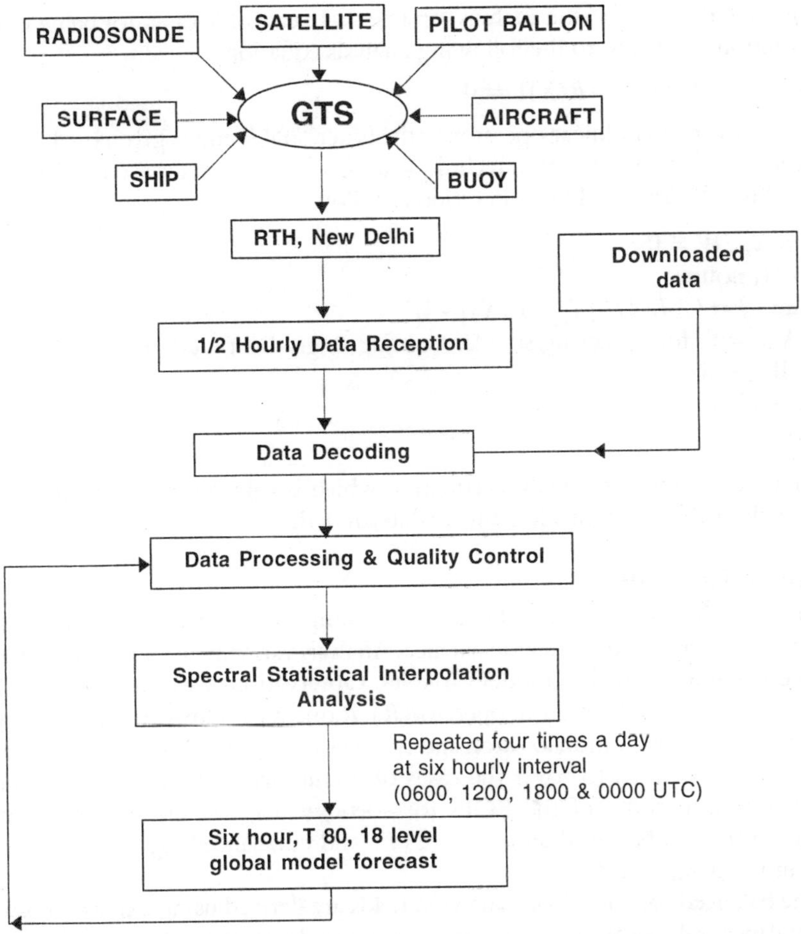

FIG. 1 Schematic diagram of data assimilation system operational at NCMRWF

Where,

X is N component vector of analysis variables
X_b is N component vector of first guess variables / background
Y_{obs} is M component vector of observations
B is N × N background error co-variance matrix
O is M × M observational error co-variance matrix
R is an operator from analysis to observation space
N is analysis degree of freedom
M is number of observations
T is transpose of the matrix

The two terms in J on the right hand side of above equation deal with the fit of the first guess field and the observations with the analysis respectively. In order that the analysis should fit best with both the first guess and the observations, J should be minimized with respect to the analysis variable X. In case of R-operator being nonlinear minimization of **J** leads to the following analysis equation.

$$W - C^T L^T O^{-1} [\Upsilon_{obs} - R(X)] = 0 \qquad ...(2)$$

Where L is the tangent linear operator to the functional R and is given by $L = \delta R/\delta X$. Thus following the perturbation technique in this case, the analysis equation (2), may be solved for W using a double iteration as follows.

Set $X = X_b$, $W = 0$
Do k = 1, nouter
Evaluate $f = C^T L^T O^{-1} [\Upsilon_{obs} - R(X)] - W$
Solve A δ = f, for δ, using standard conjugate gradient method
$W = W + \delta$
$X = X + C\ \delta$
End do

Equation (2) is the basic analysis equation which is solved for W, using conjugate gradient algorithm and thus X is finally determined.

2.2 Analysis Variables

Analysis variables are suitably chosen to simplify background term, to achieve speedy convergence and to include balance. Analysis variables are so chosen that those can be easily transformed to model variables. Observations used in GDAS are linked with analysis variable through an operator(R). R operator transforms analysis variable to the individual observation location and type via a suitable gaussian grid. R operator has also been extended to cover the cases of nonlinearity. Analysis variables in this scheme are, amplitudes of the EOFs for vorticity (δ), fast or unbalanced part of divergence (D), unbalanced temperature (T) and unbalanced log of surface pressure ($\ln P_s$) and mixing ratio (q).

The balanced part of various analysed fields are derived using a quasi-geostrophic linear balance relationship. H operator is used to derive the balanced field.

$$H = \nabla^{-2} (\nabla . f \nabla) \nabla^{-2}$$

Background error covariance statistics and the constants in the balance equations are derived from 48 – 24 hour forecast differences. The term J_c in equation (1) contains a penalty on the difference in the divergence tendency from the background divergence tendency.

Recently a high-resolution data assimilation-forecast system (T170 / L28) has been developed (Kar, 2002). Forecast error co-variance matrices have been computed at this resolution. This data assimilation system is run on a regular basis and medium-range weather forecasts are currently being prepared using a global model at T170/L28 resolution.

3. OBSERVING SYSTEM EXPERIMENTS USING SATELLITE DATA

Several observing system experiments were conducted at NCMRWF using various types of satellite data to improve the initial analysis. Empirical techniques similar to that used at JMA and BMRC for deriving synthetic vertical moisture profiles were obtained using INSAT IR data (Rajan *et al.* 2001; Rajan *et al.* 2002). This moisture data had marginal positive impact in some case studies. Later two other empirical methods (Basu *et al.*, 1997) were also tested to derive vertical moisture profile from satellite derived total column water vapor and also from layer moisture of NOAA polar satellites. Assimilation of these moisture data showed positive impact on rainfall forecasts. Procedures for nudging observed rainfall (reverse Kuo and reverse similarity computations) data during pre-data assimilation phase were developed and tested for a version of the global model. It was felt that good quality observed rainfall data is required for rainfall nudging. In another experiment, the divergent part of the wind during monsoon period was improved by using INSAT OLR information as input in analysis system (Kulkarni *et al.*, 1997). Inclusion of this divergent part of wind showed a positive impact on analysis and forecast. Inclusion of divergence part of wind only, without proper moisture information limited the impact upto two days. Experiments were also carried out to see the impact of the temperature profiles from NOAA both at coarse (500 km) and fine resolution (85 km HRPT). Objective scores showed that the satellite temperature data bring about a general improvement in the prediction over most part of the region (Bohra *et al.* 1998; Prasad *et al.*, 1999).

An attempt was made to improve the quality of INSAT derived cloud motion vectors (CMV) through the height re-assignment by employing first guess fields (short range forecasts) of the global forecast model operational at the NCMRWF. This method is found to be useful in improving the quality of CMV retrieval. At present, India Met. Department (IMD) has included this height assignment method by using NWP forecasts from regional and global forecast models in their real time CMV derivation procedure. Surface wind vectors from scatterometer (ERS) were found to be useful in improving the initial condition and positive impact was also noticed in forecasts (Das Gupta *et al.*, 2002). Results from the experiments with the QuikScat surface wind data are encouraging and it will be made operational at NCMRWF shortly. Many satellite operators have developed automated wind-extraction algorithms to derive atmospheric motion vector winds (AMWs) at very high density and the same are being transmitted through GTS and also through their ftp servers. NCMRWF has started receiving these data through GTS aas well as ftp access. These winds are deduced from the sequential imageries in IR, VIS and WV channel bands. Before using these winds in the operations, experiments were conducted with the real time quasi-operational high-resolution winds available from EUMETSAT as well as from Cooperative Institute for Meteorological Satellite Studies (CIMSS), Wisconsin, USA. These winds are derived for all operational geo-stationary satellites (IR, VIS and WV channels), using automated wind-extraction algorithms. At NCMRWF efforts are being made to study the representativeness of the data and their impact on global data assimilation forecast system.

Microwave instrument based geophysical parameters (viz., total precipitable water vapour, ocean surface wind speed, cloud liquid water, rainfall rates) from SSM/I have become available on near real time basis. Similar parameters from MSMR of Indian satellite IRS-P4 (Oceansat), have also become available to NCMRWF. To assimilate these new parameters (surface wind speed and moisture), SSI system of NCMRWF was modified and the new scheme has been tested (Rizvi, *et al.*, 2000). Inclusion of microwave information has improved the analysis and forecast. Experiments with the inclusion of DMSP SSM/I wind speed and moisture information has shown that the strength of the monsoon flow is maintained better (Kar, *et al.* 2003a). Various experimental analyses have been carried out at T80 resolution to examine the impact of wind speed and total precipitable water content (TPWC) from TRMM (TMI), DMSP (SSM/I) and IRS-P4 (MSMR) satellite data on the analyses and predictions of the Orissa super cyclone case of October 1999(Kar, *et al.* 2003b). It has been found that the best analysis for this cyclone was obtained by using TMI/TRMM data.

4. RESULTS FROM SOME RECENT EXPERIMENTS

Experiment was conducted with the high-density winds from METEOSAT-5 located at 63°E for the period June 1999 to assess its impact basically on monsoon flow. Henceforth

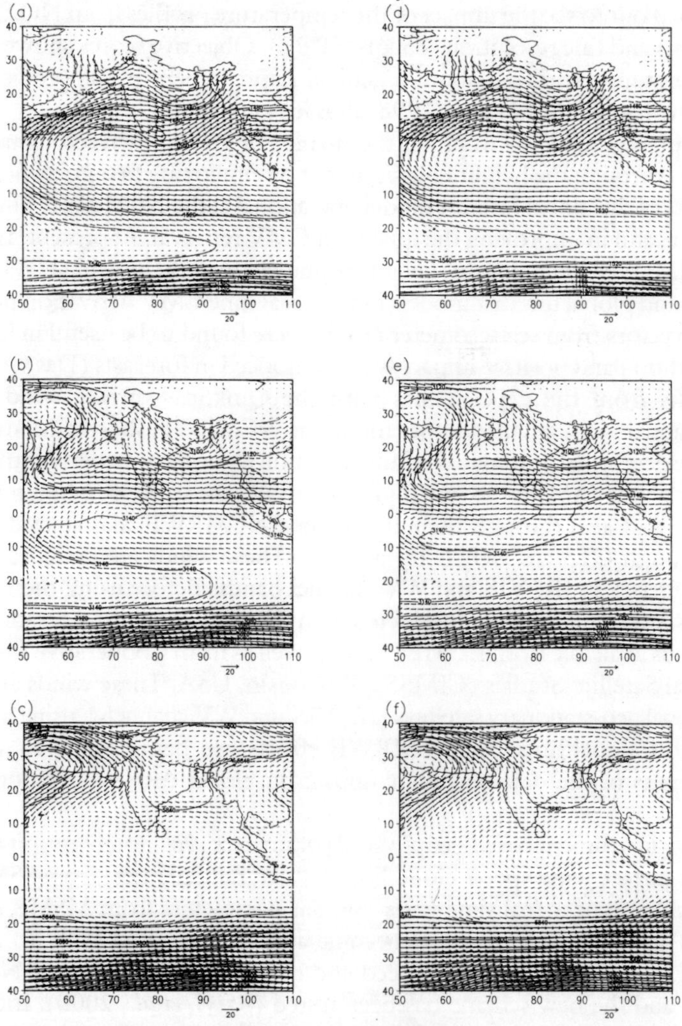

FIG. 2 Mean monthly analysed wind and height fields for June 1999 (a) 850 hPa (b) 700 hPa (c) 500 hPa for AMV and (d) 850 hPa (e) 700 hPa (f) 500 hPa for CTL

this experiment will be refereed as AMV. Control runs (CTL) for this period were carried out with conventional and satellite data from operational archive without high density winds from METEOSAT-5. Inclusion of the high density METEOSAT-5 winds in the experiment (AMV) has helped to improve the initial analysis of monsoon flow. Figure 2 shows the flow patterns at different vertical levels(850, 700, 500 hPa) for AMV and CTL respectively. The inclusion of the AMVs improved the cross-equatorial flow and the low-level westerly jet of the Arabian Sea at 850 hPa (Figure 2a). At 700 hPa the depth of monsoon flow was better defined with the AMVs, as a result the westerly system that was penetrating towards the equator was realistically pushed northward and the monsoon winds could reach the Indian sub-continent even at 700 hPa(Figure 2b). Even at 500 hPa the unrealistic penetration of westerly in the Arabian Sea was not seen in AMV(Fig 2c). The difference plot (AMV-CTL) in Figure 3 clearly shows the strengthening of the cross-equatorial flow, low-level westerly jet over the

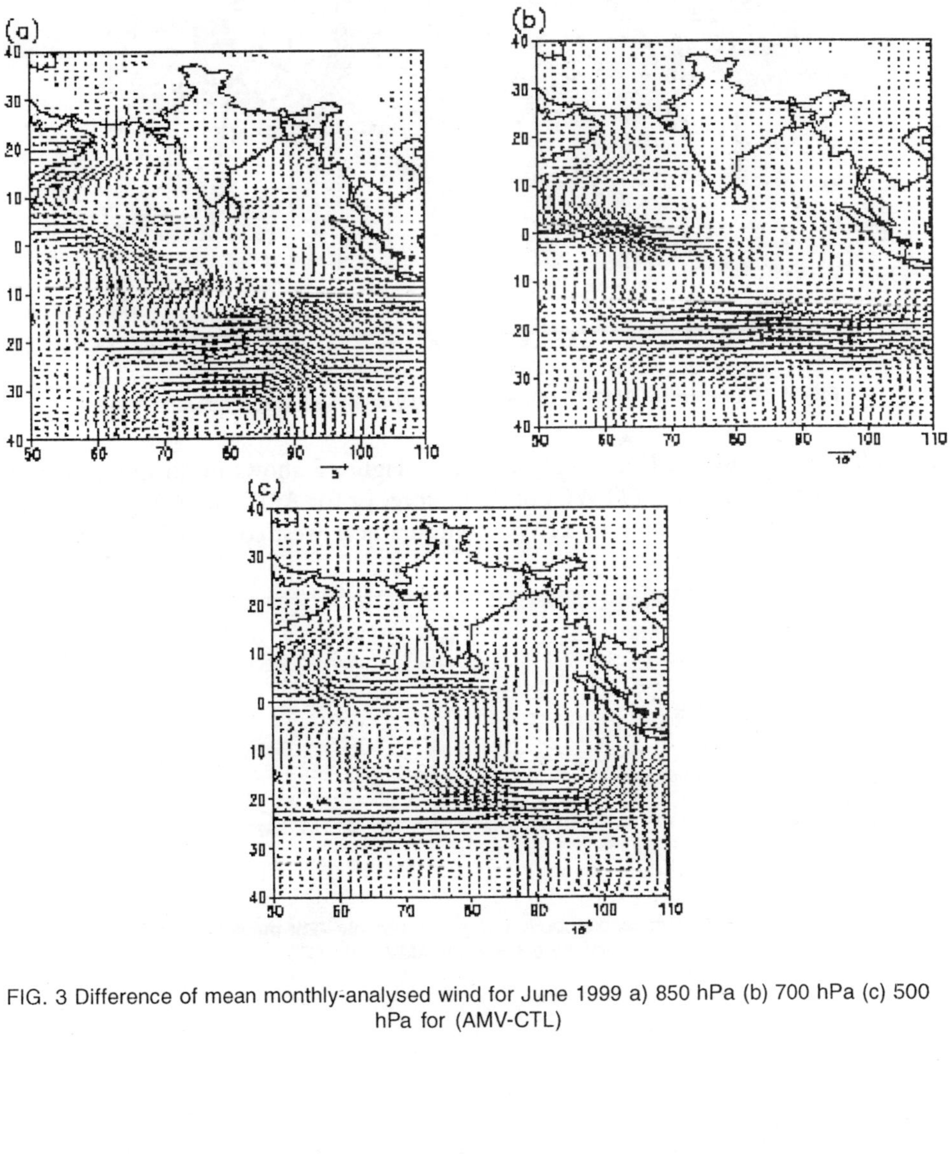

FIG. 3 Difference of mean monthly-analysed wind for June 1999 a) 850 hPa (b) 700 hPa (c) 500 hPa for (AMV-CTL)

Arabian Sea along with the realistic reduction of westerly flow regime over north Arabian Sea. In addition, inclusion of AMVs strengthens the intensity of the Mascarene High. Figure 4 depicts the difference (AMV-CTL) of mean monthly-analysed mean sea level pressure (MSLP). Difference plot clearly indicates the strengthening of Mascarene High, which could be contributing towards strengthening of cross equatorial flow and low level jet. Since a large part of the moisture over the Indian region is generally advected through monsoon currents in the month of June, an impact is likely

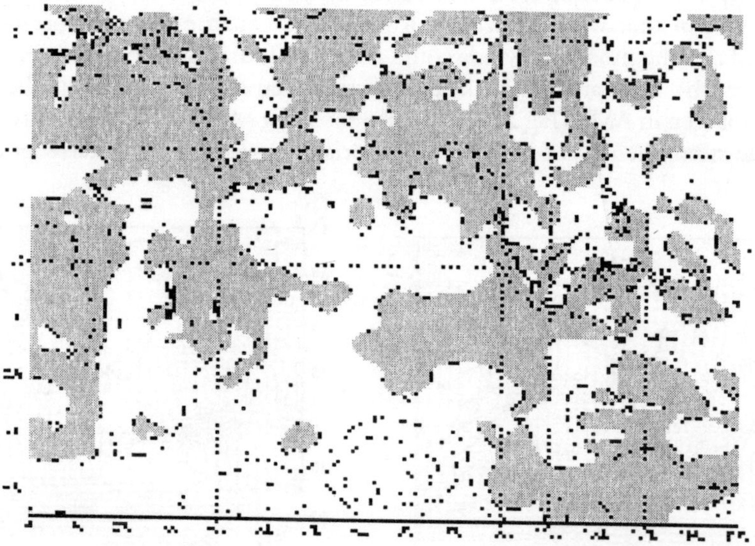

FIG. 4 Difference (AMV-CTL) of mean monthly analysed MSLP for June 1999

to be felt on the analysed moisture filed also. Figure 5 shows mean monthly total precipitable water content (TPWC) of 72 hr. forecast for AMV and CTL. It is seen from the plot that spreading of TPWC contours near west coast of India reduced

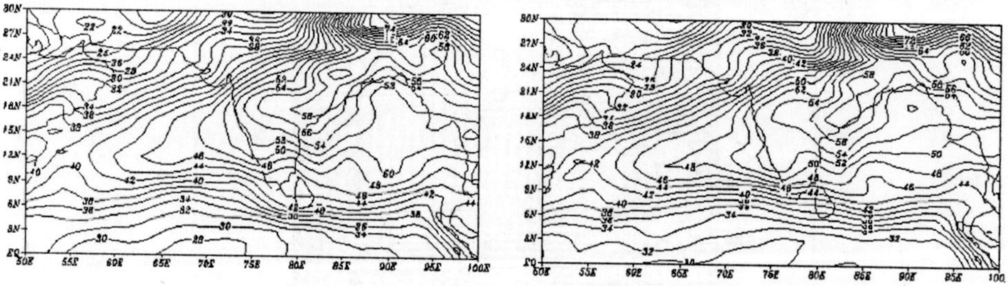

(a) (b)

FIG. 5 Geographical distribution of mean monthly 72hr predicted TPWC for June 1999 (a) AMV (b) CTL resulting in more realistic rainfall prediction up to day four. Figure 6 depicts day-3 rainfall forecasts (total for June 1999) for CTL and AMV. In CTL, rainfall maxima over the west coast is located south of the observed maxima, whereas in AMV the rainfall maxima has been realistically shifted north ward. The spreading of the rainfall over the Arabian Sea is also reduced in AMV. Rainfall prediction over Bay of Bengal and east coast of India is also seems to be more realistic in AMV compared to CTL.

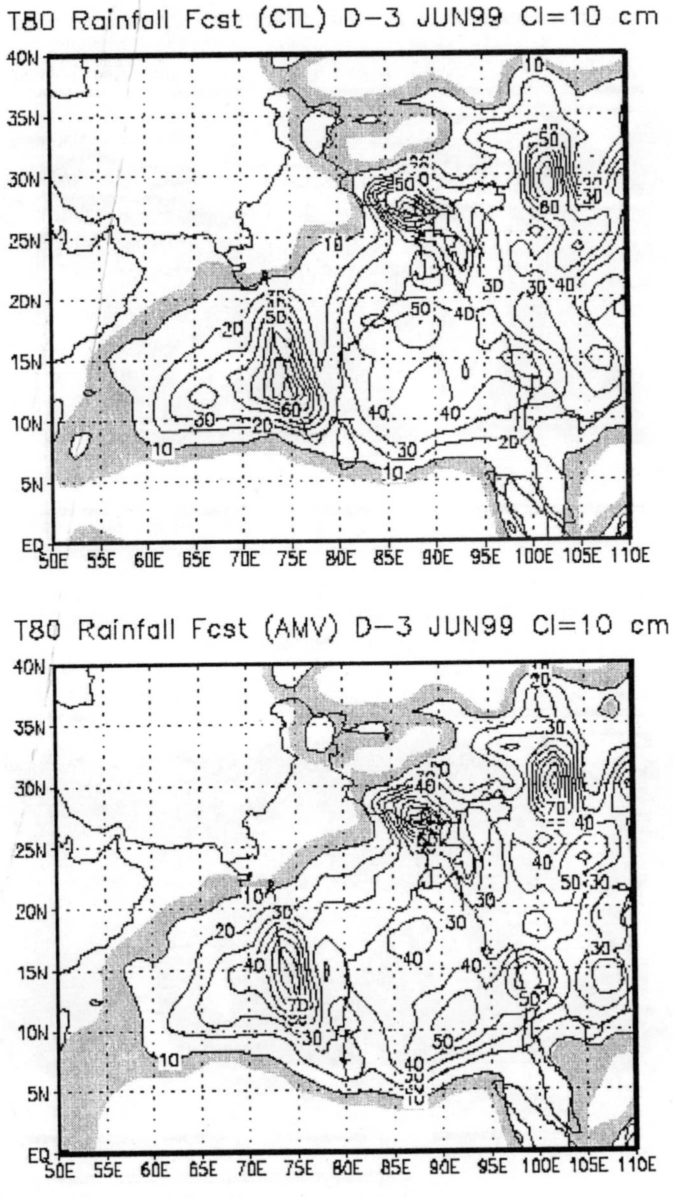

FIG. 6 Cumulative rainfall for June 1999 in Day-3 forecasts for CTL and AMV

NESIDS derived temperature and moisture information from ATOVS (NOAA-15&16) is available through GTS and ftp server at 120 km resolution. Few experiments have been conducted to study the impact of these data sets on NCMRWF's analysis–forecast system during summer monsoon of 2002 for the period, 25th August- 30th September. Figure 7 shows global distribution of TPWC derived from DMSP SSM/I sensor for 0000UTC of 29 September 2001, downloaded at NCMRWF on near real time from ftp site of 'Remote Sensing Systems'. Figure 8 is similar to as Figure 7 but from the NOAA-15 &16 ATOVS. It is clearly seen from plots that SSM/I TPWC over

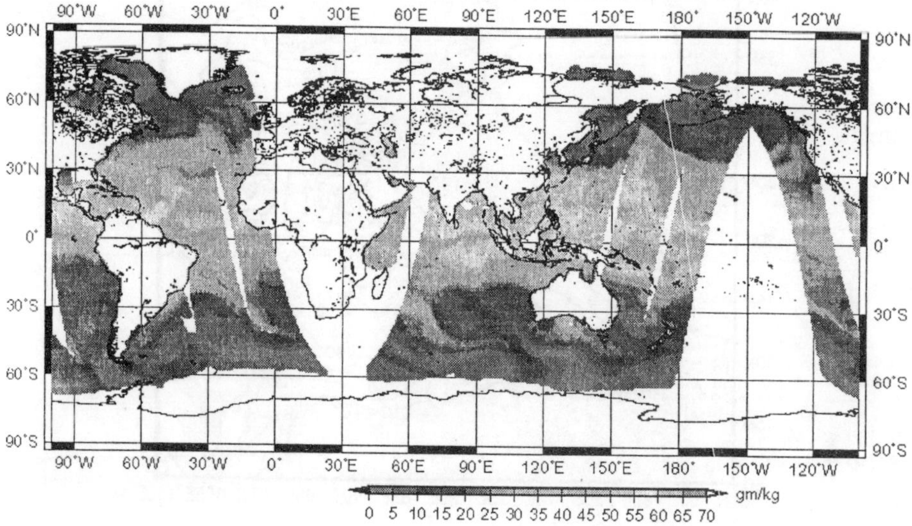

FIG. 7 TPWC from SSM/I data received at NCMRWF
(~ 0000UTC 29th Sept. 2001)

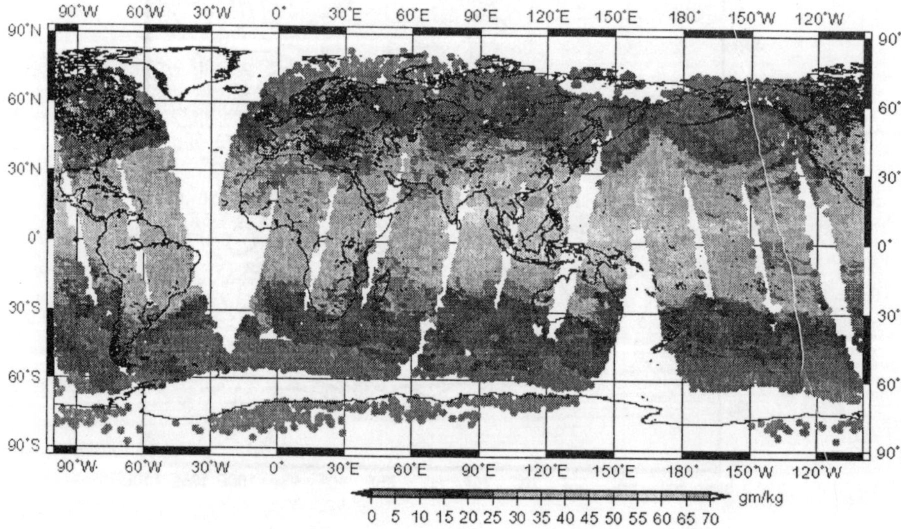

FIG. 8 TPWC from NOAA data received at NCMRWF
(~ 0000UTC 29th Sept. 2001)

Indian seas and adjoining oceans is more than that of ATOVS. Several observing system experiments were carried out with different combinations of meteorological parameters from ATOVS and SSM/I data sets in order to find out the most effective way of using these data sets for improving operational analysis and subsequent forecasts. For brevity, only three experiments out of total eight that were carried out have been discussed here. Exp-3 was conducted using conventional data along with temperature profiles at 16 standard isobaric levels and TPWC from NOAA ATOVS both over land and ocean. Exp-6 used conventional data along with SSM/I TPWC and ocean surface wind speed from SSM/I. In Exp-8 ATOVS TPWC and temperature profiles (both over ocean only) and SSM/I ocean surface wind speeds were assimilated along with all other conventional data. Results from these experiments have shown that the combination used in Exp-8 in general produces best result amongst all other combinations. Figure 9 shows cumulative monthly-analysed rainfall for September 2001 along with the same in day-1 forecasts from Exp-3, Exp-6 and Exp-8 valid for the same period. As seen in Exp-3, inclusion of ATOVS TPWC and temperature over land produces unrealistically high rainfall over northwest part of India. Assimilation SSM/I over ocean also produced unrealistic rainfall over the equatorial Indian Ocean near 66^{0}E longitudes (Exp-6). It is noticed that the combination of temperature and TPWC over ocean from ATOVS and

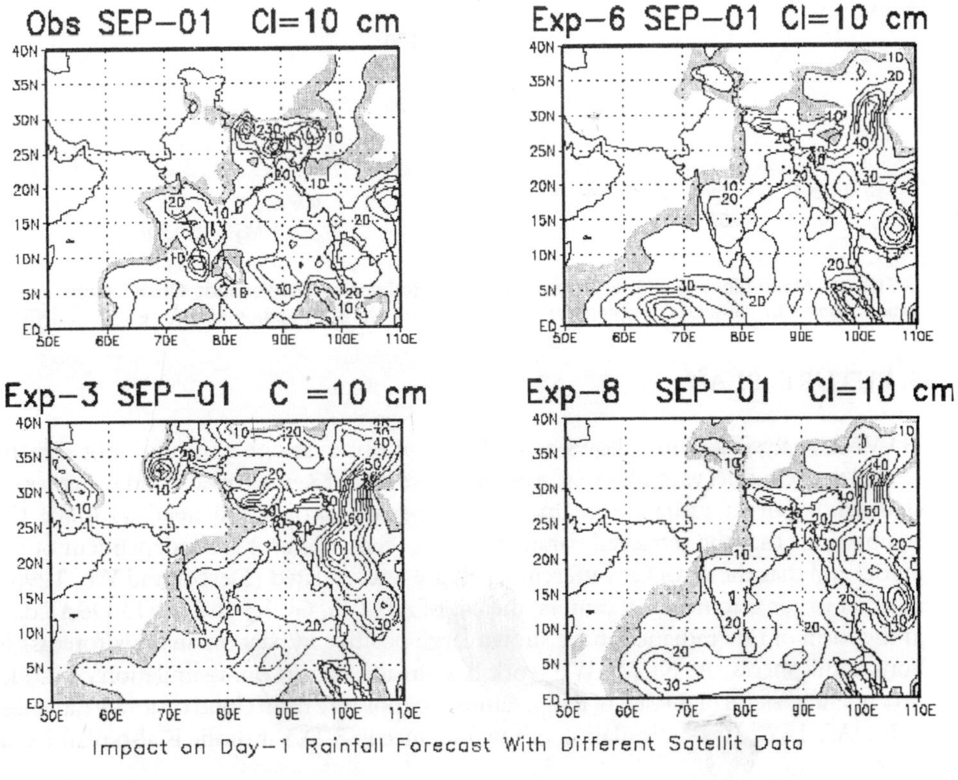

FIG. 9 Cumulative monthly observed rainfall for September 2001 along with cumulative rainfall in Day-1 forecasts for Exp-3, Exp-6 and Exp-8

the surface wind speed from SSM/I produces the best rainfall forecast for the period of study. Although day-1 prediction is only shown here the same trend is seen upto day-3 forecasts. Beyond 72 hours there is hardly any noticeable impact of data on rainfall prediction. Various statistical measures have also been computed for all these sets of experiments. Fig 10 shows mean monthly root mean square error of temperature and moisture analysis (0000UTC) at different vertical levels, computed against observations (radiosonde) over Indian region for Exp-6 and Exp-8. RMSE of moisture and temperature for Exp-8 is less than that of Exp-6 specially in lower and middle levels.

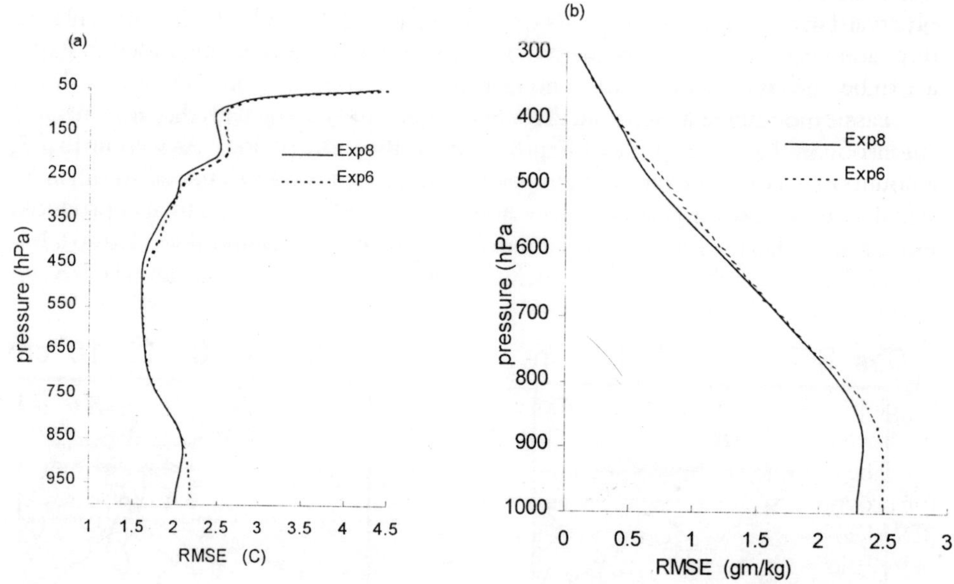

FIG. 10 Mean Monthly RMSE of Analysis Computed against Radiosonde Observations over Indian Region for 0000UTC September 2001 (a) temperature (b) sp.humidity for Exp-6 and Exp-8

5. FUTURE PLAN

3D-VAR analysis scheme allows assimilation of observations which are not directly related to the model variables (e.g. raw radiances from IR, microwave and water vapor channels). Direct radiance assimilation approach avoids the need to make the assumptions that the retrieved parameters (e.g., temperature) are independent of the model and that the errors at different levels are uncorrelated (Derber and Wu, 1998). In other major international centers, the use of ATOVS, GOES and METEOSAT data in the form of raw radiances have shown large positive impact on analysis/forecast in both hemispheres. At NCMRWF work has already started on assimilation of model dependent retrieval (1-DVar) of temperature and moisture from cloud clear raw radiances of NOAA-15 & 16. Assimilation of raw radiance in 3-D Var mode is also planned in future.

To assimilate the high density (in space and time) satellite derived winds from IR, VIS and WV channels effectively, methods are to be developed for their quality control.

Observation system experiments have to be carried out to find out the optimum thinning of these data. Speed biases in these winds are related to height assignment problems. Statistical scheme are to be developed to correct biases in these winds through the use of model short range forecast. Scatterometer wind data are very valuable because they are available in the data sparse regions of tropical oceans and also in cloudy as well as rainy conditions. The spatial representation and direction ambiguity of scatterometer wind (SeaWinds on QuickSCAT, ADEOS II) should also be addressed in a realistic way for different weather systems to be used operationally. Appropriate thinning of scatteromter data to be used in a global model has to be worked out carefully, as the observations may be spatially correlated. Quality monitoring, rain contamination, wind direction noise characteristics and wind direction ambiguity are few other factors that are to be investigated before their assimilation.

Passive microwave sensors (SSM/I, MSMR) in extremely moist conditions (monsoon environment) reduce the accuracy of derived surface wind speed. Large amount of moisture also degrade the moisture estimation as the brightness temperature signal saturates. Different types (1-D Var and radiative transfer model) of retrieval using SSM/I brightness temperatures and background field from short range forecast from NWP model has to be tried for a better retrieval of wind speed and moisture information. Another alternative is to use neural network based retrieval algorithms (used at NCEP). Assimilation of rain are to be carried out to improve temperature and moisture profiles for producing better rainfall forecast which in turn requires the development of tangent linear and the adjoint code of cloud physics and convection parameterisation schemes.

The national and international agencies are going to launch new type of instruments for providing better atmospheric/ocean observations (INSAT-3A, METSAT, INSAT-3D, Megha-Tropiques, GPM). Simultaneous assimilation of these high-resolution satellite observations to produce the best-desired analysis is the real challenge to NWP community. Close collaboration between data providers and NWP community will be needed to make effective use of these data.

REFERENCES

Basu Sujit, R. M. Gairola, C. M. Kishtawal, P. C. Pandey, A. K .Bohra, D. Rajan and A.K.Mitra, 1997: Retrieval of moisture profiles using satellite measurements of precipitable water study of their impact on NWP models. *Indian J of Radio and Space Physics*, **26**, 49-76.

Bohra, A. K., K. J. Ramesh, V. S. Prasad and G. R. Iyengar, 1998: Impact of satellite derived temparature profile data in global data assimilation and forecasting system: A case study. *The Global Atmos. and Ocean System* , **5** , 345-365.

Das Gupta, M., S. R. H. Rizvi and A. K. Mitra, 2002: Impact of ERS-2 Scatterometer wind on Global analysis-Forecast System. *Mausam*, **53**, 153-164.

Derber, J. C. , D. F. Parrish and S. J. Lord, 1991: The new global operational analysis system at NMC. *Wea. Forecasting,* **6** , 538-547.

Derber, J.C., and W. S. Wu, 1998 : The use of TOVS cloud cleared radiances in the NCEP's SSI analysis system. *Mon. Wea. Rev.*, **126**, 2287-2299.

Kar, S.C., 2002: Description of a high resolution global model (T170 / L28) developed at NCMRWF. *NCMRWF Technical Report No. 1 / 2002*, 28 pp.

Kar, S.C., S. Pattanaik, K. Rupa V.S. Prasad and S. V. Singh, 2003a: Impact of surface wind & moisture data from microwave sensors in the analysis & prediction of weather in the Indian ocean region. *Mausam* **54**, 1-12.

Kar, S.C. , K. Rupa, M.Das Gupta and S. V. Singh, 2003b: Analyses of Orissa Super Cyclone using TRMM(TMI), DMSP(SSM/I) and Oceansat-I(MSMR) derived data. *Glob. Atmos. Ocean Sys.* (In Press)

Kulkarni, P.L., A.K.Mitra, S. G. Narkhedkar, A. K. Bohra and S. Rajamani, 1997: On the impact of divergent part of the wind computed from INSAT OLR data on global analysis and forecast fields. *Meteor and Atmos Phys*, **64**, 61-82.

Parrish, D.F. and J. Derber, 1992: The NMC's spectral and statistical-interpolation analysis System. *Mon. Wea. Rev.* , **120** , 1747-1763.

Parrish, D. F., J. Derber, J. Purser, W. Wu and Z. Pu, 1997: The NCEP global analysis system: Recent improvements and future plans. *J Met. Soc.* Japan, **75**, 18359 - 18365.

Prasad V.S., K. J. Ramesh, A. K. Bohra and R. C. Bhatia, 1999: Assessment of quality and impact of full resolution TOVS temperature profile data on the operational global data assimilation-forecast system of India. *Meteor. Atmos. Phys.*, **68**, 197-212.

Rajan. D., A. K. Mitra, S. R. H. Rizvi, R. K. Paliwal, A. K. Bohra and V. B. Bhatia, 2001: Impact of Satellite Derived Moisture in a Global Numerical Weather Prediction Model. *Atmosfera* ,**14**, 203-209.

Rajan, D., A. K. Mitra, A. K. Bohra and V. B. Bhatia 2002: Moisture profiles from satellite data over Indian Ocean area . *Int. J Remote Sens*.**23**, 2959-2969.

Rizvi, S. R. H., R. Kamineni and U. C. Mohanty, 2000: Assimilation of MSMR Data in NCMRWF Global data assimilation system. *NCMRWF scientific report no. 1/2000*, (August 2000), 44 pp.

LIST OF ACRONYMS

ATOVS	Advanced Television Infrared Observation Satellites
BMRC	Bureau of Meteorology Research Centre
CIMSS	Cooperative Institute for Meteorological Satellite Studies
CMV	Cloud Motion Vectors
DMSP	Defense Meteorological Satellite Program
EUMETSAT	Europe's Meteorological Satellite Organization
GOES	Geostationary Operational Environmental Satellites
GMS	Geostationary Meteorological Satellite
GTS	Global Telecommunication System
HRPT	High Resolution Picture Transmission Project
INSAT	Indian Satellite
IR	Infrared
IRS-P4	Indian Remote sensing Satellite (Polar-4)
JMA	Japan Meteorological Agency
MSMR	Multifrequency Scanning Microwave Radiometer
NCEP	National Centers for Environmental Prediction
NCMRWF	National Centre for Medium Range Weather Forecasting
NESDIS	National Environmental Satellite, Data, and Information Service
NOAA	National Oceanic & Atmospheric Administration
OLR	Outgoing Longwave Radiation
RTH	Regional Telecommunication Hub
SSI	Spectral Statistical Interpolation
SSM/I	Special Sensor Microwave Imager
TRMM	Tropical Rainfall Measuring Mission
TMI	TRMM Microwave Imager
VIS	Visible
WV	Water Vapour

Almost Anything Can be the Leading Singular Vector

JON AHLQUIST
Department of Meteorology, Florida State University
Tallahassee, FL 32306–4520, Submitted to J. Atmos. Sci.
Email: ahlquist@met.fsu.edu

ABSTRACT

Lorenz singular vectors perform.
To optimize growth with a norm.
But new metrices portray.
Other vectors that may
Have quite different growth rate and form.

1. INTRODUCTION

For a given linear transformation, the first singular vector is that initial condition that grows more than any other. Growth is measured as the ratio of the output norm to the input norm. The second singular vector is that initial condition that maximizes growth among the vectors orthogonal to the first singular vector. Subsequent singular vectors maximize growth while being orthogonal to preceding singular vectors. Singular vectors are unique up to multiplication by a scalar.

Without using the name "singular vectors," Lorenz (1965) was the first in meteorology to compute singular vectors to study variations in forecast sensitivity to initial conditions. Since then, many other meteorologists have applied this approach. Important singular vectors studies (which include references to earlier work) include those by Farrell and Ioannou (1996a, b) on non-normal mode stability analysis; Buizza and Palmer (1995) and Gelaro *et al.* (1998) on generation of perturbations of ensemble forecasting; and Palmer *et al.* (1998) on adaptive observations. To adjust to various circumstances, the norms used to measure input and output vectors have been varied; see, *e.q.*, Palmer *et al.* (1998) and Ehrendorfer and Tribbia (1997). Indeed, different norms arise quite naturally (Toth 1991). Also, if disturbances evolve long enough under linearized dynamics, they become so-called Lyapunov vectors (Trevisan and Legnani, 1995, Trevisan and Pancotti, 1998).

In this study, we ask how much singular vectors depend on the choice of the inner product that is applied to the input vectors and the norm used to measure the output vectors. The answer is that the singular vectors and their corresponding growth amounts depend *entirely* on these choices. In fact, any vectors that transform to linearly independent vectors can be singular vectors with whatever growth factors one wants, provided one chooses an appropriate inner product for the input vectors and an appropriate norm for the output vectors. We prove this assertion by detailing the steps needed to construct the inner product for the input vectors and the norm for the output vectors.

In no way do we wish to suggest that singular vectors are unimportant. The study of perturbation growth on an evolving basic state is extremely important in meteorology, oceanography, and many other applications. Once an input inner product and an output norm are chosen, the corresponding singular vectors provide one way of decomposing arbitrary perturbations to analyze their evolution. Lyapunov vectors are another possible basis.

We wish to emphasize two points. First, one has great flexibility when choosing an input inner product and an output norm. Because different people are interested in different variables and different regions, different people can justifiably measure things in different ways, *i.e.*, there is no inherebtly best inner product or norm. Second, because anything not in the null set (defined in section) can be a singular vector—even the leading singular vector—singular vectors do not automatically have interpretations as physical modes. They are solutions to an optimization problem for particular choices of an input inner product and an output norm. Any physical meaning must come from some additional aspect of the problem.

2. MATHEMATICAL BACKGROUND

The mathematical context for this work is that of linear algebra, in particular, Hilbert spaces. We refer the reader to Herstein (1964) and Stakgold (1979) for background material on linear algebra and Hilbert spaces, respectively, but many other texts on linear algebra and/or partial differential equations contain similar material.

Let C_m denote the space of complex-valued m-vectors *i.e.*, vectors containing m complex-valued components. C_m forms a linear space under standard vector addition and multiplication by complex-valued scalars (Stakgold, pp. 225-226). For our application, an m-vector contains all quantities needed to specify the initial conditions for a weather forecast model. In some applications, the m-vector components are real-valued quantities at grid points. In other applications, they are complex-valued spectral coefficients. We shall regard all quantities as complex-valued with the understanding that real-valued vectors are just a special case. Model parameters may or may not be included among the initial conditions without restriction to what follows.

Let lower-case Greek and English letters denote scalars and vectos, respectively, and let an overbar denote the complex conjugate, e.g., $\overline{\alpha}$ and $\overline{u}$. Upper case letters will represent matrices. For arbitrary matrix E, Let E_{ij} indicate the element in the i-th row and j-th column. Let E_{*j} represent the j-th column vector of matrix E. Let "iff" stand for "if and only if," $\ni$ stand for "such that," and $\forall$ denote "for all." "QED" will mark the end of every proof.

For any given inner product, we shall always use the natural norm generated by an inner product,

$$\|u\| = \sqrt{\langle u, u\rangle} \tag{1}$$

and the natural metric generated by the norm,

$$d(u, v) = \|u - v\| \tag{2}$$

Following standard practice, if u and v are equal, written $x = y$, then every component of x equals every component of y. The classical definition of an inner product requires that $\langle u, u \rangle = 0$ iff $u = 0$.

We begin by generalizing the definition of an inner product so that even if $u \neq 0$ it may still be possible that

$$\langle u, u \rangle = 0 \tag{3}$$

This also generalizes the definition of a norm so that perhaps $\|u\| = 0$ even when $u \neq 0$ and the definition of a metric so that perhaps $d(u, v) = 0$ even when $u \neq v$. We shall say that u and v are congruent, denoted $u \cong v$, if $d(u, v) = 0$ according to a generalized metric. Therefore, it is possible that $u \cong v$ even if $u \neq v$.

As an example where congruency is relevant, let f_1 and f_2 be two vectors each containing global weather forecasts. If $f_1 \cong f_2$, then the two forecasts are the same as measured by the generalized metric, which may involve only selected variables over a limited region. Congruency has been used implicity in earlier studies when multi-level baroclinic forecasts are rated only in terms of selected variables, such as the 50 kPa height; *i.e.*, not all variables are scored or even can be scored reliably.

Mathematically, vector congruence defines an *equivalence relation* (Herstein 1964, pp. 6-9), which requires $u \cong u \; \forall \; u$; $v \cong u$ if $u \cong v$; and $u \cong v$ if $u \cong w$ and $w \cong v$. Verification of the first two equivalence relation requirements follows directly from standard requirements for a metric. To verify the third requirement, suppose $u \cong w$ and $w \cong v$, and examine $d(u, v)$ to see whether $u \neq v$. By metric non-negativity and the triangle inequality, $0 \leq d(u, v) < d(u, w) + d(w, v) = 0 + 0 = 0$, so $d(u, v) = 0$ and $u \neq v$.

Let Z denote the set of vectors in C_m that are congruent to the zero vector through a generalized norm, *i.e.*,

$$Z \equiv \{z \in C_m \ni \|z\| = 0\}. \tag{4}$$

If f represents any forecast and z is any element of Z, then $f + z \cong f$, since $\|(f + z) - f\| = \|z\| = 0$. Forecasts indistinguishable from a forecast f can be gathered into an *equivalence class* defined as $\text{cl}(f) \equiv \{f + z \ni z \in Z\}$. See Herstein (1964, pp. 6-9).

2.1 An Inner Product Consistent with 3 Congruency

The most common standard inner product is the vector dot product, which for complex-valued vectors may be written as

$$\langle u, v \rangle_m \equiv \bar{v}^T u = \sum_i^m \bar{v}_i u_i, \tag{5}$$

where u and v are treated as column vectors and where $\bar{v}^T$ denotes the conjugate transpose, also known as the Hermitian transpose. We shall call (5) the "conjugate dot product." To avoid confusion, we include the integer subscript m on the inner product brackets to indicate the dimensionality of the vectors in the dot product. Thus, $\langle u, u \rangle_m$ denotes the standard conjugate dot product of vectors u and v in C_m. Similarly, we shall write $||u||_m$ and $d(u, v)_m$ to indicate the natural norm and metric generated by the conjugate dot product in C_m.

Any generalized inner product on C_m that is consistent with congruency can be written in the form

$$\langle u, v\rangle_{A_m} \equiv \bar{v}^T A_m u = \langle A_m u, v\rangle_m \tag{6}$$

where A_m is an $m \times m$ positive semidefinite matrix, also called nonnegative definite. Matrix A_m is positive semidefinite if $\bar{u}^T A_m u \geq 0 \; \forall \, u \in C_m$. On the left hand side of (6), the positive semidefinite matrix is a subscript on the inner product brackets to clarify the inner product's definition. Similarly, we shall write

$$||u||_{A_m} \equiv [\langle u, u\rangle_{A_m}]^{1/2} \tag{7}$$

to indicate the natural norm generated using (6). We can even write $u \overset{A_m}{\cong} v$ to indicate $||u - v||_{A_m} = 0$.

Positive semidefinite matrices are easy to construct. Let $\Phi = [\Phi_{i,j}]$ be an arbitrary complex-valued matrix having m rows and an arbitrary number of columns, say s. Let Δ be an $s \times s$ matrix having non-negative values $\{\Delta_{jj}\}$ along the main diagonal and zeros elsewhere. Then the $m \times m$ matrix

$$A_m = \Phi\Delta\bar{\Phi}^T \tag{8}$$

is positive semidefinite since

$$\bar{u}^T A_m u = \sum_{j=1}^{s} \Delta_{jj} \sum_{i=1}^{m} |\bar{\Phi}_{i,j} u_i|^2 \geq 0 \tag{9}$$

for any column vector u with components u_i.

The classical definition of an inner product requires that A_m in (6) be strictly positive definite, *i.e.* $\bar{u}^T A_m u > 0 \; \forall \, u \neq 0_m$.

3. PROPERTIES OF A LINEAR TRANSFORMATION

Let L be an arbitrary complex-valued $n \times m$ matrix, where n and m denote the number of rows and columns, respectively. Then L defines an arbitrary complex-valued linear transformation

$$y = Lx \tag{10}$$

from x in C_m to y in C_n. In a meteorological context, x may contain the initial conditions for a linearized (perturbation) forecast in which case y would contain the (perturbation) forecast. Often a linear transformation is used to approximate the initial growth of forecast errors, which may remain small enough to be governed by linearized dynamics even whenthe forecast itself is governed by nonlinear dynamics. Often, $n = m$, but that restriction is not necessary here, where m and n may be arbitrary positive integers.

The *range* of L consists of all possible linearized forecasts,

$$L(C_m) \equiv \{Lx \ni x \in C_m\} \tag{11}$$

no matter how outrageous the initial conditions.

We now set the stage for our main result in the next section through the singular value decomposition (SVD) theorem.

Theorem 1 *An arbitrary complex-valued matrix, L, can be written as*

$$\underbrace{L}_{n\times m} = \underbrace{P}_{n\times n}\ \underbrace{\Lambda}_{n\times n}\ \underbrace{\overline{E}^T}_{m\times m} \tag{12}$$

where E is an $m \times m$ matrix whose columns are orthonormal with respect to the conjugate dot product in C_m. P is an $n \times n$ matrix whose columns are orthonormal with respect to the conjugate dot product in C_n, and Λ is an $n \times m$ matrix that is everywhere zero except along the main diagonal, where the real-valued elements are nonnegative and sorted as $\Lambda_{11} \geq \Lambda_{12} \geq \ldots \Lambda_{qq} \geq 0$ for $q \equiv \min\{m, n\}$.

Proof: This theorem makes the simple and reasonable statement that there are convenient orthonormal bases for the input to and output from L such that, for $j \leq q$, input vector E_{*j} maps to an output vector pointed in the direction of P_{*j} having length Λ_{jj}. This uses the standard conjugate dot product (5) and the norm it generates.

This is a slightly generalized version of singular value decomposition found in Preisendorfer and Mobley (1988, p. 36) or Press *et al.* (1992, pp. 51-56). Here E and P are complex-valued and contain complete bases for C_m and C_n, respectively.

We prove this singular value decomposition theorem by explaining how to compute the decomposition, which is equivalent to a principal component analysis, also known as an empirical orthogonal function (EOF) analysis. For a full numerical prediction model, this calculation may not be practical because of the size of the matrices involved, but it is possible in principle.

We treat $n \times m$ matrix L as a data matrix and compute the $m \times m$ "covariance matrix" $C = \overline{L}^T L$. Then we consider the eigenvalue problem,

$$C E_{*j} = \mu_j E_{*j}. \tag{13}$$

Since C is Hermitian, *i.e.*, $C_{i,j} = \overline{C}_{ji}$, its eigenvalues are real. Further, given the covariance form of this Hermitian matrix, its eigenvalues are nonnegative. If C is complex-valued, its eigenvalues and eigenvectors can be found using subroutine CH from EISPACK (Smith *et al.* 1976); subroutine RS applies if C is real. Newer software includes the commercial software package IMSL (see www.imsl.com) and the public domain package LAPACK (see www.netlib.org/lapack), both of which offer more than one possible routine, depending on what is desired.

If an eigenvalue μ_j is not equal to any other eigenvalue, then its eigenvector when normalized as $\langle E_{*j}, E_{*j}\rangle_m = 1$ is unique up to multiplication by $e^{\sqrt{-1}\theta}$ for any angle θ. (When C is real-valued, an eigenvector is unique up to multiplication by ± 1.) If two or more eigenvalues are equal, then their eigenvectors are not unique, and any other orthonormal vectors that span the same subspace are just as good. Renumbering the eigenvalues so that $\mu_1 \geq \mu_2 \geq \ldots$, the rank r of L is the number of nonzero eigenvalues; *i.e.*,

$$r = \text{rank of } L \equiv \max j \ni \mu_j \neq 0. \tag{14}$$

It is always true that $r \leq q \equiv \min\{m, n\}$. If $r < m$, then one must choose additional eigenvectors, E_{*j} for $j = r + 1, ..., m$. To do this, select an arbitrary m-vector and apply Gram-Schmidt orthogonalization (Stakgold 1979, pp. 268-269) to convert it into E_{*r+1}, a unit vector orthogonal by (5) to all the eigenvectors E_{*j} of lower index, $j = 1, ..., r$. If, by rare chance, the randomly chosen vector contains no component orthogonal to E_{*j}, for $j = 1, ..., r$, then randomly select new vectors until one does have an orthogonal component. Once one has E_{*r+1}, one can proceed in the same way of randomly selecting vectors and applying Gram-Schmidt orthogonalization to define E_{*r+2}, *etc.*, through E_{*m}.

To complete the proof, the matrices Λ and P must be defined. For $j = 1, ..., r$, set $\Lambda_{jj} = \sqrt{\mu_j}$ where μ_j is the j-th eigenvalue of (13), and $P_{ij} = \langle L_{i*}, E_{*j} \rangle_m / \Lambda_{ij}$. For any and all j such that $r < j \leq n$, set $\Lambda_{jj} = 0$ and construct P_{*j} by applying Gram-Schmidt orthogonalization to randomly chosen n-vectors just as we did for E_{*j} for $r < j \leq m$. QED

As a side note, a positive semidefinite matrix written as (8) is not is general a singular value decomposition, (12), because the column vectors of Φ need not be orthonormal by the conjugate dot product. Even so, the verbal interpretation at the start of the SVD proof clarifies the *physical* reason why a matrix given by (8) is positive semidefinite. $A_m x = \Phi\Lambda(\overline{\Phi}^T x)$ is a linear tranformation which projects input vector x onto column vectors Φ_{*j} and which outputs stretched or compressed versions of those components in the same direction, Φ_{*j}. Thus, no component of x could ever be reversed by $A_m x$, making it impossible for $\langle A_m x, x \rangle_m$ to be negative. Back now to core issues.

Define the *null space* of L as the set of all vectors in C_m that map to the zero vector in C_n; i.e.,

$$N \equiv \{x \in C_m \ni Lx = 0_n\} \tag{15}$$

N is a linear manifold, *i.e.*, subspace, since αu and $u + v$ are also in N for any scalar α and any vectors u and v in N. The elements of N can be viewed as the homogeneous solutions to $y = Lx$. If L describes the linearized evolution of forecast error from an initial error x to a final forecast error y, the null space contains all those initial conditions which result in no final forecast error, no matter what metric is used. For example, N would contain all the perturbations that damp out by linear processes during the model's spin-up time.

Define $N^{\perp}$ as the linear manifold of C_m which is orthogonal to N by the standard conjugate dot product, i.e.,

$$N^{\perp} \equiv \{x \in C_m \ni \langle x, z \rangle_m = 0 \ \forall z \in N\} \tag{16}$$

Then $N^{\perp}$ is spanned by (*i.e.*, has for a basis) the column vectors $\{E_{*1,...,}E_{*r}\}$, and N is spanned by column vectors $\{E_{*r+1,...,}E_{*m}\}$, where E_{*j} is the j-th eigenvector of (13). $N^{\perp}$ and N are subspaces with dimensions r and $m - r$, respectively. We shall refer to $N^{\perp}$ as the orthogonal complement to N.

Define $L(C_m)^{\perp}$ as the linear manifold which is orthogonal to $L(C_m)$ by the standard conjugate dot product in C_n. Then $L(C_m)$, the range of L, is spanned by column vectors

$\{P_{*1}, P_{*2}, ..., P_{*r}\}$, and $L(C_m)^{\perp}$, the rest of C_n, is spanned by column vectors $\{P_{*r+1,...,}P_{*n}\}$. $L(C_m)$ and $L(C_m)^{\perp}$ are subspaces with dimensions r and $n-r$, respectively.

By definition, a finite or infinite set of vectors $\{y_{*1}, y_{*2},...\}$ is *linearly independent* if the only solution to $\Sigma_j\, \alpha_j y_{*j} = 0$ is $\alpha_j = 0\, \forall j$.

Theorem 2 *Suppose that an inner product is given, either traditional (such as the conjugate dot product) or generalized, and let the norm be the natural norm generated by the inner product. If $\{p_{*1}, p_{*2}, \cdots\}$ is a set of orthogonal vectors with nonzero norms, then the set is linearly independent.*

Proof : Suppose $\Sigma_j\, \alpha_j p_{*j} = 0$. To solve this equation for any α_k, take the inner product with p_{*k} and apply orthogonality, yielding $\alpha_k \langle p_{*k}, p_{*k} \rangle = 0$. Since the inner product norm of each p_{*k} is nonzero and since k is arbitrary $\alpha_k = 0\ \forall\ k$. Hence, $\{p_{*1}, p_{*2}, \cdots\}$ is a linearly independent set. QED

Theorem 3 *L is one-to-one from $N^{\perp}$ onto $L(C_n)$; that is, if $a, b \in N^{\perp}$ and $La = Lb$, then $a = b$.*

Proof : Choose any $a, b \in N^{\perp}$, and write

$$a = \sum_{j=1}^{r} \alpha_j E_{*j}, \quad b = \sum_{j=1}^{r} \beta_j E_{*j} \tag{17}$$

where $\alpha_j = \langle a, E_{*j} \rangle_m$ and $\beta_j = \langle b, E_{*j} \rangle_m$. Suppose $La = Lb$. Now $LE_{*j} = \Lambda_{jj} P_{*j}$, so

$$\text{if} \quad 0_n = La - Lb \tag{18}$$

$$\text{then} \quad 0_n = \sum_{j=1}^{r} (\alpha_j - \beta_j) \Lambda_{jj} P_{*j} \tag{19}$$

Since $\{P_{*j}\}$is an orthonormal set with respect to the conjugate dot product, it is a linearly independent set by theorem 2, so the only solution to (19) is $(\alpha_j - \beta_j)\Lambda_{jj} = 0\ \ \forall\ j$. Since $\Lambda_{jj} \neq 0$, this implies $\alpha_j = \beta_j \forall j$, so $a = b$. QED

4. SINGULAR VECTORS OF AN ARBITRARY LINEAR TRANSFORMATION

With the usual singular vector problem, one chooses an inner product, $\langle\ \rangle_{A_m}$, for the input vectors in C_m and a norm $|\ \ ||_{B_n}$, for the output vectors of L in C_n. Usually, the norm for the output is generated by an inner product. That need not be the case, but it will be for this study. If $m = n$, the inner product used to evaluate orthonormality of the input vectors to L may also generate the norm used to meausre the output from L, but different measures may be used for the input and output; see, *e.g.*, Ehrendorfer and

Tribbia (1997). Singular vectors are the input vectors x_{*j} for $j = 1, ..., r$ which maximize the output to input ratio, $||Lx_{*j}||_{B_n} / ||x_{*j}||_{A_m}$, while being orthogonal to all preceding singular vectors, *i.e.*, $\langle x_{*j}, x_{*k}\rangle_{A_m} = 0 \; \forall \; k < j$. The output to input ratio can be called the singular value. The largest of the singular values is the norm of L, defined as (Stakgold 1978, pp. 295-296)

$$||L|| \equiv \max_{x \in C_m} \frac{||Lx||\,B_n}{||x||_{A_m}} = \max_{\|x\|_{A_m} = 1} ||Lx||\,B_n. \tag{20}$$

Here, we look at the problem in reverse. Just how much do the singular vectors and singular values depend on the inner products chosen to measure the input to L and the output from L? As we shall see, the dependence is extreme. With the appropriate choice of inner products in C_m and C_n, any vector not in the null space can develop by whatever amount one wants. This will be shown in theorems 6 and 7. First, we must prove background theorems 4 and 5.

Let L be an arbitrary $n \times m$ complex-valued matrix of rank r, and arbitrarily choose r vectors $\{x_{*j}\}$ in C_m under the restrictionthat $\{Lx_{*j}\}$ is a linearly independent set in C_n. Set $y_{*j} = Lx_{*j}$.

Theorem 4 *If* $\{Lx_{*j}\}$ *is a linearly independent set in* C_n, *then* $\{x_{*j}\}$ *is linearly independent in* C_m.

Proof: Suppose $0_m = \sum_{j=1}^{r} \alpha_j x_{*j}$ Applying L, we have $0_m = \sum_{j=1}^{r} \alpha_j Lx_{*j}$. By the linear independence of $\{Lx_{*j}\}$, the only solution is $\alpha_j = 0 \; \forall \; j$. QED

Theorem 5 *If* $\{Lx_{*j}\}$ *is a linearly independent set in* C_n, *then the projection of* $\{x_{*j}\}$ *onto* $N^{\perp}$ *is linearly independent in* $N^{\perp}$.

Proof: Since $\{Lx_{*j}\}$ is linearly independent in C_n and since L is one-to-one from $N^{\perp}$ onto $L(C_n)$ by theorem 3, linear independence carries back from $L(C_n)$ to $N^{\perp}$. QED

We are now ready to present our main finding in terms of the next two theorems. Theorem 6 states that, given a fairly arbitrary collection of linearly independent input vectors, an inner product can be defined that makes the set orthonormal. Theorem 7 states that we can design an "output norm" to make those vectors singular vectors with whatever singular values we want. These theorems are followed by three corollaries that apply these two theorems.

Theorem 6 *Let* $\{x_{*1}, ..., x_{*r}\}$ *be an arbitrary set of vectors such that* $\{Lx_{*1}, ..., Lx_{*r}\}$ *is linearly independent in* C_n. *Then an inner product exists that makes* $\{x_1, ..., x_r\}$ *orthonormal.*

Proof: To construct the appropriate generalized inner product given by (6), we seek a positive semidefinite matrix $A_m = \Phi\Delta\overline{\Phi}^T$ such that

$$\langle x_{*i}, x_{*j} \rangle_{A_m} = \overline{x}_{*j}^T \Phi\Delta\Phi^T x_{*i} = \delta_{i,j} \tag{21}$$

where δ_{ij}, is the Kronecker delta, *i.e.*, $\delta_{ij} = 1$ if $i = j$ and $\delta_{ij} = 0$ if $i \neq j$.

This will be satisfied if Δ is the identity matrix in C_r, $\Delta = I_r \equiv [\delta_{ij}]$, and if

$$\langle \Phi_{*i}, \Phi_{*j} \rangle_m = \overline{x}_{*j}^T \Phi_{*i} = \sum_k \overline{x}_{kj} \Phi_{ki} = \delta_{ij} \tag{22}$$

for $1 \leq i, j \leq r$. This defines the equations that must be solved to find the so-called ***dual or reciprocal basis*** (Stakgold 1979, p. 293). Spiegel (1959, p. 171), refers to these as conjugate or reciprocal tensors. To guarantee a unique reciprocal basis, we restrict the columns of Φ to elements of $N^{\perp}$, so we write

$$x_{*j} = \sum_{k=1}^{r} \chi_{jk} E_{*k} \tag{23}$$

$$\Phi_{*i} = \sum_{l=1}^{r} \phi_{il} E_{*l} \tag{24}$$

where $\chi_{jk} = \langle x_{*j}, E_{*k} \rangle$ and where the coefficients ϕ_{il} are for the moment unknown. Using the orthonormality of the column vectors of E,

$$\langle \Phi_{*i}, x_{*j} \rangle_m = \sum_{k=1}^{r} \chi_{jk} E_{*k} \tag{25}$$

The linear independence of $\{x_{*j}\}$ projected onto $N^{\perp}$ guarantees that $\det([\chi_{ik}]) \neq 0$. Therefore, there is a unique solution for $[\phi_{ik}]$ to $[\chi_{ki}]^T [\phi_{ik}] = I$ identity matrix in C_r.

By restricting the column vectors of Φ to $N^{\perp}$, we are imposing $|x||_{A_m} = 0$ iff $x \in N$; that is, $x \equiv 0_m$ iff $x \in N$. In words, we have made the norm of an input vector 0 if it maps to the zero vector, 0_n. QED

Theorem 7 *Let $\{x_{*1}, \ldots, x_{*r}\}$ be an arbitrary set of vectors such that $\{Lx_{*1}, \ldots, Lx_{*r}\}$ is linearly independent in C_n. Let $\{\sigma_1, \ldots, \sigma_r\}$ denote any r arbitrary nonnegative numbers arranged so that $\sigma_1 \geq \ldots \sigma_r \geq 0$. Then there exists a norm $\|\ \|_{B_n}$ on C_n such that*

$$\| Lx_{*j} \|_{B_n} / \| x_{*j} \|_{A_m} = \sigma_j, \tag{26}$$

where B_n is an $n \times n$ positive semidefinite matrix and A_m is the matrix from theorem 6.

Proof: Recalling (8), we seek to express B_n as

$$B_n = \Psi\Gamma\overline{\Psi}^T \tag{27}$$

where Ψ is an $n \times r$ complex-valued matrix, and Γ is an $r \times r$ real-valued matrix which is zero except along its main diagonal where it is nonnegative. With $y_{*j} = Lx_{*j}$, it follows that

$$||Lx_{*j}||_{B_n} = [\overline{y}_{*j}^T \Psi\Gamma\overline{\Psi}^T y_{*j}]^{1/2} \tag{28}$$

We achieve our goal if we set

$$\Gamma_{ij} = \sigma_i^2 \delta_{ij} \tag{29}$$

for $1 \le i,\ j \le r$ and if we require that

$$\langle \Psi_{*i}, y_{*j} \rangle = \sum_k \overline{y}_{kj} \Psi_{ki} = \delta_{ij} \tag{30}$$

for $1 \le i,\ j \le r$. We can solve (30) for the column vectors of Ψ as a reciprocal basis for $\{y_{*i}\}$ restricted to $L(C_m)$ in the same way that we solved earlier for the column vectors of Φ as a reciprocal basis for $\{x_{*i}\}$ restricted to $N^{\perp}$. With this norm,

$$||Lx_{*j}||_{B_n} = \sigma_i \tag{31}$$

$$||y||_{B_n} = 0 \text{ if } y \notin L(C_m). \tag{32}$$

By (31), we now have the singular vectors and values we want, since $\langle x_{*i}, x_{*j} \rangle_{A_m} = \delta_{ij}$ by (21). QED

A number of corollaries follow immediately.

Corollary 1 *Choose any inner product* $\langle\ \rangle_{A_m}$ *on* C_m. *If vectors* $x_{*1}, x_{*2}, \ldots, x_{*r}$ *are orthonormal with respect to* $\langle\ \rangle_{A_m}$ *and if* $Lx_{*1}, Lx_{*2}, \ldots, Lx_{*r}$ *are linearly independent in* C_n, *then* $x_{*1}, x_{*2}, \ldots, x_{*r}$ *can be singular vectors with whatever norm values one wants.*

Proof: Because we are given an inner product and orthonormal set on C_m, we can use these for theorem 7 without needing theorem 6. QED

As an application of the first corollary, suppose that $x_{*1}, x_{*2}, \ldots, x_{*r}$ are singular vectors with respect to inner products $\langle\ \rangle_{A_m}$ and $\langle\ \rangle_{B_n}$ and that $\alpha_1, \alpha_2, \ldots, \alpha_r$ are the corresponding singular values, *ie.*, $\alpha_j = ||Lx_{*j}||_{B_n} / ||x_{*j}||_{A_m}$. By corollary 1, one can design a new inner product $\langle\ \rangle_{B_n'}$ such that $x_{*1}, x_{*2}, \ldots, x_{*r}$ are still the singular vectors but the singular values, w_j, are reversed compared to those with $\langle\ \rangle_{B_n}$, *i.e.*, $w_j = \alpha_{r+1-j}$. That is, the first shall be last, and the last shall be first. B_n' is closely related to B_n. If $B_n = \Psi\Gamma\overline{\Psi}^T$, then $B_n' = \Psi\Gamma'\overline{\Psi}^T$, where Γ' is the same as Γ except that the main diagonal elements are reversed in order. Of course, the diagonal elements of Γ' could be rearranged in any order, which would pair the singular values with the singular vectors in any way one wants, or other non-negative diagonal elements could be used, and their square roots would be the singular values.

Corollary 2 *If x is any vectory such that $Lx \neq 0_n$, then a norm for C_n can be designed to make x the most important singular vector.*

Proof: If an inner product $\langle\ \rangle_{A_m}$ is specified, then we have a special case of the first corollary, as long as $\|x\|_{A_m} \neq 0$. If no inner product $\langle\ \rangle_{A_m}$ on C_m is specified, then we have a special case of theorems 6 and 7. In either case, we can usually construct additional singular vectors along with x. QED

Corollary 3 *If $y \in C_n$ is any vector in the range of L, they y can be viewed as the final state of the most important singular vector.*

Proof: Since L is one-to-one from $N^{\perp}$ onto $L(C_m)$, there is a unique "preimage" x in $N^{\perp}$ such that $y = Lx$. Now we apply corollary 2 to make x the most important singular vector. QED

The last two corollaries serve as a caution to those doing diagnostic studies. If one has a certain initial perturbation or final forecast error in mind, one can construct an output norm to make that state appear to be the most important of all. In some cases, it might be useful to know the form of such a norm. It is just as easy, though, to design an output norm to make x the least important of all vectors not in the null set.

5. DISCUSSION

The previous section proved that inner products for the domain (input) and range (output) of any linear transformation L can be designed to make almost anything a singular vector, even the leading singular vector, with any desired nonnegative singular value. Without a doubt, many inner products wiould generate unusual norms, but two points need to be made.

One, the coefficients for positive semidifinite matrices A_m and B_n can be varied continuously, giving rise to a continuous variation in the singular vectors and singular values. Therefore, one can go from a standard norm to a weird norm by continuously varying the coefficients of $A_m = \Phi\Delta\overline{\Phi}^T$ and/or $B_n = \Psi\Gamma\overline{\Psi}^T$, and there is no magic point at which a norm suddenly becomes weird. For example, the comments following the first corollary point out that one can totally change the relative significance of the singular vectors simply by moving around the main diagonal elements of matrix Γ in $B_n = \Psi\Gamma\overline{\Psi}^T$.

Two, applications give rise to a wide variety of norms. A norm that rates global forecasts in terms of performance over the United States might not pick the same forecast as a norm that emphasizes performance over India. A norm for an airline company that emphasizes flight-level winds between two cities would be far different from an agricultural norm emphasizing surface conditions. Many business would probably prefer a norm that had a direct interpretation in dollars or rupees (Katz and Murphy 1997).

The trouble with using any norm to score a weather forecast is that it characterizes a three-dimensional multivariate global field by a single number. The world's population is interested in too many different regions and too many different weather variables to make one norm the university accepted "best" way to score forecasts. If the norm is arbitrary, then the results here indicate that singular vectors are arbitrary, too. Without an inherent norm, singular vectors are solutions to an optimization problem, but they are not physical modes because almost anything can be a singular vector. Therefore, any physical meaning for singular vectors must come from some other aspect of the problem.

6. SUMMARY

Singular vectors represent an important generalization of the way in which stability analysis can be viewed. The basic state can evolve nonlinearly in time, and the singular vector perturbations can evolve in both time and space via linearized dynamics. Classical stability analysis is much more restrictive in that it considers only oscillating or exponentially growing perturbations of constant spatial structure on a typically steady basic state.

In the present study, we have shown that any disturbance not in the null space of a transformation can be viewed as the singular vector of greatest development if we choose to measure the initial and final disturbances in an appropriate way. In general, the inner products and their associated norms may have unusual weighting coefficients in order to make an arbitrary disturbance the fastest growing disturbance. Nonetheless, the variety of human interests gives rise to a wide variety of norms.

The calculations outlined here may not be practical for operational weather forecasting models because the matrix problems would be so large, but the primary point is that the solutions exist in principle. This fundamental lack of uniqueness for singular vectors means that, although a set of singular vectors is a solution to a particular optimization problem, that optimization does not bestow any physical significance on the singular vectors. Nature responds to an initial value problem based onthe initial conditions, independent of whatever norm may be used to analyze the problem.

The results of this study in no way invalidate any earlier studies regarding the mathematics of singular vectors. If one has decided how to measure forcast error, *i.e.,* if one knows which inner products and norms one wishes to use, then singular vectors are an entirely appropriate way to decompose arbitrary initial conditions, plan adaptive observations, *etc.* But singular vectors carry no intrinsic meaning as physical modes because of their complete dependence onthe inner products chosen. Any physical meaning must come from some aspect beyond the fact that singular vectors are the solution to an optimization problem.

Acknowledgements: This research was sponsored by NOAA grant NA07WA0104. The author thanks Zoltan Toth and Eugenia Kalnay for useful discussions and comments on a draft of the manuscript.

REFERENCES

Buizza, R., and T. Palmer, 1995: The singular-vector structure of the atmospheric general circulation. *J. Atmos. Sci.,* **52**, 1434-1456.

Ehrendorfer, M., and J.J. Tribbia, 1997: Optimal error covariances through singular vectors, *J. Atmos, Sci.,* **54,** 286-313.

Farrell, B.F., and P.J. Ioannou, 1996a: Generalized stability theory. Part I: Autonomous operators. *J. Atmos. Sci.,* **53,** 2025-2040.

Farrell, B.F., and P.J. Ioannou, 1996b: Generalized stability theory. Part II: Nonautonomous operators. *J. Atmos. Sci.,* **53,** 2025-2040.

Gelaro, R., R. Buizza, T.N. Palmer, and E. Klinker, 1998: Sensitivity analysis of forecast errors and the construction of optimal perturbations using singular vectors. *J. Atmos. Sci.,* **55,** 1012-1037.

Herstein, I.N., 1964: *Topics in Algebra.* Xerox College Publishing, Lexington, MA. 342 pp.

Katz, R. W., and A.H. Murphy, eds., 1997: *Economic Value of Weather and Climate Forecasts.* Cambridge University Press, Cambridge, England, 222 pp.

Lorenz, E.N., 1965: A study of the predictability of a 28-variable atmospheric mode. *Tellus,* **17,** 321-333.

Palmer, T., R. Gelaro, J. Barkmeijer, and R. Buizza, 1998: Singular vectors, metric, and adaptive observations. *J. Atmos. Sci.,* **55,** 633-653.

Smith, B.T., and others, 1976: *Matrix Eigensystem Routines—EISPACK Guide,* 2nd ed., vol. 6 of Lecture Notes in Computer Science, Springer-Verlag, New York.

Spiegel, M., 1959: *Vector Analysis.* Schaum's Outline Series, McGraw-Hill, New York. 225 pp.

Stakgold, I., 1979: *Green's Functions and Boundary Value Problems.* Wiley, New York, 638 pp.

Toth, Z., 1991: Intercomparison of circulation similarity measures. *Mon. Wea. Rev.*, **119,** 55-64.

Trevisan, A., and R. Legnani, 1995: Transient error growth and local predictability: a study of the Lorenz system. *Tellus,* **47A,** 103-117.

Trevisan, A., and F. Pancotti, 1998: Periodic orbits, Lyapunov vectors and singular vectors in the Lorenz system. *J. Atmos. Sci.,* **55,** 390-398.

Challenges of Mesoscale Weather Prediction in North Carolina

SETHU RAMAN

North Carolina State University, Raleigh, NC 27695-8209

E mail : *sethu_raman@ncsu.edu*

ABSTRACT

A high-resolution non-hydrostatic mesoscale numerical model (MM5) is used to investigate local circulations caused by three different geographical features in North Carolina, United States. One is the land - ocean contrast along the coast. The second is the "Sandhills" region marked by a change in the soil type from clay to sand. The third region is along the mountains in the western part of the state. Numerical simulations were performed during a three-day period in the summer when synoptic forcing is minimal so that mesoscale features can be studied. The model simulated three mesoscale circulations - the sea and land breezes along the coast, a shallow circulation over the Sand hills, and a lee circulation in western North Carolina. Model simulations are compared with the surface observations provided by the NC ECONet (North Carolina Environment and climate Observing Network). As expected, the results were the best in the eastern part of the state where there is no complex topography. The mesoscale circulations caused by the surface heterogeneities and the complex topography has the potential of triggering intense convection when the conditions are favorable through scale interaction.

1. INTRODUCTION

Mesoscale processes, dictated by surface characteristics, play a dominant role on the circulation patterns and regional scale weather predictability. Large geographical variability in North Carolina (NC) provides a wide variety of weather events and climatological regimes. There are many challenges for weather prediction in North Carolina, United States due to the heterogeneity in topography, land use, and soil type, presence of the ocean and the Gulf Stream just offshore. These heterogeneous attributes and the availability of surface data from a network of automated weather stations (NC ECONet) feature North Carolina as an excellent location for simulations and validation of a regional model such as the MM5 numerical modeling system with observations of high spatial and temporal resolution.

MM5 is utilized for a non-precipitation case and a convective case including precipitation with a 5-km domain centered over the Carolinas (North and South Carolina). Model integration is for 72 hours from 0000UTC August 15, 2000 to

0000UTC August 18, 2000. Simulated hourly surface and sub-surface values are evaluated against in-situ surface observations. These simulations consist of real case studies involving MM5 Version 3 with the MRF PBL scheme coupled to the Oregon State University (OSU) land surface model (LSM). The OSU LSM uses 1 km resolution land-use and soil data as input into MM5 for capturing the dynamics of land-surface forcing.

The acquisition and combination of different agro-meteorological data across NC provides high-resolution observations for validation at multiple model grid points. For the case studies, these data incorporate hourly observation sites throughout North Carolina including 19 ASOS (Automated Surface Observing Sites / owned and operated by the NWS and FAA) sites and 15 ECONet (Environmental and Climate Observing Network: maintained by the State Climate Office of North Carolina) sites. Multiple parameters compared and investigated using this network of observations include: ECONet: Air Temperature (2m), Relative Humidity (2m), Wind Speed (10m), Wind Direction (10m), Soil Temperature (10 cm), Soil Moisture (10cm), Hourly Precipitation rate; ASOS: Air Temperature (2m), Dewpoint (2m), Wind Speed (10m), Wind Direction (10m), Hourly Precipitation, Weather Conditions, and Cloud Layers.

The approach involves examination of simulated horizontal and vertical wind patterns across the coast, piedmont, and mountainous regions in North Carolina. These simulated wind fields are used to investigate local land-sea interactions near the coast, effects of land surface processes in the piedmont, and pollution transport potential over complex terrain in the mountains of North Carolina. Precipitation patterns generated by the model are also compared with daily observations of precipitation amounts in conjunction with the hourly stations across the central portion of the 5 km domain, particularly North and South Carolina.

Following a graphical comparison of 2D and 3D fields, statistical methods are applied to provide quantitative relationships for errors and biases in the simulations. The analyses give additional insight into model performance. This is especially important when validating complex and comprehensive interactions and processes that occur in North Carolina. Statistical measures used include: absolute correlation, root mean square error (RMSE), bias, normalized mean square error (NMSE), weighted normalized mean square error of the normalized ratios (WNNR), normalized mean square error of the distribution of the normalized ratios (NNR), and the index of agreement. Typical results of the selected simulations are presented here for brevity.

2. NUMERICAL SIMULATION OF MESOSCALE CIRCULATIONS

The period, August 15, 2000 0000UTC (1900 LST) through August 18, 2000 0000UTC, is a summer case with no precipitation. Throughout this period, synoptic winds were typically light and variable, allowing mesoscale features to dominate the local weather. For most of the duration of the case study, high pressure dominated the southeast United States. Cloud cover was minimal over the Carolinas during this case study. Numerical simulation of mesoscale processes and boundary layer structure is performed. The primary focus is directed towards North Carolina (NC). The combination of topography in the west, land-use pattern variations, and close proximity to the coast in the east can cause various mesoscale interactions, which are complex in nature. Analysis of modeled boundary layer processes and interactions are evaluated and validated with observational data.

North Carolina can be divided into distinct climate divisions, Mountain, Piedmont, Coastal Plain, and Tidewater as indicated in Figure 1. Each of these regions has unique geographical characteristics with reference to proximity to water bodies, soil and vegetation type, and topography. Coastal plain and piedmont have different soil types and hence vegetation.

A well-defined sea breeze circulation was simulated on all days. This feature and the circulations in the lee of the mountains and over the Sand Hills region of the piedmont are apparent from the horizontal variations of the 10 m wind vectors shown in Figure 2. The simulated Appalachian mountain surface winds are more variable and have spatial discontinuity at some locations. The general flow pattern in the mountains.

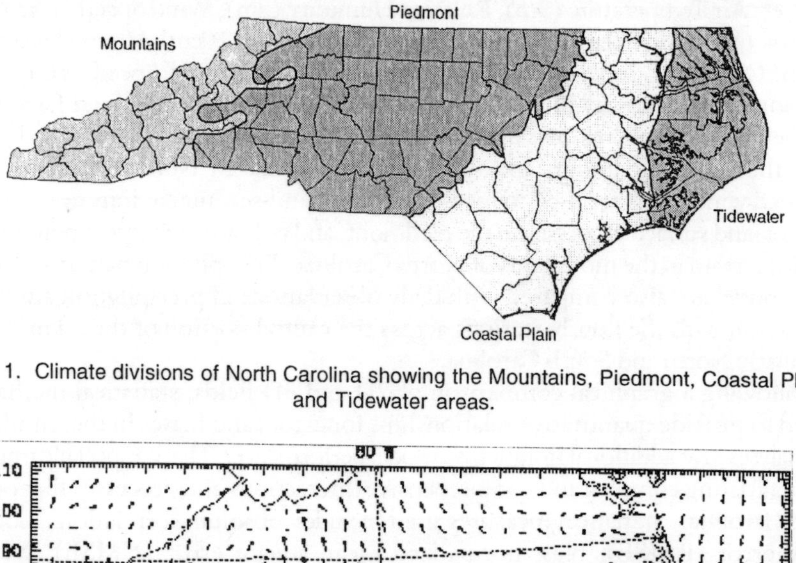

FIG. 1. Climate divisions of North Carolina showing the Mountains, Piedmont, Coastal Plain, and Tidewater areas.

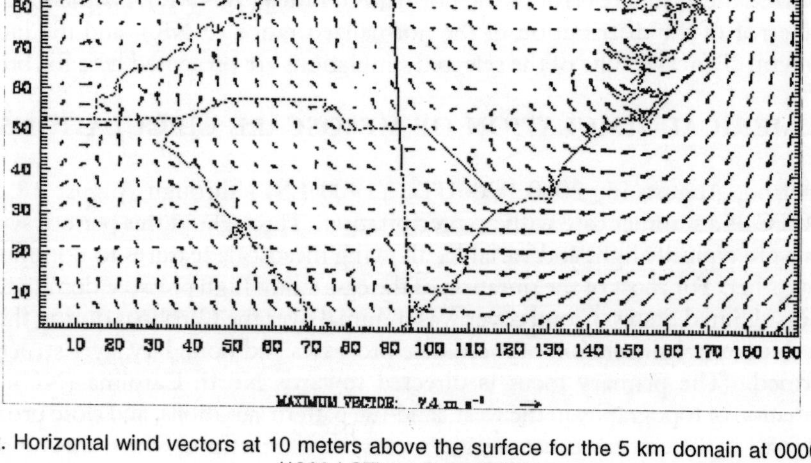

FIG. 2. Horizontal wind vectors at 10 meters above the surface for the 5 km domain at 0000UTC (1900 LST) on August 18,

for much of the simulation, was from the northwest, although there appears to be an area of convergence on the leeward side of the mountains. An area of converging winds east of and parallel to the Appalachians was simulated during afternoons. This feature, as simulated by the model, shows strong flow, perpendicular to the mountain range on the lee side of the mountains that is most likely induced by the mountainous terrain. This flow appears to converge with the overall light, ambient flow from the southwest. To examine the vertical extent of this simulated feature, a cross section of the wind, temperature, and moisture fields were analyzed across the axis of the mountains (not shown). Complex flow patterns, including terrain induced convergences and mesoscale circulations, were evident. An accelerated flow was simulated over the top of the mountain range, about 5 to 8 ms^{-1}. This accelerated flow converges with slower moving air on the leeward side of the mountains, approximately 1 to 3 ms^{-1}.

On the leeward slope of the mountains, a return flow circulation develops during the late afternoon hours. This circulation extends 50 km in the horizontal and from the surface to approximately 1 km. The well-mixed layer, mountain waves, and circulation patterns are more pronounced during the daytime and get eroded as the night progresses.

There also appears to be an area of convergence in the Sand Hills region near the Piedmont / Coastal Plains intersection (Figure 1) where a strong southerly flow of about 6m s^{-1} suddenly ceases in a narrow area of near calm winds. This convergence zone extends well into South Carolina and has an orientation of northeast to southwest.

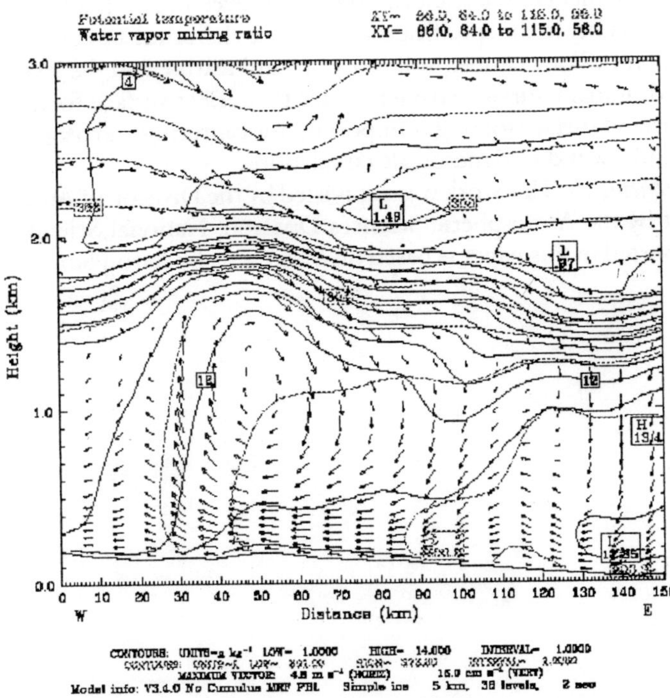

FIG. 3. Vertical cross-section showing circulation vectors (m/s), potential temperature (K), and water vapor mixing ratio (g/kg) across the Sand Hills region at 0000UTC (1900 LST)

This region consists of a narrow band of sandy soil that borders the clay-based soils in the piedmont. Sharp contrasts in soil characteristics can result in differential heating across this region. These heat flux gradients apparently initiate local mesoscale circulations.

The center of the simulated Sand Hills circulation is elevated to an altitude of approximately 1 km. The horizontal extent stretches over 70 km and has a vertical extent of 2 km (Figure 3) There is a strong surface flow, about 4.6 ms^{-1}, associated with this circulation, with a strong upward component on the westward side of the circulation. This circulation becomes even stronger when the sea breeze circulation along the coast merges with it.

Evolution of this feature over time shows a westward migration of this circulation feature where the center travels 35 km in the horizontal from east to west over a span of 6 hours. These circulations are not evident in the nighttime. The simulated flow in the evening is fairly uniform and is from the west at a maximum of 4.3 m s^{-1}.

3. CONCLUSIONS

During weak synoptic conditions, mesoscale processes can have significant impact on regional weather. Examples of these processes include local surface heat flux gradients caused by differences in evaporation and transpiration from the earth's surface, and circulations caused by topography. Surface characteristics can also play a dominant role in the development of the planetary boundary layer and in cloud and precipitation patterns. Uncertainty in the specification of surface characteristics can impact regional scale predictability. For example, soil moisture and texture and the land use itself are some of the dominant surface characteristics that affect surface forcing. Correctly treating these land surface forcing is important for capturing and properly simulating terrain and land-use induced mesoscale circulations.

Acknowledgements. This work was supported by the Division of Air Quality, North Carolina and by the Atmospheric Sciences Division, National Science Foundation. Research reported in this paper is based on M.S.thesis by Aaron Sims (2001).

Mesoscale Modeling Research and Development at NCAR.

JIMY DUDHIA
MMM Division, National Center for Atmospheric Research, Boulder, Colorado, USA
E mail: dudhia@ucar.edu

ABSTRACT

NCAR has several research programs relating to mesoscale modeling. Here focus is on two particular models at NCAR, namely the Pennsylvania State University / NCAR Mesoscale Model (MM5) and the new Weather Research and Forecast Model (WRF).

1. INTRODUCTION

In the following sections history and uses of MM5 are outlined along with motivation for the development of WRF. Current status of WRF are also discussed. Finally potential areas of collaboration are addressed in the context of current research and development interests at NCAR.

2. MM5

2.1 History

MM5 is the fifth generation of the Penn State / NCAR Mesoscale Model, that originated from a hurricane model and mesoscale model developed by Anthes, Warner and co-workers at Penn State University in the 1970's. The model has survived largely owing to keeping up with computational and scientific advances during the last twenty years. During the 1980's the model was hydrostatic, had few physics options, was capable of using limited datasets for initialization, and only ran on the Cray computer at NCAR. The 1990's saw the development of a nonhydrostatic version with multiple nests and data assimilation together with many more physics options, and portability onto almost all the current computer platforms, including Linux PCs and distributed-memory platforms. The data ingest system was also generalized to allow easy initialization with any gridded datasets and observations, and a land-surface model was added. Concurrently with the increased model portability, the MM5 modeling community grew to its current size of over 700 users worldwide.

NCAR has the role of supporting the modeling system programs, holding tutorials for new users twice per year, and hosting an annual workshop for model users to present results and development related to MM5. The code continues to be developed and the latest release (Version 3.5) was in December 2001.

2.2 Uses

As a full-physics nonhydrostatic mesoscale model, MM5 has a wide variety of applications. Many universities use MM5 as a real-time forecast model, providing local weather guidance, as well as a tool for teaching and research. The Air Force Weather Agency (AFWA), Taiwanese Civil Aviation Authority and Korean Meteorological Administration use MM5 operationally to provide forecasts at the national scale, and in other regions of the world. NCAR runs nested 30 km and 10 km MM5 forecasts over the United States and Antarctica, posting results on the Web. We also have an experimental global version of MM5 that has been running 5-day forecasts with a 120 km grid since 1999.

As a research tool, MM5 has been used to study a variety of phenomena, including mid-latitude ocean cyclones, tropical storms and hurricanes, polar lows, severe convective outbreaks, winter storms, fronts and topographic effects. A large number of papers have been published with MM4 and MM5 research.

MM5 also serves as a framework for research into numerical weather prediction methods, primarily physics parameterizations, initialization techniques and sensitivities, data assimilation research, verification and grid resolution studies.

Other diverse applications include air quality modeling, regional climate studies, and idealized studies.

3. WRF

3.1 Motivation

The Weather Research and Forecast Model (WRF) is the result of a collaborative effort between NCAR, operational centers (the National Centers for Environmental Prediction, Forecast Systems laboratory and AFWA), government laboratories and universities to produce a mesoscale model that can be used both operationally and for research in meteorology and other applications.

This joint effort is motivated by the value foreseen in closer ties between the operational and research communities in mesoscale modeling. Using a common modeling framework would allow easier transfer of ideas and personnel between these communities that are currently separated by using very different modeling frameworks and techniques. Furthermore, both in research and operations, the existing models are becoming outdated with regard to overall code design, numerical techniques, and their flexibility to work down to cloud-resolving scales.

WRF is designed with the 1-10 km grid scales in mind, as these resolutions will soon be typical in national scale forecast models. The Eta model of NCEP is already being run at 12 km over the contiguous United States.

In addition to capabilities as a forecast model, it is a priority for WRF to be suited to idealized research, such as storm dynamics, orographical flows, ideal baroclinic waves, etc. This allows researchers to study basic dynamical problems within the same framework as the full forecast model by setting up their own initial conditions, and using minimal physics.

The goal is that in a few years WRF will take over the roles of MM5 and NCEP's Eta model and FSL's MAPS model, as a new state-of-the-science model. WRF development started in 1997-1998, and currently there is a basic version of WRF that runs with either idealized or real initial conditions. The following sections will outline the current status and plans for WRF in a few general aspects.

3.2. Code Design

WRF is written in standard Fortran 90, and designed to work on a variety of computing architectures ranging from single-processor machines, to distributed clusters of shared-memory machines without code modification by using the MPI and Open MP standards. It achieves this by a method called two-level domain decomposition, whereby the model domain is divided first into an arbitrary number of patches, and these are further divided into tiles (Figure 1). Patches are subdivisions of the memory for distributed-memory architectures, so a shared-memory machine or single processor would have only one patch covering the entire domain. Tiles are a subdivision of the computing task for parallel threads on a shared-memory machine. Each processor executes one tile at a time, and, for instance, each call to a physics subroutine executes one tile of calculations. The tiles determine the size of the horizontal loops and local memory in the low-level subroutines. The inner loop index was chosen to be a horizontal index to take advantage of vector processors where possible.

The code has driver-layer, interface, and model-layer modules of subroutines. The driver-layer handles architecture-specific tasks such as memory allocation, array initialization, and calling input/output functions, in addition to handling potentially multiple domains when nesting is implemented. The model layer modules are architecture

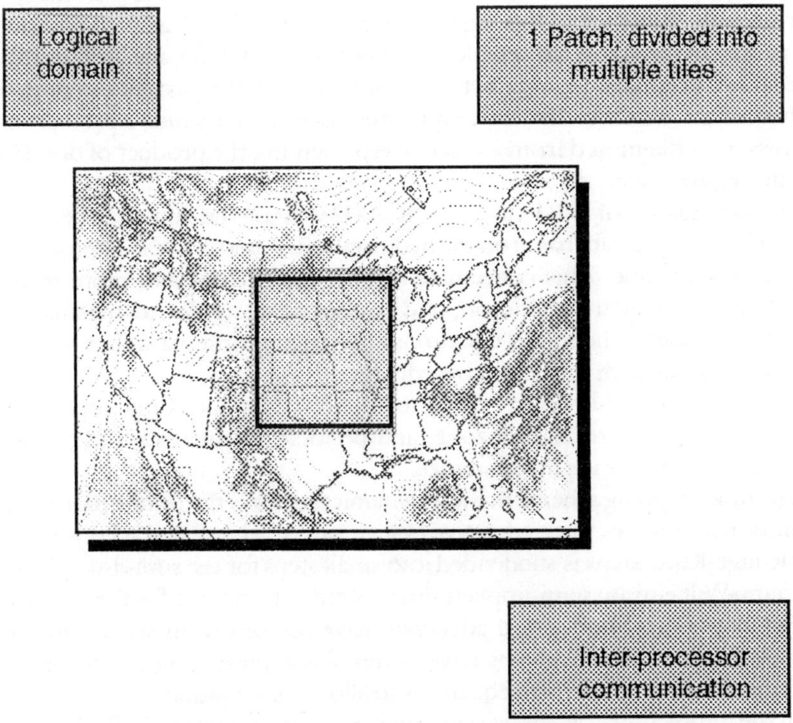

FIG.1. Domain decomposition in WRF

independent, and use basic Fortran 90 to do physics and dynamics calculations. The interface layer for the model is the main solver that advances one domain by one timestep, and calls the physics and dynamics subroutines of the model layer. It has some code to handle tiles and distributed-memory data transfers.

A unique feature of the code is the Data Registry, an ASCII file that contains information about all the state arrays that define the state of the model domain at a given time, in addition to other arrays required at the solver level such as tendencies and temporary arrays. This Registry also defines namelist variables, including those that select physics packages. The Registry allows code modification to be made easily even when new arrays are required, since adding an array to the Registry automatically makes it available at the solver level without changes being required in the driver layer. Thus WRF is written with code development by scientists in mind, which is a priority for such a model to be a research tool.

Input and output is also designed to be modular with respect to the format. Currently the released code has only NetCDF format files, but other formats are being developed.

3.3 Dynamics and Numerical Techniques

The initial release of WRF, made in November 2000, uses a terrain-following height coordinate and a C-grid staggering of the velocity components. The basic equations solved are the full unapproximated nonhydrostatic equations with Coriolis and curvature effects relevant to a map projection on a sphere. It can also use a Cartesian grid for idealized cases. A new feature in WRF is that it solves the continuity equation for dry density directly, and this together with its treatment of all scalars allows a strict conservation of the scalars to numerical round-off accuracy. Only anelastic and hydrostatic models have been able to achieve this conservation in the past, not unapproximated non-hydrostatic models. The basic equations also do not include a pressure equation, but pressure is diagnosed from a variable representing the product of dry density and potential temperature.

The next release of WRF due in early 2002 will use the same basic equations, but with a vertical coordinate based on the hydrostatic dry pressure (the Laprise coordinate) that varies with time. This coordinate imposes a different and more realistic free-surface upper boundary condition, and is also more similar to that used in many hydrostatic models. The model equations also strictly conserve scalars while allowing acoustic modes as with the height coordinate.

A variety of lateral-boundary condition options exist, including imposing real-data boundary values in a relaxation zone, and periodic, symmetric, and open-radiative conditions suitable for idealized studies.

The time-stepping scheme in these dynamical cores is the third-order Runge-Kutta technique that involves three advective evaluations to achieve this accuracy. Each of the three Runge-Kutta steps is subdivided into small steps for the sound-wave terms using the Klemp-Wilhelmson semi-implicit time-splitting technique for the acoustic modes.

The horizontal and vertical advection have options from second to sixth order accuracy. Higher order schemes have better shape preserving properties for scalar advection. These are flux-form equations to allow conservation.

In the future, other dynamical cores will be tested within the WRF framework and compared to the current two cores. These will include a semi-implicit semi-Lagrangian core with sophisticated numerics being developed at NCEP.

3.4 Physics

WRF is designed as a full-physics numerical model, and will provide multiple options for each physics category ranging from faster schemes for operational applications to more sophisticated schemes for research purposes. The primary categories are (i) microphysics, (ii) cumulus parameterization, (iii) planetary boundary layer, (iv) land surface, (v) radiation, (vi) sub-grid turbulence. So far physics packages have come from a variety of sources including MM5 and the Eta model. These have been converted to Fortran 90 modules, and also must obey strict rules to fit into the WRF plug-compatible physics interface. The physics interface defines which arrays are passed to the physics, which indices the loops must cover, the dimensions of local and state arrays, and the tendency arrays to be updated by the physics. Each physics module contains all the necessary subroutines to initialize and run a particular physics option, together with necessary local constants and work arrays.

To date, the physics options are:

Microphysics: Passive water vapor, Kessler, Lin *et al.*, NCEP 3-class, NCEP 5-class, Zhao-Carr.

Longwave radiation: Rapid Radiative Transfer Model, GFDL.

Shortwave radiation: Dudhia, Goddard, GFDL.

Surface Layer: MM5, Eta.

Land Surface: 5-layer MM5 soil thermal model, Oregon State University LSM.

Planetary Boundary Layer: Mellor-Yamada-Janjic, Hong-Pan,

Cumulus Parameterization: Kain-Fritsch, Betts-Miller-Janjic.

Sub-grid Turbulence: Constant diffusion, Smagorinsky-type, TKE.

Work continues to update these schemes and add new options.

3.5 Initialization

WRF currently can be run either as an idealized model or with a real-data analysis. To provide initial and boundary conditions for real-data applications, a pre-processor has been developed to read and interpolate Grib-format files to the chosen WRF domain, and to provide topographic and map information for the model. For idealized runs several modules have been developed as options. These include two-dimensional flow over a hill and squall lines, and three-dimensional supercell thunderstorms and baroclinic waves. These can be adapted by the user to any other idealized set-up.

Currently the real-data capability includes no method to improve the analysis with observational data, but a 3DVAR program is being developed as part of the WRF system that will allow ingest of a variety of data types.

3.6 Testing and Validation

WRF is being run at several centers already as a test real-time forecast model. Figure 2 shows a sample forecast of Tropical Storm Barry in 2001. Its precipitation scores are routinely being compared to those of the existing operational model at NCEP, and other models being run over the US (Figure 3). These indicate that WRF's skill already matches the other models by the standard measures. Additionally these

FIG. 2. Simulation of Tropical Storm Barry near Florida in real-time tests of WRF as a forecast model. 48 hr forecast verifying at 0000 UTC, August 6, 2001

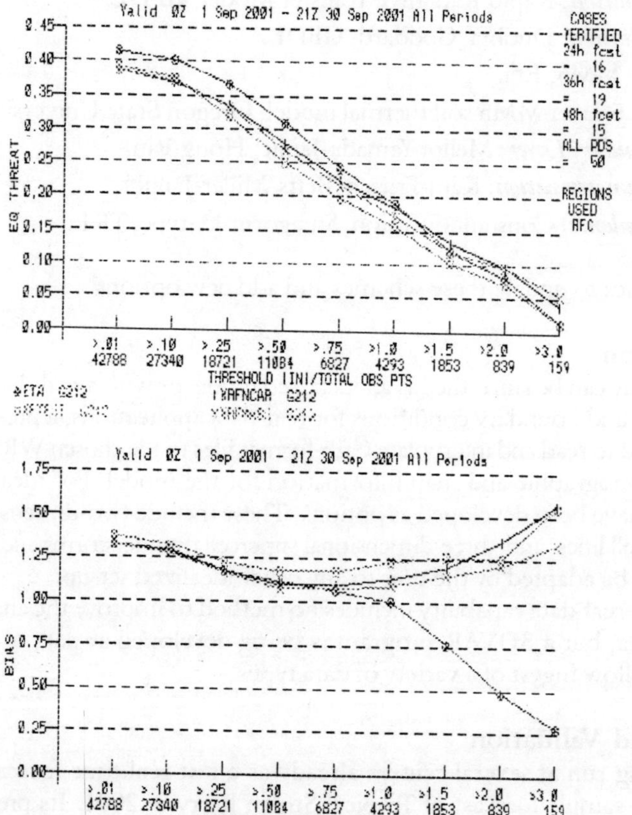

FIG. 3. Threat (top) and bias (bottom) scores for test WRF forecasts during September 2001 for the US domain. Thresholds in inches per 24 hrs.

tests have been a valuable tool to isolate initial bugs and allow the model's robustness to be tested. At NCAR WRF is being run on a 22 km US grid with both the height and mass vertical coordinate and different physics options twice per day. Additionally a 10 km run with finer vertical resolution is being run over the central US once per day.

Plans for validation include choosing extended periods of about a month that will serve as testbeds for improvements and various versions of WRF. The validation datasets will include precipitation and other observations, and benchmark scores of existing models.

3.7. Long-term Plans

The current version of WRF, released in November 2000, and updated in May 2001, is regarded as a "Basic Version" of WRF, and has been undergoing continual changes. In early 2002, the height coordinate will be replaced by the mass coordinate described above, and that will be released as the newest basic version. During 2002, primary goals are to include two-way nesting in WRF, and 3DVAR. There will also be additional physics options, and possibly new plotting capabilities in addition to the simple ones currently available.

Around the end of 2002, a "Research Version" is planned for release. This should be a fully tested robust code that can be used for research in a competitive manner with existing research models. NCAR will continue to be the center for user support for WRF, similarly to its current handling of MM5 support. Annual workshops and tutorials for WRF have already started, and a Web site is maintained for WRF information and to enable users to register themselves and download the code, which is freely available.

Dynamics cores will be continually evaluated within the WRF framework, and physics will be updated with newer packages as they become available. The physics interface and WRF code hierarchy separating the model layer from the driver layer, will make clean evaluations of different dynamical cores with the same physics possible.

In the longer term, the operational centers will evaluate WRF as a replacement for their existing models, and it is possible that by 2008 they will be running WRF as their primary forecast models.

4. POTENTIAL AREAS OF COLLABORATION

As described above, NCAR is playing an active role in mesoscale modeling research, development, and support in the United States. We have had several collaborations in recent years that have allowed us to gain experience with tropical environments, and these include major projects with Hong Kong and Taiwan in which local forecast systems were developed based on MM5. The primary issues in these collaborations have been the data void associated with the Pacific Ocean, and the initialization and forecasting of typhoons. Similar issues would pertain to the Indian forecasts. While some of the problems have been overcome, and there are now many published papers using MM5 successfully for tropical cyclone simulations, this is still an interesting area of research, particularly if applied to a new environment, such as the Indian Ocean and Bay of Bengal.

Heavy rainfall episodes associated with the monsoon are also of interest, and NCAR collaborations with Chinese and Korean scientists have focussed on their regional flooding cases. Again, India presents some unique problems in forecasting such events, and it is possible that much could be learned from Indian meteorologists with experience in these areas, and that a mesoscale model could be used to verify theories about conditions required for heavy rainfall.

Colorado includes a high-plains region and part of the Rocky Mountains, and the population is concentrated near the foothills of this range. Similarly India has a high population near the Himalayan foothills, that may be affected by similar weather, in particular high wind events and hailstorms. A model can be used to understand the dynamics associated with mountainous topography and its nearby plains, an area that has already been much studied in the context of Colorado using MM5. It would be interesting to compare and contrast the conditions for severe weather in northern India where the primary ranges are oriented west to east.

Acknowledgments. Many staff contribute to NCAR's mesoscale model development for MM5 and WRF and their support. These include Ying-Hua (Bill) Kuo, Joe Klemp, John Michalakes, Bill Skamarock, Dave Gill, Wei Wang, Al Bourgeois, Kevin Manning, Dale Barker, Chris Davis, Jim Bresch, Yong-Run Guo, Cindy Bruyere and others. Collaborations with researchers at The Pennsylvania State University, NOAA's Forecast Systems Laboratory, Air Force Weather Agency, and various other laboratories and universities have also contributed to these modeling systems.

Mesoscale Modeling of Convective Systems over India: Status and Scope

U.C. MOHANTY, M. MANDAL, A.K. DAS & A.P. DIMRI
Centre for Atmospheric Sciences, Indian Institute of Technology, Delhi
Hauz Khas, New Delhi – 110016, INDIA.
E-mail: mohanty@cas.iitd.ernet.in

ABSTRACT

This paper presents the status and future scope of non-hydrostatic regional/mesoscale modeling used for the simulation of intense convective processes over India. A number of leading institutes/organizations in India are presently using different non-hydrostatic mesoscale models for simulation of important convective weather events over the Indian region. Among these, MM5 model is the most common. Some of the results obtained at the Centre for Atmospheric Sciences, Indian Institute of Technology, Delhi (IITD) are presented to show the simulation / forecast skill of the model over the Indian region. Model shows reasonably good skill in simulating the tropical cyclones, western disturbances and heavy precipitation events over the region. In future, the model forecast skill could be increased significantly by improving/modifying the physical parameterization schemes and the model initial conditions.

1. INTRODUCTION

The atmosphere is a continuous fluid medium and is governed by the hydro-dynamical and thermo-dynamical equations (laws). In Numerical Weather Prediction (NWP), these equations are solved using numerical techniques. In a NWP model, the hydro-dynamical and thermo-dynamical equations are integrated in 3-dimension to determine the prognostic variables at some future time. Dynamic (dynamical framework), numerics (numerical techniques) and physics (parameterization of physical processes) are the major components of a NWP model. The last component is needed to take into account the effect of sub-grid scale (a scale smaller than model grid size) physical processes. In "parameterization", the effects of these processes are approximated in terms of large-scale (resolved) variables based on theory and experimental observations. Almost all NWP models include parameterization of atmospheric radiation, convection, boundary layer, surface processes and cloud microphysics. In addition to these, the global climate models (GCMs) parameterize very slow atmospheric processes as well viz., land surface changes, slow chemical processes etc. Further, as pointed out by Bjerknes (1904) the NWP is an initial value problem. The simulation / prediction of future state with NWP models depend on accurate and consistent representation of initial state of the atmosphere.

The weather events in tropics are developed through complex interaction of convective clouds. Physical forcing plays dominant role (compared to dynamical forcing) in development / modulation of these complex convective systems. In addition, the simple geostrophic balance does not hold good in equatorial tropics, causing the boundary layer friction and transients to play active role in the balance of the forces. Furthermore, providing accurate initial condition to NWP models is also a problem in tropics as it is a data sparse region with almost about 60% covered by oceans, mountains and desert. The above-mentioned aspects make NWP in tropics very challenging and are some of the reasons for the lack of forecast skill of the NWP models in tropics as compared to mid-latitude. The development of non-hydrostatic regional/mesoscale models that can be integrated at a very high resolution with physical processes either resolved explicitly or parameterized through sophisticated schemes has increased the forecast skill of NWP models to a large extent. Improvement of model initial condition through assimilation of new datasets from meteorological satellites also plays very important role in improving the forecast skill of the models.

Delineation of the fine structure of small scale / mesoscale phenomena are very important to improve the forecasting skill of NWP models used in tropics. For this purpose, the resolution of regional models have been increased to few kilometers. The hydrostatic approximation that is being used in large-scale global and most of the regional models may not hold good at this resolution. In addition, to represent small-scale phenomena such as thunderstorms, or convective clouds that have vertical accelerations that are not negligible compared to buoyancy forces, it is not valid to use hydrostatic approximation. A number of non-hydrostatic models have been developed in order to simulate intense mesoscale convective systems. Some of the widely used non-hydrostatic mesoscale models include the Advanced Regional Prediction System (ARPS) of Oklahoma University (Xue *et al.*, 1995), the Penn State University (PSU) / National Center for Atmospheric Research (NCAR) Mesoscale Model (MM5) (Dudhia, 1993), the National Center for Environmental Prediction (NCEP) Regional Spectral Model (RSM) (Juang et al, 1997), the Coupled Ocean-Atmosphere Mesoscale Prediction System of US Navy (COAMPS) (Hodur, 1997), the Mesoscale Compressible Community (MCC) model (Laprise *et al.*, 1997), the Colorado State University Regional Atmospheric Model (RAMS) (Pielke *et al.*, 1992). These models have been used at a very high resolution for the forecast/simulation of wide range of weather events in different parts of the world.

In this paper, section 2 presents a brief status of mesoscale modeling in India. Some of the results obtained using MM5 model for simulation of tropical cyclone, heavy precipitation and western disturbances are presented in section 3. A brief conclusion and future scope of improving forecast skill of mesoscale models are presented in section 4.

2. MESOSCALE MODELING ACTIVITIES IN INDIA

India is a tropical country and its economy is mainly based on agriculture. The southwest summer monsoon is the lifeline of Indian agriculture. Although, there are extensive studies on various observational aspects of Indian summer monsoon, the modeling studies have been limited. Requirement of huge computing resources have restricted the use of global models for coarser resolution.

Sub-grid scale physical processes and land surface heterogeneity play important role in the evolution and modulation of the weather events over the tropical Indian region. As mentioned earlier, a high-resolution non-hydrostatic regional/mesoscale model has shown higher forecast skill in simulating such weather events in other part of tropics and is expected to be so in the Indian region as well. The Regional Climate Models (RCM), developed with non-hydrostatic mesoscale as the base and includes the effect of urbanization and changes in surface characteristics can also be a useful tool for simulation of climate over the Indian region. To use these models, one needs to give special attention to the following two aspects. Firstly, the physical processes are to be parameterized / represented explicitly depending upon the resolution of the model and nature of the atmospheric processes. The physical parameterization schemes could be suitably modified / improved depending on the observational datasets obtained from field experiments and extensive sensitivity experiments. Secondly, the land surface characteristics are to be provided in greater detail as lower boundary condition to the model so as to have complete information about heterogeneity of soil type, vegetation cover, vegetation type etc. Further, inhomogeneous and rugged topography of the region, which has a large influence on the nature of distribution of monsoon rainfall over India, can be better represented in a high-resolution mesoscale model. A number of leading research institutes in India including Centre for Atmospheric Sciences, IITD, Indian Institute of Tropical Meteorology (IITM), Pune; National Centre for Medium Range Weather Forecasting (NCMRWF), New Delhi are using mesoscale models recently. The mesoscale/regional models used by some of the leading institutes/ organizations are provided in Table1. However, it is to be noted that the list in the table is prepared based on the informations provided by these organizations and therefore, it may not be a complete and extensive one to reflect all organizations / institutes involved in mesoscale modeling studies in India. In general, PSU/NCAR mesoscale model MM5 has been most widely used in simulating important weather events over India viz., tropical cyclones, monsoon depressions, heavy rainfall events, western disturbances etc. Besides MM5, other non-hydrostatic models like ARPS, ETA and RAM are also used. The NCMRWF has started using MM5 model for providing semi-operational forecast over mountainous northwest India from 1st January 2002. MM5 model has already showed its skill in simulating weather system like tropical cyclone, western disturbance and heavy rainfall. Use of physical parameterization schemes suitable for the region and improved initial condition through high-resolution data analysis can further increase the forecast skill of the model. A large number of sensitivity experiments are performed at IITD to find a suitable combination (among the schemes already implemented in the model) of cumulus convection, planetary boundary layer and radiation parameterization scheme in simulating tropical cyclone and western disturbances over the Indian region (Azadi et. al., 2001; Mandal et. al., 2003). Attempts are also made to improve the model initial condition through high-resolution analysis at IITD (Das *et al.*, 2003; Mandal *et al.*, 2002a) and Space Application Centre (SAC), Ahemdabad. Some experiments are performed at IITD and NCMRWF to study the impact of model horizontal resolution in simulation of different weather events over Indian region (Mandal *et al.* 2002b). NCMRWF have also explored the use of MM5 model at cloud resolving scale (at 1 km resolution) (Das, 2002)

3. RESULTS AND DISCUSSIONS

As mentioned in Table1, important weather events like tropical cyclone, heavy precipitation and western disturbances over the Indian region are simulated at IITD using MM5 model. A few of these results are presented in this section.

TABLE 1. Mesoscale modeling activities in India

Organisation	Mesoscale model(s)	Area of R&D
Center for Atmospheric Sciences, IIT Delhi	MM5 and ARPS	Tropical cyclone, monsoon depression, western disturbance, heavy rainfall events
National Centre for Medium Range Weather Forecasting, New Delhi	MM5 and ETA	Mountain meteorology, Semi--operational run
India Meteorological Department, New Delhi	High resolution LAM, FSU	Operational NWP forecast
Indian Institute of Tropical Meteorology, Pune	MM5 and RAMS	Tropical cyclone, monsoon depression and heavy rainfall events
Andhra University, Vishakapattnam	MM5	Tropical cyclone
Space Application Centre, Ahmedabad	MM5	Tropical convective systems
Space Physical Laboratory, Trivendrum	ARPS	Convective systems
Snow and Avalanche Study Establishment, Chandighar	MM5	Mountain Meteorology
Cochin University, Cochin	MM5	Role of Western Ghats in Modulating Indian Summer Monsoon

3.1 Tropical Cyclone

Tropical cyclones that originate from the Bay of Bengal are one of the deadliest weather events in the Indian region. Reasonably accurate prediction of intensity and track of these cyclones are of great importance and very challenging task to the meteorologists. Some of the results obtain from simulation of the super cyclone that crossed Orissa coast on 29th October 1999 are presented here. The initial vortex formed over the gulf of Thailand on 24th October 1999, intensified into a cyclonic storm at 0000 UTC on 26th October and lay centered at 13.5° N/96.5° E. Thereafter it moved almost in a straight line in the northwestward direction and crossed Orissa coast close to Paradip at 0530 UTC on 29 October. The model is integrated for 120 hours producing 5 days simulation of the storm. NCEP/NCAR reanalysis datasets interpolated to model grids are used as initial and boundary condition for model integration. A few of the results obtained from simulating the cyclone at 90 and 30 km model resolutions are presented below.

Model simulated mean Sea Level Pressure (SLP) (day-1 to day-3) are presented in Figure 1. On day-1, central pressure is simulated almost accurately by the model with central SLP of 999 hPa and 998 hPa at 90 km and 30 km resolution respectively

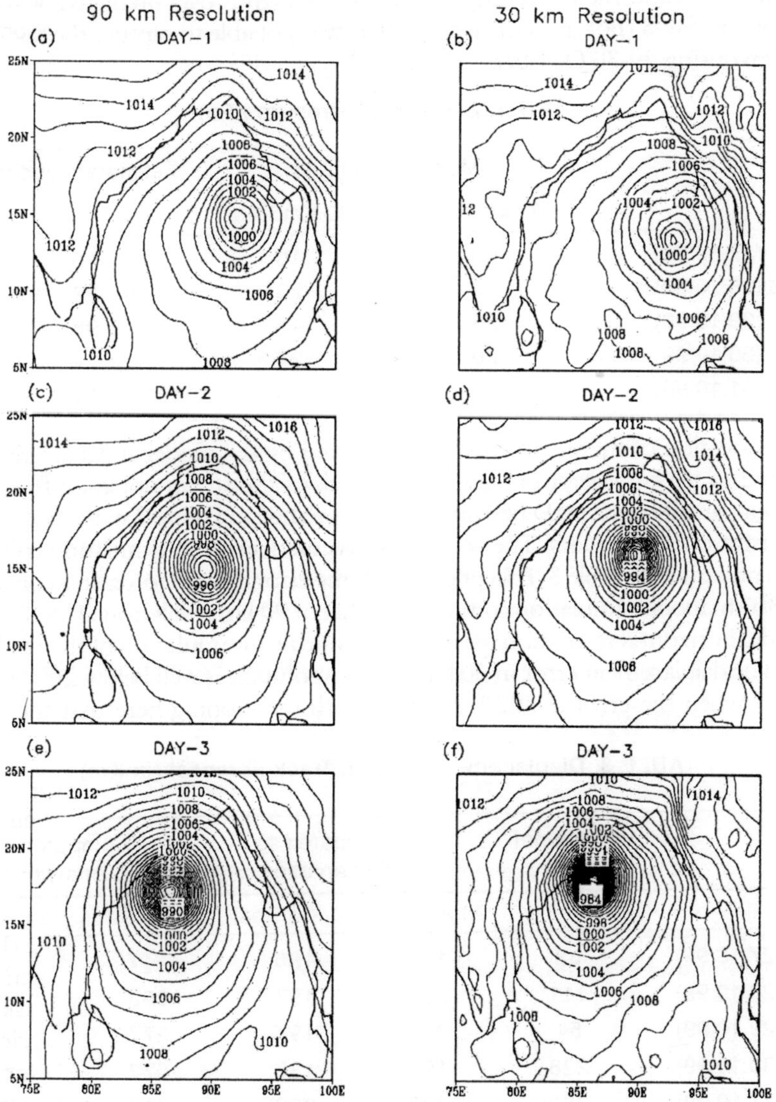

FIG. 1. Model simulated mean and sea level pressure (all at 0000 UTC)

compared to 998 hPa in the observation. On day-2, simulation shows 994 hPa and 985 hPa central SLP at 90 km and 30 km model resolution respectively compared to estimated central SLP of 986 hPa. This shows that the model horizontal resolution has notable impact in simulating the intensity of the storm. The impact of resolution becomes more significant on subsequent days simulation. On day-3, model at 30 km resolution simulates central SLP of 963 hPa compared to 983 hPa simulated by 90 km model resolution and 912 hPa estimated. Although, the intensity of the cyclone on day-3 is considerably under-predicted, the pressure drop of 45 hPa (Table 2) simulated by the model at 30 km resolution can be considered as reasonably good as far as simulation by NWP model is concerned. Intensity of the storm is slightly over predicted by the model on day-4 and day-5 (not shown), as the model was not able to capture the fast decay of the storm during 29-30 October.

TABLE 2 Pressure Drop (in hPa)

Time	Model Simulated		Estimated/Observed
	90 km Res.	30 Km Res.	
Day-1 (27.10.99)	09	10	10
Day-2 (28.10.99)	15	23	20
Day-3 (29.10.99)	25	45	98
Day-4 (30.10.99)	20	24	14
Day-5 (31.10.99)	12	15	12

The location of the storm is better simulated by the model at 30 km resolution compared to 90 km except on day-1. The track of the storm obtained from model simulation, analysis and IMD best-fit track is presented in Figure 2 with the location of the storm at every 24 hours. The vector displacement error in simulating track of the storm is presented in Table 3. Except on day-1, the error in track forecast is considerably less in 30 km model simulation compared to that of 90 km resolution. This shows that the model horizontal resolution plays important role in simulation of the track of the storm. The displacement error in 30 km model resolution is well below 200 km except on day-5 when the error is 302 km. It is worth mentioning here that the error in

TABLE 3 Displacement Error in Track Forecast (in Km)

Time	Positional Error	Compared to Observation		Compared to Analysis	
		90 km Res.	30 Km Res.	90 km Res.	30 km Res.
Day-1 (27.10.99)	63	155	196	217	247
Day-2 (28.10.99)	111	283	157	393	267
Day-3 (29.10.99)	54	272	197	277	216
Day-4 (30.10.99)	228	197	137	077	143
Day-5 (31.10.99)	350	402	302	312	318

Initial Positional Error: 122 km

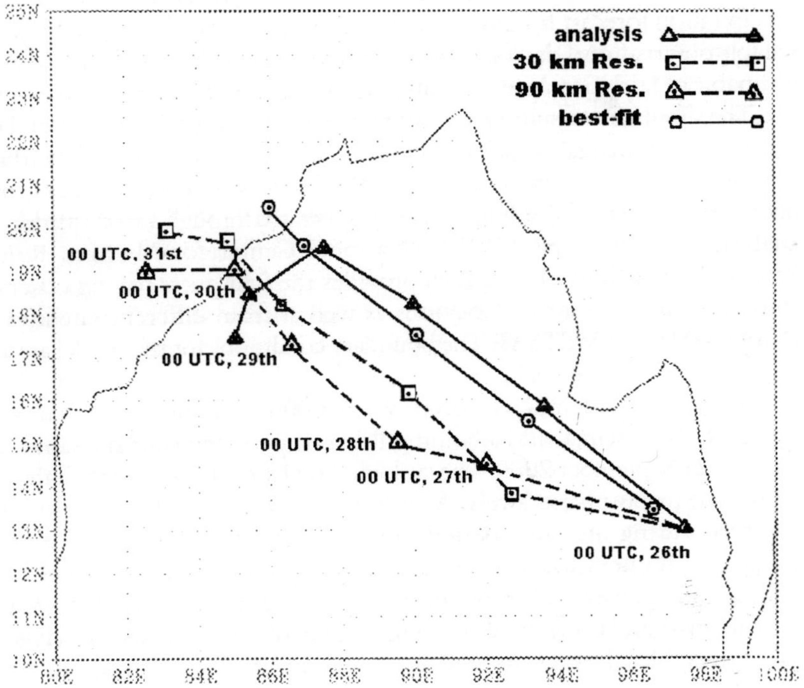

Fig. 2. Track of the storm during the period 26-30 October 1999.

specifying the initial position of the storm (at 0000 UTC on 26th October) itself is 122 km. In view of this, the track predicted by the model could be considered to be reasonable.

Model simulated precipitation over Orissa is compared with the precipitation recorded by IMD (not shown). It is found that the model resolution has significant impact in simulating precipitation as well. The magnitude and distribution of precipitation is simulated reasonably well by the model at 30 km resolution.

3.2 Heavy Precipitation Event

Rainfall activities over Indian subcontinent during summer monsoon depend on several convective systems viz. monsoon depressions, intense lows, squall lines, thunderstorms, offshore vortices and mid-tropospheric disturbances. Large-scale synoptic event like monsoon depression embedded with organized convection are instrumental in causing heavy precipitation events. But the low pressure systems less intense than the monsoon depression are also significant that are frequently responsible for substantial amount of precipitation within a short time span as well over a specific location. These low-pressure systems are complex in terms of genesis, intensification and temporal evolution. Although each event occurs in a large-scale flow that is favorable for cyclogenesis, several crucial details relating to the extreme mesoscale convective developments are to be identified and studied. Understanding of atmospheric processes and their interactions within the mesoscale convective systems that determine the size, scale, intensity and movement of the event as a whole, is necessary to provide quantitative precipitation forecast for any one location. If high-resolution geophysical details and reliable observational database are available, a mesoscale model with finer grid spacing is capable of fairly accurate simulation of certain heavy rainfall events.

The simulation results of a heavy rainfall event during 18-21 September 2000 over Gangetic West Bengal, which produced a rainfall amount of 100 cm within 3 days, are discussed here. The mesoscale model MM5 is integrated for 48 hours, after 12 hours nudging with the high-resolution analyses generated through a multiquadric interpolation analysis procedure. NCMRWF (National Center for Medium Range Weather Forecasting) analysis (1.5° x 1.5°) provides the first guess having observations from conventional upper-air observations as well as from different satellites viz. GOES, INSAT, GMS and METSAT. The boundary conditions for the model are also obtained from NCMRWF analysis.

A low pressure area was observed at 0000 UTC on 18 September 2000 over Gangetic West Bengal and adjoining Bihar. It persisted over the same region for 19 through 21 September 2000. The cyclonic circulation associated with the low extended up to mid-tropospheric levels. A cyclonic circulation in the lower troposphere was observed during the same period in southwest Bay of Bengal off and over coastal Tamilnadu. It interacted co-operatively with the circulation over Gangetic West Bengal leading and intensified. Strong south-southwesterlies of 15-17 ms^{-1} were observed in coastal Orissa and Gangetic West Bengal region on 18-19 September (Figure 3) leading to considerable advection of moisture over Gangetic West Bengal and Bihar as seen in

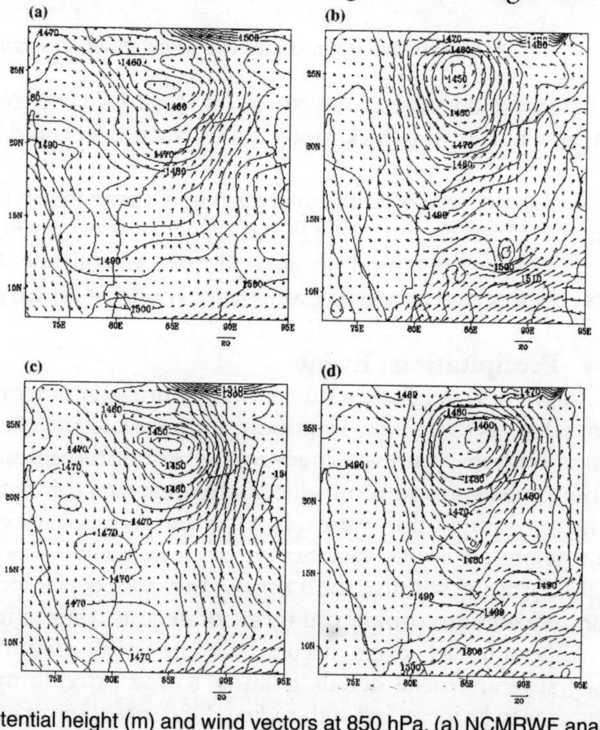

FIG. 3. Geopotential height (m) and wind vectors at 850 hPa. (a) NCMRWF analysis and (b) simulated fields after 24 hours valid on 0000 UTC 18 September, 2000 (c) NCMRWF analysis and (d) simulated fields after 48 hours valid on 0000 UTC 19 September, 2000

the verifying analysis as well as 24 and 48 hour predictions. Figure 4 shows an anti-cyclonic circulation located over North West Madhya Pradesh and adjoining Uttar Pradesh in the analysis at 250 hPa on 18th and 19th September 2000. However, as seen, the location of the center is shifted eastward in the simulated fields compared to the NCMRWF analysis. The center of the anti-cyclone gradually shifts west-northwest wards after remaining at the same position for the first 2 days. Over Gangetic West Bengal and neighborhood, this anticyclonic circulation is expected to provide considerable divergence and the northerly winds. The cold northerlies can create considerable hydrostatic thermal instability. The entrance region of quasi-stationary high-level divergence and the axis of strong low-level moisture convergence feeding the surface boundary usually determine the position of mesoscale convective system.

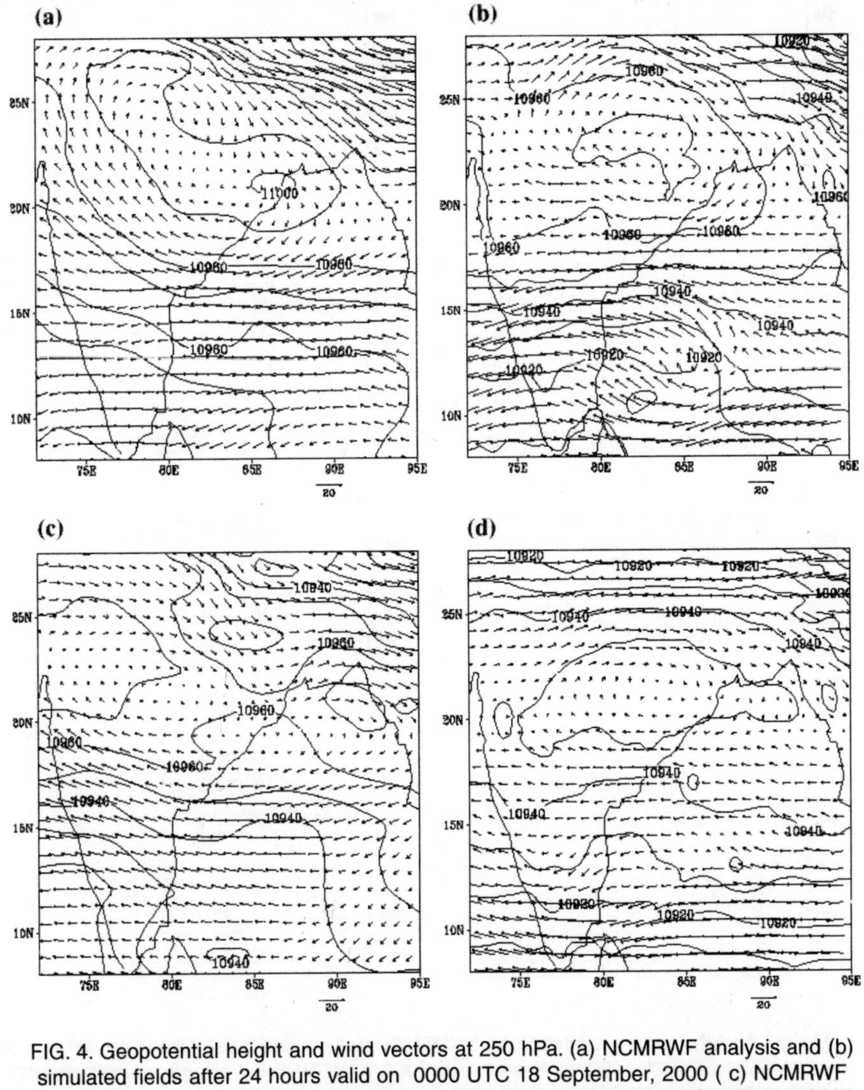

FIG. 4. Geopotential height and wind vectors at 250 hPa. (a) NCMRWF analysis and (b) simulated fields after 24 hours valid on 0000 UTC 18 September, 2000 (c) NCMRWF analysis and (d) simulated fields after 48 hours valid on 0000 UTC 19 September, 2000

Therefore, the heaviest convective rainfall usually occurs in the region of high moistur and maximum ambient or elevated instability, best lift and slow movement of the systen Accumulated precipitation amounts plotted in Figure 5 represent the model simulatio against the rainfall analysis from NCMRWF. The MM5 simulated rain shows mor features as well as produces more convective rain compared to the analysis. This coul be attributed to the fact that the rainfall analysis is done at a lower resolution (150 km) The locations of heavy precipitation and distribution of rainfall are represented i greater detail in the simulated results. The simulation brings out the two major pocket of heavy rainfall situated at the northwest and southeast sectors of the system. Whil

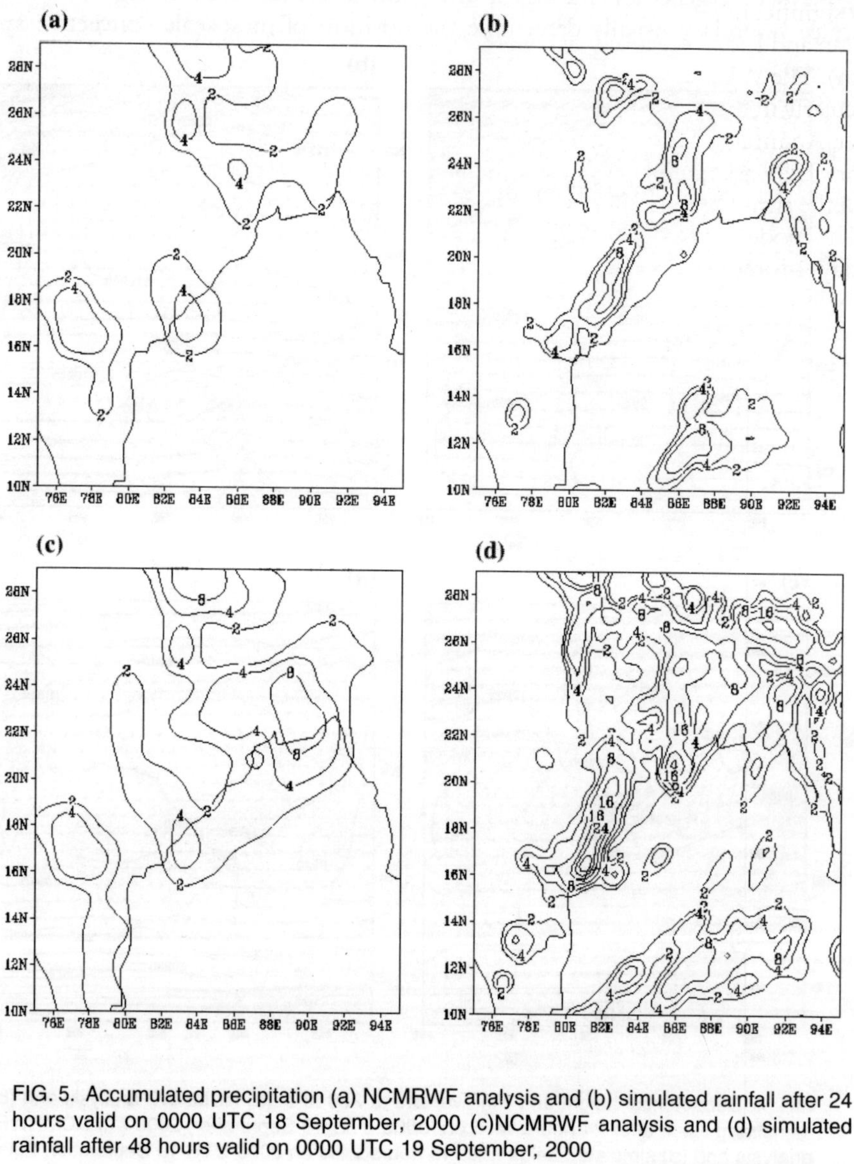

FIG. 5. Accumulated precipitation (a) NCMRWF analysis and (b) simulated rainfall after 24 hours valid on 0000 UTC 18 September, 2000 (c)NCMRWF analysis and (d) simulated rainfall after 48 hours valid on 0000 UTC 19 September, 2000

he first sector receives rain due to the thermal instability and orographic lifting, the ther sector of rain owes its existence to the strong moisture convergence that is upported by low-level mean flow. As a consequence, very heavy precipitation, ranging rom 15 to 60 cm in 24 hours, was reported from many stations in Gangetic West 3engal, Bihar and Orissa on all days during the period 18 through 21 September 2000.

3.3 Western Disturbance

Western disturbances (WDs) are eastward moving low-pressure systems in the ower troposphere originating in the Mediterranean Sea or in mid west Atlantic Ocean. WDs significantly affect weather over the countries such as Iran, Iraq, Afghanistan, Pakistan and India, especially during the post-monsoon and winter periods (October-March). The WDs cause heavy precipitation during winter months in northwest India, articularly in the western sector of the Himalayas, usually in the form of snow, during winter. An intense WD system that affected northwest India during the period 21-25 anuary 1999 associated with heavy precipitation is simulated using MM5 model and a ew of the simulation results are presented here.

The model simulated 24-hour accumulated precipitation for day-1, day-2, day-3 and day-4 forecasts and corresponding observations are shown in Figure 6. Similar

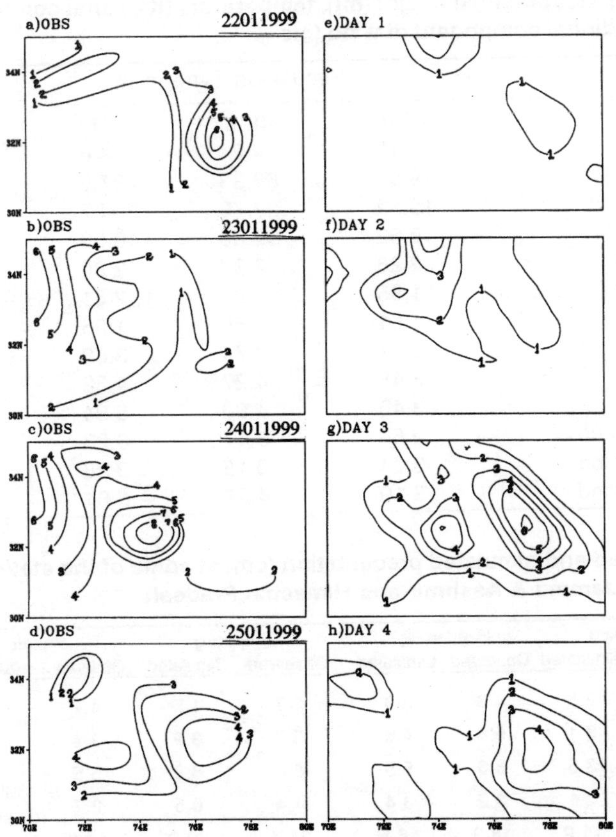

FIG. 6. Precipitation (cm) valid at 0300 UTC (a) Observed (b) Model simulated

comparison for some of the selected stations are presented in Table 4. Figure 6 shows the simulated precipitation maxima are located along the NW-SE oriented axis with few pockets of higher precipitation. Though the region under consideration has relatively fewer observations, the orientations of model simulated precipitation maxima as well as overall cellular structure of precipitation belt agrees reasonably well with the observation. The difference in magnitude of precipitation in the forecast and observation could be due to the complex topography over the said region. While comparing the precipitation amounts at some of the stations, it is seen that the model is able to simulate them reasonably well. During these four days under consideration, model simulated 24 hours accumulated precipitation amount was seen to be considerably closer to the corresponding observations (see Table 4). Although the precipitation is seen to be slightly under predicted on day 1 for, day 2 onwards the prediction comes closer to the observations.

In order to evaluate the performance of the model some skill scores are also calculated. Root mean square (rms) errors associated with some of the basic meteorological parameters are presented in Table 5. Model simulated sea level pressure shows good agreement with the verification analysis. The geo-potential height at 500

TABLE 4 Root mean square errors for the day 1 to day 4 in model simulation of sea level pressure (hPa), Geopotential height (mt), temperature (K), zonal component of wind (ms^{-1}) and meridional component of wind (ms^{-1}).

Variables	Simulation Time in hr			
	24 hr	48 hr	72 hr	96 hr
Sea Level pressure	4.13	4.24	4.6	4.04
850 hPa GPH	19.57	66.33	37.6	15.52
500 hPa GPH	15.32	22.70	20.87	19.54
200 hPa GPH	19.62	36.86	48.59	58.70
850 hPa temperature	2.26	2.37	2.86	2.87
500 hPa temperature	1.85	2.21	2.61	2.81
200 hPa temperature	1.61	1.86	1.95	1.96
850 hPa zonal wind	3.17	3.74	3.80	3.83
500 hPa zonal wind	3.48	4.27	4.52	4.80
200 hPa zonal wind	3.46	4.90	6.94	8.55
850 hPa meridional wind	2.64	3.17	3.23	3.46
500 hPa meridional wind	3.04	3.15	3.28	3.47
200 hPa meridional wind	3.19	4.29	5.81	5.96

TABLE 5. Observed and simulated precipitation (cm) at some of the station in Jammu & Kashmir and Himachal Pradesh

	Sonamarg		Haddan Taj		Gulmarg		Manali	
	Observed	Simulated	Observed	Simulated	Observed	Simulated	Observed	Simulated
Day 1	2.1	2.5	4.3	3.3	2.7	1.7	4.7	3.7
Day 2	3.9	3.1	5.8	4.6	5.7	3.9	3.4	2.7
Day 3	4.7	3.5	5.0	5.5	8.6	6.8	5.5	4.9
Day 4	2.5	2.1	3.2	3.4	4.4	6.5	2.7	2.1
Accumulated	13.2	11.2	18.3	16.8	21.4	20.9	16.3	13.4

hPa shows that model does not show much deviation from the verifying analysis as integration time progresses. At 200 hPa level rms error in forecast grows with model integration and in day 4 reaches a maximum value of 58.7 meters. At 850 hPa, forecast errors grow on day 1 till day 2 and decrease on day 3 and 4. In particular, the error on day 2 is higher. This inconsistency may be attributed to the complex topography over the region and the fact that many of the stations are above 850 hPa level. The rms errors associated with temperature, zonal and meridional component of wind are found to increase consistently with time. The magnitude of the errors show that the meteorological fields in the middle and upper troposphere are well simulated by the model. Comparatively lower forecast skill of the model in simulating fields at lower troposphere and surface is due to the complex orography over the region.

4. CONCLUSIONS AND FUTURE SCOPE

A brief review on the use of mesoscale models in India during last half decade reflects significant progress in this direction in the country. A number of leading research institutes / organizations have started using different high-resolution non-hydrostatic mesoscale models for simulation of convective weather systems over the Indian region.

PSU/NCAR mesoscale model MM5, is most widely used for the simulation of important weather events over the Indian region. The results presented in section 3, show that the model could simulate the track and intensity of the tropical cyclones, the rainfall associated with the tropical cyclones, western disturbances and monsoon low-pressure systems with reasonable accuracy. There is immense scope to improve the efficiency of the model in simulating these events.

The large-scale analysis used as initial and boundary condition for the mesoscale models can not represent the small-scale features that are inherent part of the tropical weather systems. Therefore, high-resolution analysis is very much necessary to improve the model initial condition ingesting conventional and non-conventional observations from satellites, radars, buoys, aircraft etc. Consequently, the forecast skill of the model can be improved further through 4-dimensional data assimilation using different sophisticated techniques.

There is an urgent need to setup meso-network of observations at some selected sites of specific interest. These special observations will facilitate the validation/ improvement of parameterization schemes for different physical processes. In addition, model performance will be better evaluated using these observations.

As the surface processes and the land surface inhomogeneity in terms of surface gradient plays crucial role in initiation / modulation of the tropical convective systems, the high resolution and accurate surface parameters viz., soil type, soil temperature, soil moisture, vegetation type, vegetation cover etc are to be obtained from satellites / conventional observations.

Extensive sensitivity experiments and model diagnostic studies are to be performed for better understanding of the physical processes and the role of geophysical surface characteristics. These studies will help to identify the changes to be incorporated in the model towards improving its forecast skill.

Ensemble / super ensemble approach using a number of mesoscale models will further reduce the systematic forecast errors over the Indian region.

Acknowledgements: NCAR USA is acknowledged for providing the model and large-scale analysis used in the ongoing research. The authors sincerely acknowledge NCMRWF and IMD for assistance towards data and computing facilities.

REFERENCES

Azadi M., U.C Mohanty, O.P. Madan and B. Padmanabhamurty, 2001: Prediction of precipitation associated with a western disturbance using a high-resolution regional model: Role of parameterization of physical processes. *Meteorol. Appl., 7,* 317-326.

Bjerknes V. ,1904: Das problem der wettervorhersage, betrachlet vom standpunkte der mechanik und der physik. *Meteor. Z.,* **21**, 1-7.

Das, A.K., U.C. Mohanty, S. Das , M. Mandal and S.R. Kalsi , 2003: A Study on Circulation Characteristics of Monsoon Depression during BOBMEX-99: Mesoscale Analysis. *Proc. Indian Acad. Sci., Earth and Planetary Sciences (In Press).*

Das S., 2002: Mesoscale and cloud resolving scale simulation of a heavy precipitation episode and associated cloud system using MM5 model. *INDO-US workshop on weather and climate modeling, 7-9 February, 2002. NCMRWF, New Delhi.*

Dudhia J. ,1993 : A non-hydrostatic version of the Penn State-NCAR mesoscale model. *Mon. Wea. Rev.,* **121**, 1493-1513.

Hodur, R.M.,1997: The U.S. Navy's Coupled Ocean/Atmosphere Mesoscale Prediction System (COAMPS), *Mon. Wea. Rev.,* **125**, 1414-1430.

Juang, H.-M. H., Hong, S.-Y. and Kanamitsu, M.,1997: The NCEP regional spectral model: an update. *Bulletin Amer. Metero. Soc.,* **78**, 2125-2143.

Laprise, R., D. Caya, , G. Bergeron, and M. Giguere, 1997: The formulation of Andre Robert MC2 (Mesoscale Compressible Community) Model. *Atmosphere-Ocean Special Vol. XXXV No1,* 195-220.

Mandal M., U.C.Mohanty, and S. Raman , 2003: A Study on the Impact of Parameterization of Physical Processes on Prediction of Tropical Cyclones over the Bay of Bengal with NCAR/ PSU Mesoscale Model (MM5). *Natural Hazard (In Press).*

Mandal M., U.C.Mohanty, A.K. Das and S. Behera ,2002a: Mesoscale Analysis and Tropical Cyclone Forecast: A Case Study. *Workshop on 'Mesoscale modeling with special emphasis on mountain weather forecasting',* NCMRWF, New Delhi, July 2002.

Mandal M. and U.C. Mohanty ,2002b: Simulation of Orissa super-cyclone using MM5 model at different resolution. *TROPMET-2002,* Bhubaneswar, Orissa, India, February 2002.

Pielke, R.A., W. R Cotton, R. L.Walko, C. J. Tremback, W. A Lyons, L. D. Grasso, M.E. Nicholls, M.D. Moran,D. A.Wesley, T.J. Lee, and J.H. Copeland, 1992: A comprehensive meteorological modeling system — RAMS. *Meteor. Atmos. Phys.,* **49**, 69-91.

Xue, M., K.K. Droegemeier, V. Wong, A. Shapiro and K. Brewster, 1995: ARPS Version 4.0 User's Guide. Center for Analysis and Prediction of Storms, [Available from CAPS, Univ. of Oklahoma, 100 E. Boyd St., Norman OK 73019], 380 pp.

Mesoscale Convective Systems during SCSMEX: Simulations with a Regional Climate Model and a Cloud-Resolving Model

W.-K. TAO[1], Y. WANG[2], J.-H. QIAN[3], C.-L. SHIE[4], W. K.-M. LAU[1] & R. KAKAR[5]

[1]Laboratory for Atmospheres
NASA/Goddard Space Flight Center
Greenbelt, MD
[2]*Joint Center for Earth Systems Technology*
University of Maryland, Baltimore County
Baltimore, MD
[3]*International Research Institute for climate prediction*
LDEO of Columbia University
Palisades, NY
[4]*Goddard Earth Sciences and Technology Center*
University of Maryland, Baltimore County
Baltimore, MD
[5]*NASA/Headquarters*
Washington DC
E-mail: tao@agnes.gsfc.nasa.gov

ABSTRACT

The South China Sea Monsoon Experiment (SCSMEX) conducted in 1998 was aimed towards better understanding of the role of key physical processes for the onset and evolution of the summer monsoon over Southeast Asia and southern China. A regional scale model (with grid size of 20 km) and Goddard Cumulus Ensemble (GCE) model having 1 km grid size are utilized and integrated to understand the precipitation processes during SCSMEX especially the soil precipitation interaction and feedback processes associated with the flood events that occurred around China's Yantz River during SCSMEX. In addition, several sensitivity tests pertaining to various land surface models, cumulus parametrization schemes, sea surface temperature (SST) variations and mid-latitude influences are also performed. These tests have indicated that the land surface model has a major impact on the circulation over the S. China Sea. Cloud processes can effect the precipitation pattern while SST variation can effect the precipitation amounts over both land and ocean. The exact location (region) of the flooding can be effected by the soil-rainfall feedback.

The GCE model captured many observed precipitation characteristics due to its finer resolution. For example, the model simulated rainfall temporal variation compared quite well to the sounding-estimated rainfall. The results show there are more latent heat fluxes prior to the onset of the monsoon. However, more rainfall was simulated after the onset of the monsoon.

This modeling study indicates the latent heat fluxes (or evaporation) have more of an impact on precipitation processes and rainfall in the regional climate model simulations than in the cloud-resolving model simulations. Research is underway to dertermine if the difference in the grid sizes or the moist processes used in these two models is responsible for the differing influence of surface fluxes on precipitation processes.

1. INTRODUCTION

The South China Sea Monsoon Experiment (SCSMEX) was conducted in May-June 1998. One of its major objectives is to better understand the key physical processes for the onset and evolution of the summer monsoon over Southeast Asia and southern China (Lau *et al.* 2000). Multiple observation platforms (e.g., soundings, Doppler radar, ships, wind profilers, radiometers, etc.) during SCSMEX provided a first attempt at investigating the detailed characteristics of convection and circulation changes associated with monsoons over the South China Sea region. SCSMEX also provided precipitation derived from atmospheric budgets (Johnson and Ciesielski 2003) and comparison to those obtained from the Tropical Rainfall Measuring Mission (TRMM).

In this paper, a regional climate model and a cloud-resolving model are used to perform multi-day integrations to understand the precipitation processes associated with the summer monsoon over Southeast Asia and southern China. The regional climate model is used to understand the soil-precipitation interaction and feedback associated with a flood event that occurred in and around China's Yantz River during SCSMEX. Sensitivity tests on various land surface models, cumulus parameterization schemes (CPSs), sea surface temperature (SST) variations and midlatitude influences are also performed to understand the processes associated with the onset of the monsoon over the S. China Sea during SCSMEX.

Cloud-resolving models (CRMs) use more sophisticated and physically realistic parameterizations of cloud microphysical processes with very fine spatial and temporal resolution. One of the major characteristics of CRMs is an explicit interaction between clouds, radiation and the land/ocean surface. It is for this reason that GEWEX (Global Energy and Water Cycle Experiment) has formed the GCSS (GEWEX Cloud System Study) expressly for the purpose of improving the representation of the moist processes in large-scale models using CRMs. Progress in studying precipitating convective systems in GCSS is reported in Moncrieff *et al.* (1997) and Moncrieff and Tao (1999), Moncrieff (2003). The Goddard Cumulus Ensemble (GCE) model is a CRM and is used to simulate convective systems associated with the onset of the South China Sea monsoon in 1998. The GCE model includes the same land surface model, cloud physics, and radiation scheme used in the regional climate model. A comparison between the results from the GCE model and regional climate model is performed.

2. MODEL DESCRIPTION

2.1 Regional Climate Model

A Regional Land-Atmosphere Climate Simulation (RELACS) System is being developed and implemented at NASA Goddard Space Flight Center. One of the major goals of RELACS is to use a regional-scale model with improved physical processes, in particular land-related processes, to understand the role of the land surface and its interaction with convection and radiation as well as the water and energy cycles in Indo-China, the South China Sea (SCS), China, N. America and S. America.

The Penn State/NCAR MM5 atmospheric modeling system, a state of the art atmospheric numerical model designed to simulate regional weather and climate, has been successfully coupled to the Goddard Parameterization for Land-Atmosphere-Cloud Exchange (PLACE) land surface model. PLACE allows for the effects of vegetation, and thus important physical processes such as evapotranspiration and interception are included. The PLACE model incorporates vegetation type and has been shown in international comparisons to accurately predict evapotranspiration and runoff over a wide variety of land surfaces. The coupling of MM5 and PLACE creates a numerical modeling system with the potential to more realistically simulate the atmosphere and land surface processes including land-sea interaction, regional circulations such as monsoons, and flash flood events. In addition, the Penn State/NCAR MM5 atmospheric modeling system has been: (1) coupled to the Goddard ice microphysical scheme, (2) coupled to a turbulent kinetic energy (TKE) scheme, (3) coupled to the Goddard radiation scheme, (4) modified to ensure cloud budget balance, and (5) has incorporated initialization with the Goddard EOS data sets at the NASA/Goddard Laboratory for Atmospheres. Table 1 shows the major characteristics of MM5.

RELACS is used to simulate the onset of the South China Sea monsoon in 1986, 1997 and 1998. RELACS is also used to understand the soil-precipitation interaction and feedback associated with a flood event that occurred in and around China's Yantz River during 1998. The model is typically run within a domain covering 6000-8000

TABLE 1 Characteristics of the PSU/NCAR MM5. Processes modified by the Goddard Mesoscale Modeling and Dynamics Group are marked with an *

Parameters/Processes	Penn State/NCAR MM5
Vertical Coordinate	sigma-p (z)
Explicit Convective Processes	1-2 class water & 1-2 class ice*3 ice scheme**
Implicit Convective Processes	Cumulus ParameterizationSchemes
Numerical Methods	Leapfrog in Time and 2nd Order Scheme in Advection for both Scalar and Dynamic Variables*Off-line positive definite transport**
Initialization	*Goddard DAO**, NCEP and ECMWF Analyzed (and/or Bogus) Data
FDDA	Nudging/Adjoint*Rainfall**
Radiation	Emissivity Method in LW*Goddard k-distribution and four-stream discrete-ordinate scattering (8 bands)**
Sub-Grid Diffusion	Smagorinski
Planetary BoundaryProcesses	Blackadar PBL or Dry Convective Adjustment*TKE (PSU)**
Topography	Sigma-p (z)
Two-Way Interactive Nesting	Traditional but with Monotonic Interpolation
Surface Energy Budget	Force-restore Method or*7-Layer Soil Model (PLACE)**

km by 6000-8000 km using interactive nesting techniques with multiple grid refinements and a nesting ratio of 3:1.

2.2 Cloud-Resolving Model

The Goddard Cumulus Ensemble (GCE) Model, a cloud-resolving model, has been developed and improved at NASA/Goddard Space Flight Center over the past two decades. Novel characteristics of the GCE model are the explicit representation of warm rain and ice microphysical processes, and their complex interactions with solar and infrared radiative transfer processes, and with surface processes. The GCE model is being linked with other physical models such as: passive microwave radiative transfer and spaceborne precipitation radar models for the purposes of developing and improving retrieval algorithms of precipitation and latent heat release, and a photo-chemistry model to assess the impact of vertical transport and mixing of important trace species on O3 production/reduction processes. More than 90 GCE modeling papers were published over the last two decades (Tao 2003). Table 2 shows the GCE model's major characteristics.

TABLE 2. Characteristics of the Goddard Cumulus Ensemble Model.

Parameters/Processes	GCE Model
Vertical Coordinate	z
Explicit Convective Processes	2 class water & 2 moment 4 class iceSpectral Bin Microphysics
Implicit Convective Processes	Betts & Miller or Kain & Fritsch
Numerical Methods	Positive Definite Advection for Scalar Variables;4-th Order for Dynamic Variables
Initialization	Initial Condition with Forcingfrom Observations/ Large-Scale Model
FDDA	Nudging
Radiation	k-distribution and four-stream discrete-ordinate scattering (8 bands)Explicit Cloud-Radiation Interaction
Sub-Grid Diffusion	TKE (1.5 order)
Topography	Sigma-z*
Two-Way Interactive Nesting	Radiative-Type
Surface Energy Budget	Force-restore Method7-Layer Soil Model (PLACE)TOGA COARE Flux Module

The equations that govern cloud-scale motion (wind) are anelastic by filtering out sound waves. The subgrid-scale turbulence used in the GCE model is based on work by Klemp and Wilhelmson (1978). In their approach, one prognostic equation is solved for subgrid kinetic energy, which is then used to specify the eddy coefficients. The effect of condensation on the generation of subgrid-scale kinetic energy is also incorporated in the model (see Soong and Ogura 1980; Tao and Soong 1986 for details). The cloud microphysics include a parameterized Kessler-type two-category liquid water scheme (cloud water and rain), and a parameterized Lin *et al.* (1983) or Rutledge and Hobbs (1984) three-category ice-phase scheme (cloud ice, snow and hail/graupel) (see Tao and Simpson 1993 and Tao *et al.* 2003a for a detailed description of the cloud microphysics). Shortwave (solar) and longwave (infrared) radiation parameterizations are also included in the model (Tao *et al.* 1996). The TOGA-COARE bulk flux algorithm (Fairall *et al.* 1995) is linked to the GCE model to calculate the surface fluxes (Wang *et al.* 1996, 2002). All scalar variables (potential temperature, mixing ratio of water vapor, turbulent coefficients, and all five hydrometeor classes) use forward time differencing and a positive definite advection scheme with a non-oscillatory option (Smolarkiewicz and Grabowski 1990). The dynamic variables, u, v and w, use a second-order accurate advection scheme and a leapfrog time integration (kinetic energy semi-conserving method). Details of the model description and tests can be found in Tao and Simpson (1993) and Tao *et al.* (2003a).

2.3 Water Budget Analysis

To quantify the moisture processes, the regional scale model's hydrology budget was calculated horizontally and vertically integrating the continuity equation for atmospheric moisture around a rectangular domain, which yields:

$$W = FW - FE + FS - FN + E - P - R$$

Where W is the total change of atmospheric water vapor which is generally a small term; FW, FE, FS, and FN are the vertically integrated moisture fluxes along the western, eastern, southern, and northern boundaries of the budget calculation domain, respectively; $Q = FW - FE + FS - FN$ is the total moisture convergence in the domain; E is the total surface evaporation, P is the total precipitation; and R is the residual term.

3. RESULTS

3.1 Regional Climate Model Simulations

(a) South China Sea (1986, 1997 and 1998)

Prior to SCSMEX (1998), a pilot study was performed to determine the model design (i.e., size of domain, grid size), physical processes and initial data that are needed for the successful simulation of convective activity over Indo-China and the South China Sea. The MM5 with two nested domains (60 and 20 km horizontal resolution) was used to simulate convective activity over Indo-China and the South China Sea, during the monsoon season, from May 6 to May 20, 1986. Sensitivity tests indicated that a large domain was needed for good simulations (Figure 1). The model results captured several dominant observed features, such as twin cyclones, a depression system over the Bay of Bengal, strong south-westerly winds over Indo-China before and during the onset of convection over the SCS, and a vortex over the SCS. Two additional MM5 runs with different land process models, Blackadar and MRF, were performed, and their results are compared to the run with PLACE. The results indicate that the MM5 results using PLACE and Blackadar are in very good agreement, but the results using MRF do not contain the south-westerly wind over Indo-China prior to the onset of convection over the SCS. However, no surface rainfall estimates/observations were available for model validation for the pilot case.

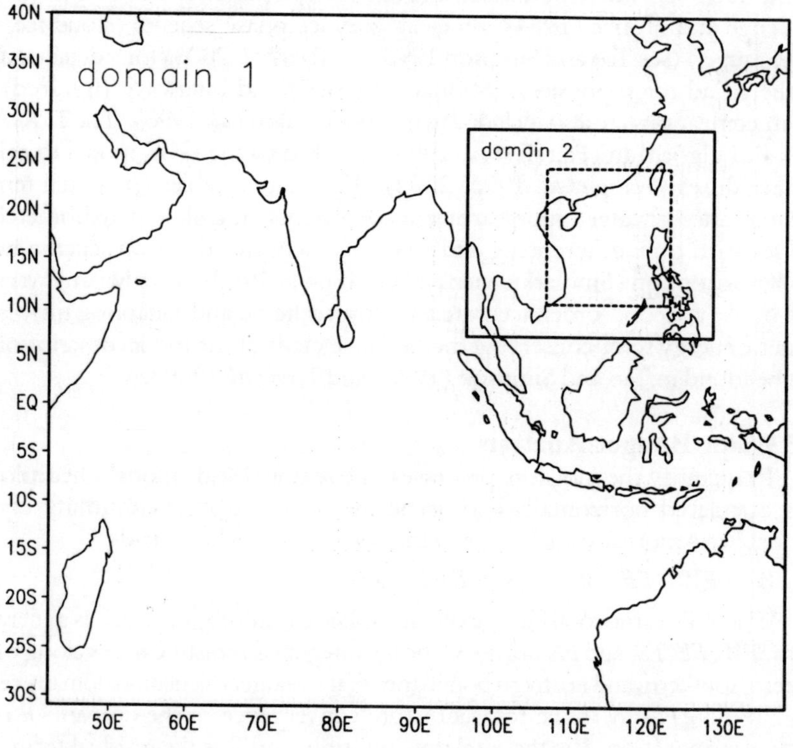

FIG. 1. Nested configuration used by the model. Horizontal resolutions for domains 1 and 2 are 60 and 20 km, respectively. The water budget is calculated over the domain marked with dashed lines.

The onset of the South China Sea monsoon (SCSM) in 1997 and 1998 has also been studied using RELACS. In both 1997 and 1998, signals for the onset of the monsoon were strong (in other cases, 1999 for example, the signal is very week). The pre-onset is also associated with the development of a "double cyclone" in the Bay of Bengal as with the pilot case (1986). The northern cyclone strengthened and the southern dissipated. When the northern cyclone reached the northwest part of the Indo-China Peninsula, it spun up a strong southwestly flow across the peninsula. The onset of the monsoon in both cases coincided with the arrival of the strong southwest winds in the South China Sea.

In 1998, beginning on May 15, strong convection occurred around the coast of southern China associated with a cold front pushing to the southeast to the northern part of the SCS. Convection intensified and rainfall peaked on May 17. The southeast circulation only penetrated into the western part of the SCS with the western part of the SCS still under the control of a West Pacific high-pressure system. By the 21st of May, the SCS was completely occupied by the southwest circulation. By the 23rd of May, a large-scale wind vortex became established over the northern part of the SCS. This vortex then propagated southeastward to the southern part of the SCS.

The onset of the 1997 monsoon is different from the one in 1998 in the following ways. First, the penetration of southwesterly winds into the SCS is much stronger.

The onset is also much more abrupt than in 1998. The strong southwestly winds completely pushed the west Pacific high-pressure system out the SCS region during the first two days of the onset (May 19-20). Secondly, the northern cold front from southern China was also stronger than that in 1998. Third, the precipitation was much more intense in 1997 during the onset period. Due to the stronger circulation and much quicker onset, the water budget for the region was also very different.

A series of two-week numerical experiments have been carried out to study the effects of local land-cover changes on convective activity over Indo-China and the South China Sea, during the monsoon season, from May 18 to May 29, 1997. The temporal variation of simulated rainfall compared well with Global Precipitation Climatology Project (GPCP)[1] estimates (Figure 2a). The maximum rainfall that occurred on May 24 and 25 was well simulated. The run without PLACE produced more rainfall compared to GPCP, and its maximum occurred on May 25. Sea surface temperature (SST) variation[2] only has a small impact on rainfall.

The Tropical Rainfall Measuring Mission (TRMM) was launched on November 1997. For 1998 case, the RELACS simulated rainfall can be compared with TRMM[3] and GPCP rainfall products. Both satellite-derived rainfall products showed several peaks in rainfall over the S. China Sea in May 1998 (Figure 2b). Both also showed

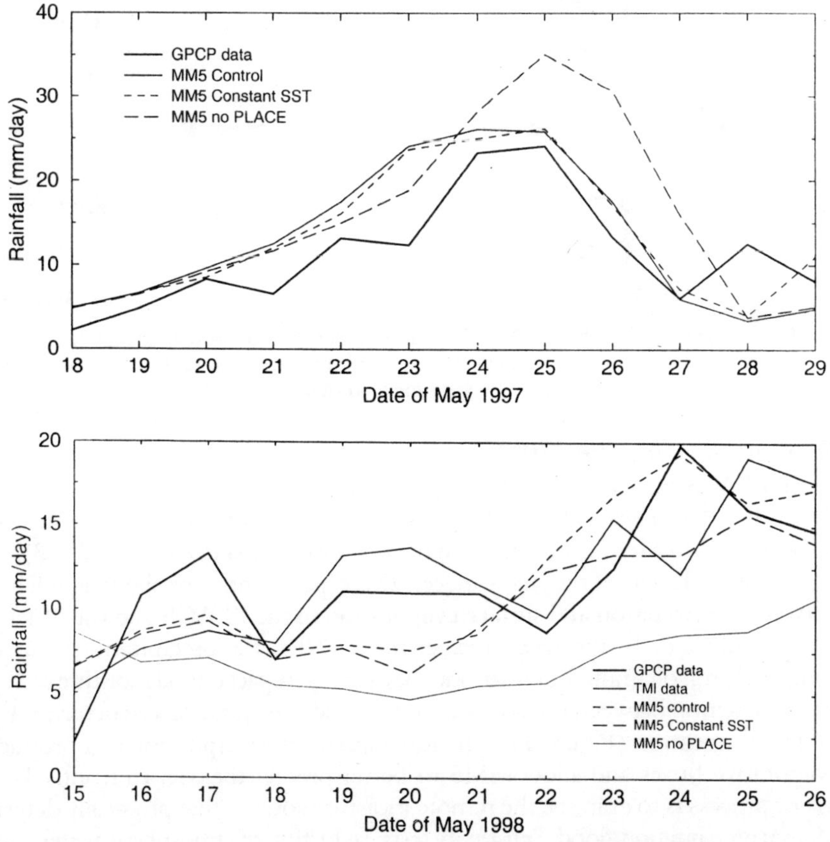

FIG. 2 **(a)** Daily rainfall (in mm day-1) averaged over the South China Sea region during May 18 to 29, 1997. **(b)** is the same as (a) expect for during 15 to 26, 1998.

daily rainfall increasing with time. In general, the RELACS system captured these observed features. For example, the model correctly simulated the onset of the SCSM on May 17, with the build up of westerlies over the northern part of the Indian Ocean, and the movement of strong convection and precipitation from the Bay of Bengal to the South China Sea. These features were also observed and simulated by MM5 in 1986. In addition, the model results are in very good agreement with TRMM rainfall during the first 4 days of integration and with GPCP during the last 5 days of integration. The run without PLACE does not capture the observed multiple rainfall peaks.

The water and energy cycles (surface precipitation and its associated latent heat release and surface fluxes) were calculated to assess the role of the atmosphere and the land-surface components responsible for the development of the monsoon for the 1997 and 1998 cases. The results indicated that there were no significant differences in terms of evaporation, rainfall and precipitation between the two cases (Figure 3). One of the major differences is that significant moisture was transported to the northern boundary (toward S. China) in 1998 but not in 1997. Another difference is that the eastern Pacific was a moisture source in 1998 but a sink in 1997. There was a greater contribution from the equatorial region in 1997.

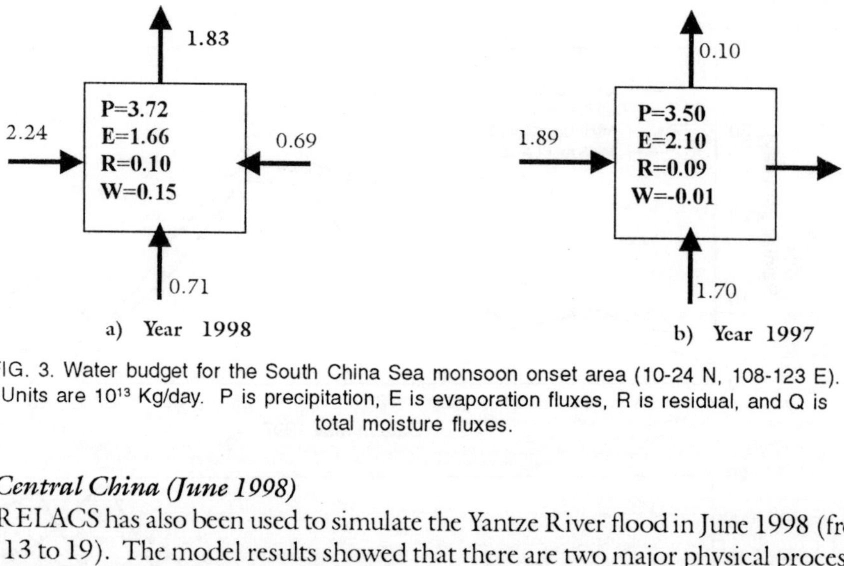

FIG. 3. Water budget for the South China Sea monsoon onset area (10-24 N, 108-123 E). Units are 10^{13} Kg/day. P is precipitation, E is evaporation fluxes, R is residual, and Q is total moisture fluxes.

(b) *Central China (June 1998)*

The RELACS has also been used to simulate the Yantze River flood in June 1998 (from June 13 to 19). The model results showed that there are two major physical processes associated with Mei-Yu precipitation over central China (Qian *et al.* 2003). The first one is the local effect of the land surface. Two experiments, one allowing full interaction between precipitation and surface evaporation in the PLACE and the other one not, were conducted. In the second experiment (NF), the precipitation is assumed to completely run off immediately without having any impact on soil moisture or vegetation. The interaction between surface evaporation and precipitation can determine the Mei-Yu front's location (Figure 4). Previous upstream precipitation can produce strong evaporative fluxes and allow MM5 to better predict the Mei-Yu front. The second major process is to examine the remote moisture sources that physically determine the Mei-yu precipitation/flood. Sensitivity tests, reduction of atmospheric water vapor fluxes from the western, southern and eastern lateral boundary, were performed. Results

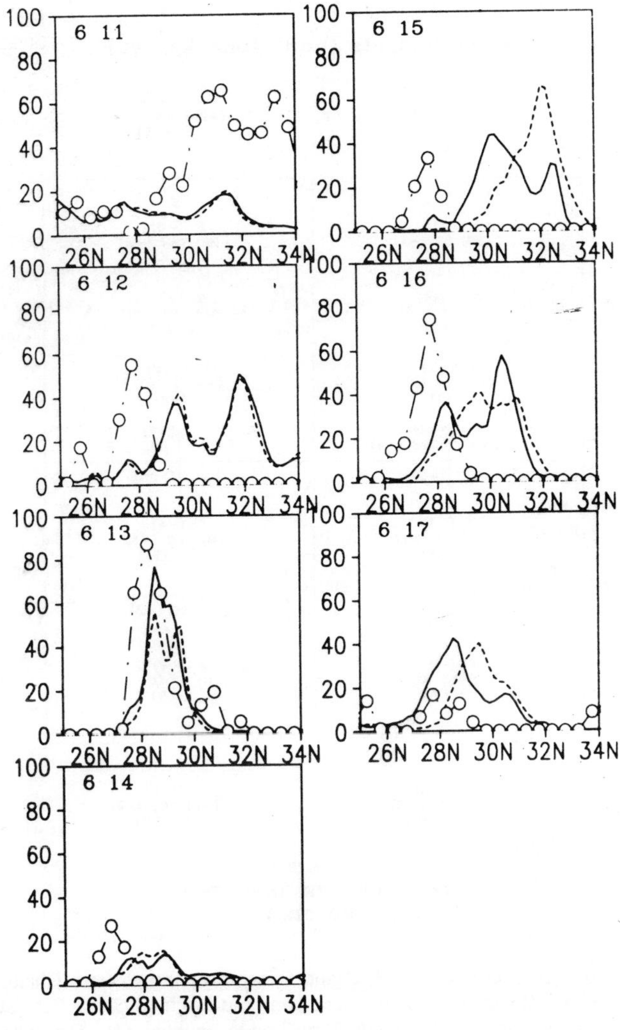

FIG. 4. Meridonal distribution of daily rain rate (mm day^{-1}), averaged from 114E to 121E for the TRMM observed (dot-dash-circle), control run (solid) and NF (dash).

indicated that the Mei-Yu precipitation is basically determined by the moisture transport of the southwesterly low level jet (LLJ) from Indo-China. There is very little impact from moisture transported from the South China Sea and the Pacific Ocean.

Water budgets were calculated over four rectangular domains A, B, C, and D, representing western China, the Yantze River Valley, southern China and the South China Sea, and the Indo-China Peninsula, respectively (Figure 5). The total atmospheric water vapor change W is small. The change in cloud water storage is included as part of the residual term R. For domain B (the Yangtze River Valley) , the total precipitation P is 144; the surface evaporation E is 45; the total moisture fluxes FS, FW, FN, and FE are 280, 450, -65, and 689, respectively. Thus the moisture convergence Q is 105. For the sensitivity test (W80) that reduced the moisture fluxes from western lateral

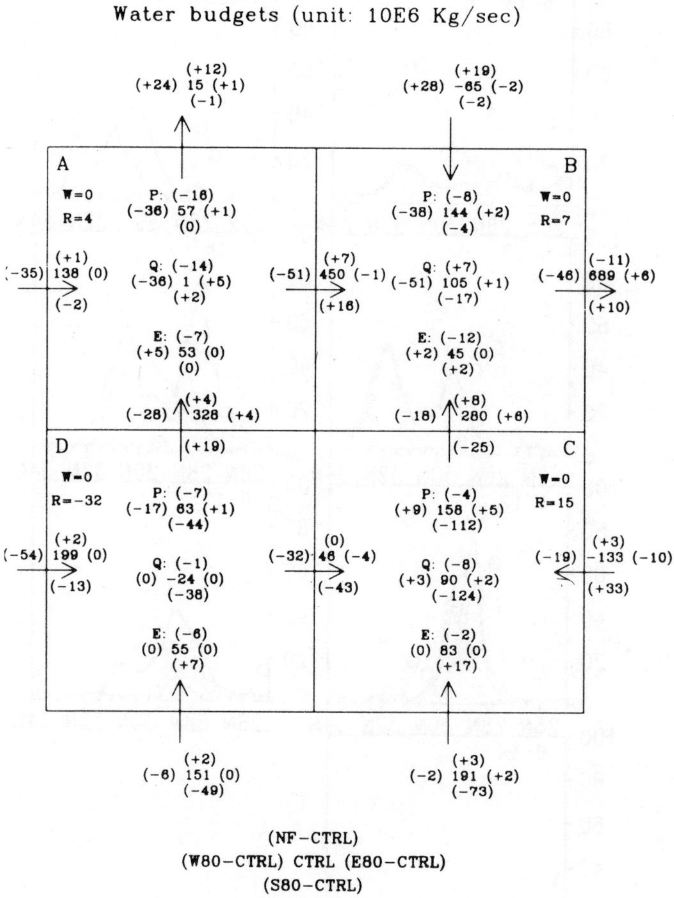

FIG. 5. Water budget in domains A, B, C and D, representing western China, the Yangtze River Vally and southern China, the South China Sea, and the Indo-China Peninsula, respectievely. Budget components P, E, Q, W and R defined in Equation (1). The units are 10^6 Kg/s

boundary by 20%, FW for domains D (the Indo-China Peninsula) and A (western China) are reduced by 54 and 35, respectively, and FW for domain is reduced by 51. The precipitation over B is reduced substantially by 38 (about 1/4 of control run). No flood was produced in the Yangtze River Valley if the moisture fluxes were further reduced to 50% from model's western boundary. For the sensitivity test (S80) that reduced the moisture fluxes from southern lateral boundary by 20%, FS for domains C (southern China and the South China Sea) and D are reduced by 73 and 49, respectively, and FS for the Yangtze River Valley is reduced by 25. But, the precipitation over the Yangtze River Valley is only reduced by 4. Thus, although the moisture transport from the tropics is quite strong, its impact on Mei-Yu precipitation is not significant.

In addition, the model results are very sensitive to the cumulus parameterization schemes used in the MM5. The Betts and Miller scheme produced too much rainfall over the East Coast of China. The Kain-Fritsch schemes, on the other hand, simulated rainfall in good agreement with observations.

3.2 Cloud-resolving model simulations

The GCE model is used to simulate convective systems for two SCSMEX convective periods, May 18-26 (prior to and during the monsoon) and June 2-11 (after the onset of the monsoon) 1998 (Tao *et al.* 2003b,c). Observed large-scale advective tendencies (or forcing) for temperature, water vapor mixing ratio, and horizontal momentum are used as the main large-scale forcing in governing the GCE model in a semi-prognostic manner (Soong and Ogura 1980; Soong and Tao 1980; Tao and Soong 1986). The main feature of this approach is that ensembles of clouds can be generated by the "*observed-prescribed forcing*". This type of cloud-resolving modeling was used in many recent modeling studies to study TOGA COARE and GATE convective systems (Das *et al.* 1999, Grabowski *et al.* 1998, Wu and Randall 1996, Wu *et al.* 1998, Donner *et al.* 1999, Johnson *et al.* 2002, Tao 2003). Please see reviews by Tao (2003) and Johnson *et al.* (2002).

The temporal variation of the GCE model simulated rainfall is in very good agreement with that estimated from soundings (Figure 6). The probability distributions of simulated

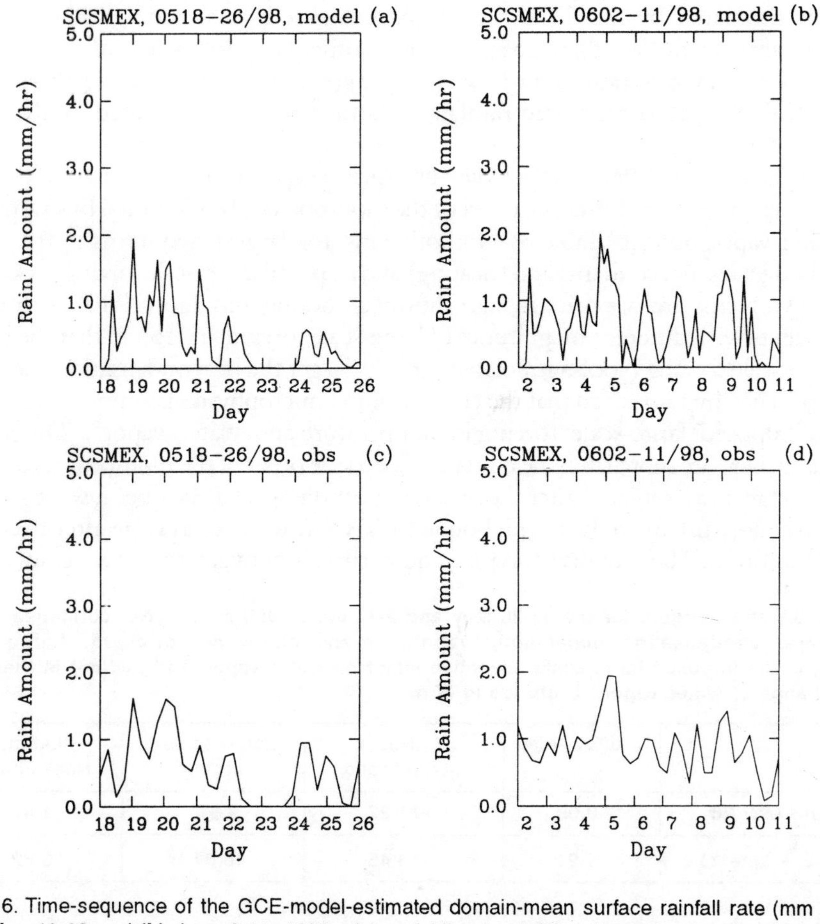

FIG. 6. Time-sequence of the GCE-model-estimated domain-mean surface rainfall rate (mm h-1) for **(a)** May 18-26 and **(b)** June 2-11, 1998. **(c)** and **(d)** are rainfall estimates using the Q2 budget.

rainfall for these SCSMEX cases and that estimated by soundings are also in good agreement for both cases. For example, a bimodal distribution of rainfall for the May case is well simulated by the model. In contrast, there is only a single peak for both simulated and observed distributions for the June case. The good agreement is mainly caused by the fact that the GCE model was forced by large-scale tendencies in temperature and water vapor computed from the sounding network. When the imposed large-scale advective forcing cools and moistens the environment, the model responds producing clouds through condensation and deposition. The fall out of large precipitation particles produces rainfall at the surface. The larger the advective forcing, the larger the microphysical response (rainfall) the model will produce (Soong and Tao 1980; Tao and Simpson 1984). On the other hand, the model will not produce any cloud or rainfall when the imposed large-scale advective forcing heats and dries the atmosphere. The rainfall amounts simulated by the GCE model and estimated by the TRMM PR, the TMI and from soundings are quite different. But, they all indicate that less rainfall occurs in the May case than in the June case. Surprisingly, the model under-estimated rainfall by 14% and 20%, respectively, for the May and June cases compared to that calculated based on soundings. In other previous GCE model simulations, the rainfall amount simulated by the GCE model and estimated by soundings is usually in excellent agreement (less than 0.5%) with each other (i.e., Johnson *et al.* 2002). The GCE-model-simulated rainfall for June is in very good agreement with the TRMM PR and GPCP. The TMI-estimated rainfall is about twice that estimated by the GCE and PR.

Similarities and differences between precipitation processes that occurred prior to and that which occurred after the onset of the monsoon can be identified by examining the water vapor budget (Table 3). In both runs, the largest two terms in the water vapor budget are net condensation (heating) and imposed large-scale forcing (cooling). These two terms are opposite in sign, however. Soong and Tao (1980) performed experiments with different magnitudes of large-scale forcing and found that the larger the large-scale forcing (cooling/moistening), the larger the net condensation (heating/drying). They hypothesized that the effect of cloud microphysics is simply a response to the "imposed large-scale forcing in temperature and water vapor". The results also show there are more latent heat fluxes prior to the onset of the monsoon. However, more rainfall was simulated after the onset of the monsoon. This is because large-scale advective moisture from the lateral boundaries is almost twice as large after the onset of the monsoon. The radiative forcing is quite similar between these two cases.

TABLE 3. Water budgets for the 18-26 May and 2-11 June 1998 cases. Net condensation is the sum of condensation, deposition, evaporation and sublimation of cloud. Large-scale forcing is the imposed large-scale advective effect on water vapor, and $L_v d(q_v)$ is the local time change of water vapor. Units are in W m^{-2}.

	d(Qv) /dt	Net Condensation	Large-scale Forcing	Latent Heat Fluxes
May18 - May 26	0.05	-11.23	9.81	1.47
June 2 – June 11	1.21	-16.45	16.84	0.82

3.3 Precipitation Efficiency

The large-scale precipitation efficiency in the simulations can be defined as (1) the ratio of the total rainfall to the large-scale water vapor transported into the domain plus latent heat flux (L-PE-1), or (2) the ratio between the total rainfall to the large-scale water vapor transported into the domain (L-PE-2). The larger the difference between these two large-scale precipitation efficiencies, the more latent heat flux (or evaporation flux) contributes to the rainfall. The definition of these precipitation efficiencies is very different from several definitions of storm-scale precipitation efficiency [see Ferrier *et al.* (1996) for a detailed review on precipitation efficiency associated with storms].

TABLE 4. Large-scale precipitation efficiencies calculated from regional climate model simulations and cloud-resolving model simulations.

	D(Qv)/dt	Precipitation (Rainfall)	Large-scale Forcing	Latent Heat Fluxes (Evaporation)	L-PE-1 L-PE-2
June 1998 China		144	105	45	0.96/1.37
June 1998 S. China Sea		158	90	83	0.91/1.76
May 1997 S. China Sea	-0.01	3.50	1.31	2.10	1.03/2.67
May 1998 S. China Sea	0.15	3.72	1.81	1.66	1.07/2.06
May 1998 CRM	0.024	10.55	9.01	1.56	0.99/1.17
June 1998 CRM	1.678	15.26	16.13	0.81	0.90/0.95

Table 4 shows the two large-scale precipitation efficiencies from the regional climate model simulations for 1997 and 1998, and the cloud-resolving model simulations for 1998. The results indicated that the L-PE-1 is close to one for all cases and for both models. This implies that the local change of water vapor over S. China and the S. China Sea does not vary much before and after convective events. However, the L-PE-2 is larger in the regional climate simulations than in the cloud-resolving model simulations. This result indicates the latent heat fluxes (or evaporation) have more of an impact on precipitation in the regional climate model simulations. The results also indicate that the impact of latent heat fluxes on precipitation processes is smaller for S. China than over the S. China Sea in the regional climate simulations. This is because the latent heat fluxes over land have a smaller contribution to precipitation than those from the ocean.

4. SUMMARY

RELACS has been used to simulate the onset of the South China Sea monsoon in 1986, 1997 and 1998. Sensitivity tests on various land surface models, cumulus parameterization schemes (CPSs), sea surface temperature (SST) variations and midlatitude influences have been performed. These tests have indicated that the land surface model has a major impact on the circulation over the S. China Sea. CPSs can effect the precipitation pattern while SST variation can effect the precipitation amounts over both land and ocean. RELACS has also been used to understand the soil-precipitation interaction and feedback associated with a flood event that occurred in and around China's Yantz River during 1998. The exact location (region) of the flooding can be effected by soil-rainfall feedback.

The GCE model was used to simulate convective systems for two SCSMEX convective periods, May 18-26 (prior to and during the monsoon) and June 2-11 (after the onset of the monsoon), 1998. The GCE-model-simulated rainfall captured the temporal variation quite well in both cases compared to the sounding estimated rainfall. A bimodal distribution of rainfall for the May case and a single peak for the June case are well simulated by the GCE model. The rainfall amounts simulated by the GCE model and estimated by the TRMM PR, the TMI and from soundings are quite different. The GCE-model-simulated rainfall is in closer agreement with the TRMM PR than the soundings and the TMI. TMI-estimated rainfall is about twice that estimated by the GCE and PR. A detailed study to determine the reasons for the differences in rainfall amounts between the different measurements is underway through the TRMM project. The results show there are more latent heat fluxes prior to the onset of the monsoon. However, more rainfall was simulated after the onset of the monsoon. Small differences in the precipitation efficiency between the two periods is noted.

This modeling study indicates the latent heat fluxes (or evaporation) have more of an impact on precipitation processes and rainfall in the regional climate model simulations than in the cloud-resolving model simulations. Research is underway to determine if the difference in the grid sizes or the moist processes used in these two models is responsible for the differing influence of surface fluxes on precipitation processes.

Acknowledgements. The authors are supported by the NASA Headquarters physical climate program, the NASA TRMM, and the interdisciplinary program of EOS. The authors thank Mr. S. Lang for reading the manuscript. We also thank Dr. M. Moncrieff for his comments on the paper. Acknowledgment is also made to NASA/Gddard Space Flight Center for computer time used in the research.

REFERENCES

Das, S., D. Johnson and W.K. Tao, 1999: Single Column and cloud ensemble model simultations of TOGA COARE connective systems. *J. Meteor. Soc.* Japan, 77, 803-826.

Donner, L. J., C. J. Seman abd R. S. Hemler, 1999: Three-dimensional cloud-system modeling of GATE convection, *J. Atmos. Sci.*, **56**, 1885-1912.

Fairall, C., E. F. Bradley, D. P. Rogers, J. B. Edson, and G. S. Young, 1996: Bulk parameterization of air-sea fluxes for TOGA COARE. *J. Geophy. Res.*, **101**, 3747-3764.

Ferrier, B. S., J. Simpson and W.-K. Tao, 1996: Factors responsible for different precipitation efficiencies between midlatitude and tropical squall simulations. *Mon. Wea. Rev.,* **124**, 2100-2125.

Grabowski, W. W., X. Wu, M. W. Moncrieff, and W. D. Hall, 1998: Cloud-resolving modeling of cloud systems during Phase III of GATE. Part II: Effects of resolution and the third spatial dimension. *J. Atmos. Sci.*, 55, 3264- 3282.

Johnson, R. H., and P. E. Ciesielski, 2003: Characteristic of the 1998 summr monsoon onset over the Northern South China Sea, *J. Metro Soc.,* Japan. Submitted.

Johnson, D., W.-K. Tao, J. Simpson, and C.-H. Sui, 2002: A study of the response of deep tropical clouds to large-scale processes, Part I: Model set-up strategy and comparison with observation, *J. Atmos. Sci.*.**59**, 3492-3518.

Klemp, J. B., and R. Wilhelmson, 1978: The simulation of three-dimensional convective storm dynamics. *J. Atmos. Sci.*, **35**, 1070-1096.

Lau, K. M., Y. Ding, J.-T. Wang, R. Johnson, T. Keenan, R. Cifelli, J. Geriach, O. Thiele, T. Rickenbach, S.-C. Tsay, and P.-H. Lin, 2000: A report of the field operations and early results of the South China Sea Monsoon experiment (SCSMEX). *Bull. Amer. Meteor. Soc.*, **81**, 1261-1270.

Lin, Y.-L., R. D. Farley and H. D. Orville, 1983: Bulk parameterization of the snow field in a cloud model. *J. Clim. Appl. Meteor.*, **22**, 1065-1092.

Moncrieff, M.W., S.K. Krueger, D. Gregory, J.-L. Redelsperger and W.-K. Tao, 1997: GEWEX Cloud System (GCSS) Working Group 4: Precipitating convective cloud systems. *Bull. Amer. Meteor. Soc.,* **78**, 831-845.

Moncrieff, M.W. and W.-K, Tao: 1999: *Cloud-resolving models.* Global Water and Energy Cycles, Ed. by K. Browning and R.J. Gurney, Cambridge University Press, 200-209.

Moncrieff, M.W., 2003: Simulation and super-parameterization of tropical cloud systems. (This publication)

Qian, J.-H., W.-K. Tao, and K.-M. Lau, 2003: Mechanisms of torrential rain associated with the Mei-yu development during SCSMEX-98. *Mon. Wea. Rev.* (Accepted).

Rutledge, S.A., and P.V. Hobbs, 1984: The mesoscale and microscale structure and organization of clouds and precipitation in midlatitude clouds. Part XII: A diagnostic modeling study of precipitation development in narrow cold frontal rainbands. *J. Atmos. Sci.*, **41**, 2949-2972.

Smolarkiewicz, P. K., and W.W. Grabowski, 1990: The multidimensional positive advection transport algorithm: nonoscillatory option. *J. Comput. Phys.*, **86**, 355-375.

Soong, S.-T., and Y. Ogura, 1980: Response of tradewind cumuli to large-scale processes. *J. Atmos. Sci.* , **37**, 2035-2050.

Soong, S.-T., and W.-K. Tao, 1980: Response of deep tropical clouds to mesoscale processes. *J. Atmos. Sci.*, **37**, 2016-2036.

Tao, W.-K., and J. Simpson, 1984: Cloud interactions and merging: Numerical simulations. *J. Atmos. Sci.,* **41**, 2901-2917.

Tao, W.-K., and S.-T. Soong, 1986: A study of the response of deep tropical clouds to mesoscale processes: Three-dimensional numerical experiments. *J. Atmos. Sci.*, **43**, 2653-2676.

Tao, W.-K., and J. Simpson, 1993: The Goddard Cumulus Ensemble Model. Part I: Model description. *Terrestrial,Atmospheric and Oceanic Sciences*, **4**, 19-54.

Tao, W.-K., J. Simpson, S. Lang, C.-H. Sui, B. Ferrier, and M.-D. Chou, 1996: Mechanisms of cloud-radiation interaction in the tropics and midlatitudes. *J. Atmos. Sci .,* **53**, 2624-2651.

Tao, W.-K., J. Simpson, D. Baker, S. Braun, M.-D. Chou, D. Johnson, B. Ferrier, A. Khain, S. Lang, B. Lynn, C.-L. Shie, D. Starr, C.-H. Sui, Y. Wang and P. Wetzel, 2003a: Microphysics, Radiation and Surface Processes in a Non-hydrostatic Model. *Meteorology and Atmospheric Physics*, 82, 97-137.

Tao, W.-K., 2003: Goddard Cumulus Ensemble (GCE) model: Application for understanding precipitation processes, AMS Meteorological Monographs - Symposium on Cloud Systems, Hurricanes and TRMM, 107-138.

Tao, W.-K., C.-L. Shie, D. Johnson, R. Johnson, S. Braun, J. Simpson, and P. E. Ciesielski, 2003b: Convective Systems over South China Sea: Cloud-Resolving Model Simulations, *J. Atmos. Sci.* (Accepted).

Tao, W.-K., C.-L. Shie, D. Johnson and J. Simpson, 2003c: Surface energy budget and precipitation efficiency of convective systems developed during TOGA COARE, GATE, SCSMEX and ARM: Cloud-resolving model simulations. *J. Atmos. Sci.* (Accepted).

Wang, Y., W.-K. Tao and J. Simpson, 1996: The impact of ocean surface fluxes on a TOGA COARE cloud system. *Mon. Wea. Rev.* **124**, 2753-2763.

Wang, Y., W.-K. Tao, and J. Simpson, and S. Lang, 2003: The sensitivity of tropical squall lines (GATE and TOGA COARE) to surface fluxes: Cloud resolving model simulations, *Quart. J. Roy. Meteor. Soc.* (In Press).

Xu, K.-M., and D. A. Randall, 1996: A semiemprical cloudiness parameterization for used in climate models. *J. Atmos. Sci.*, **53**, 3084-3102.

Wu, X., W. W. Grabowski and M. W. Moncrieff, 1998: Long-term behavior of cloud systems in TOGA COARE and their interactions with radiative and surface processes. Part I: Two-dimensional modeling study. *J. Atmos. Sci.,* **55**, 2693-2714.

Cloud-Resolving Simulation and Super-parametrization of Tropical Cloud Systems

MITCHELL W. MONCRIEFF
Mesoscale & Microscale Meteorology Division, NCAR
Boulder CO 80307-3000
E mail : moncrief@ucar.edu

ABSTRACT

The need to understand the large-scale effects of organized convection is emphasized. This need is evinced by recent progress in the numerical realization of tropical convection using the cloud-system-resolving modeling approach and its novel offshoot, super-parametrization. While super-parametrization is preparatory to eventual global cloud-resolving models, conventional parametrization will be required for the foreseeable future. But organized convection is not represented by any parametrization scheme. How to represent the dynamics of organization is an intellectual challenge. It is shown how cloud-system-resolving modeling is advancing our knowledge of the organization and the attendant scale-interaction and feedback among the physical processes. Specifically, recent progress on the following will be summarized: hierarchical modeling of oceanic tropical convection; surrogate convection in high-resolution global numerical weather prediction models; convectively coupled tropical gravity waves; and, the role of explicit convection in the tropical intraseasonal oscillation. Finally, a strategy to examine convective cloud systems in the Indian Ocean locale is outlined.

1. INTRODUCTION

Quantifying the effects of moist convection on the large scales of motion is requisite not only to the accurate numerical predictions of weather and climate but also to understanding the physics of the atmospheric water and energy cycles and the role of clouds in climate. This class of problem, in which the collective effects of small scales on large scales is the key issue, occurs throughout the physical sciences. In atmospheric science this practical (sub-grid-scale parametrization) aspect is an impediment to accurate prediction.

A major challenge lies in understanding the Earth's water and energy cycles. While the water and energy cycles are products of multiscale interactions among cloud-microphysics, radiation, convection, and boundary layer and surface hydrology, none of these processes is dominant and, moreover, all are sub-grid scale. The dynamical interactions embodied in the attendant parametrizations are not well understood and, as a result, are inadequately represented in large-scale models. The fact that weather prediction models, even the most advanced ones, have difficulty in forecasting precipitation underscores this point. Examples of major components of the water and energy cycles include the tropical intraseasonal oscillation, the inter-tropical convergence zone/Hadley circulation complex and the monsoon systems of the world.

Global observing systems (e.g., the Tropical Rainfall Measuring Mission, TRMM), along with accompanying field campaigns, are quantifying precipitation distribution and its variability. But measurement alone is not a complete answer: quantities such as latent heating are not directly available from either active sensors (radar) or passive sensors (radiometers). Cloud-system-resolving models (CSRMs) have successfully been used to derive latent hearing profiles (Tao *et al.* 1993). Herein, it will be shown how CSRMs have addressed a variety of convection-related aspects involving a range of scales.

In regard to convection, it is difficult to escape the issue of parametrization. After decades of vexing attempts to improve parametrizations of convection and cloud-radiative interaction, it is widely recognized that a physically based methodology is necessary. Cloud-resolving modeling is promising and international advocacy for it began about a decade ago under the auspices of the GEWEX Cloud System Study (GCSS), which was established in recognition of the substantial progress in simulating convective processes and cumulus ensembles using nonhydrostatic models and the implications this has for parametrization. But just as it took decades for climate models to get fully recognized as an important facility for climate research, so cloud-resolving models are slowly bridging the scientific and cultural gaps between process studies and parametrization (GCSS 2000).

Why is convective parametrization resistant to solution? Convection is a highly selective processor of water: dynamical instabilities generate precipitation of varying intensities and amounts, and the attendant precipitation patterns can be remarkably coherent. So the problem is complicated. The mass-flux approach to parametrization pioneered by Arakawa and Schubert (1974) assumes the principle of scale separation. Specifically, the dynamical scale of the parameterized process should be small compared to the resolution of the model in which it is applied. But scale separation does not truly exist because of the mesoscale organization (dynamical coherence) of convection. The key point is that convection, especially in the tropics and its environment are a continuum rather than separable entities. Fundamentally, this explains why convection is resilient to parametrization.

Mesoscale organization is also a basic reason why CSRMs realize cloud systems and their collective effects without unduly fine resolution (Moncrieff 1995). Note that CSRMs do not assume scale separation but, instead, deal with scale selection explicitly. CSRMs couple the moist processes at scales more realistic than possible in global models, whose resolution ranges from about 30 km (weather prediction) to about 300km (climate). This is justification for using CSRMs in the development of physically based parametrization (Moncrieff *et al.* 1997; GCSS 2000).

This begs the question: why does convection organize on mesoscales (~1-1000km)? Powerful instabilities and scale-interactions occur on mesoscales, and phase changes among the states of water are highly interactive with dynamics. Convection is one of few atmospheric phenomena capable of self-organization and accompanying upscale development, a behavior well documented in field experiments and satellite observations. Organized systems have long been known to be persistent and travel great distances, generating coherent patterns of rainfall that can extend up to continental scales (Laing and Fritsch 1997; Carbone *et al.* 2002). Because coherent sequences of precipitation are a product of organized convection, organization is the reason for its existence. Figure 1 implies that the global effects of organization depends on relationships among orography, mean flow and mesoscale organization. Note that the concept of mesoscale organization remains to be built into a parametrization theory.

Dynamical models show that organization is a function of wind shear, convective available potential energy and large-scale forcing (Moncrieff 1981); the action of mesoscale gravity waves (Tripoli and Cotton 1989; Mapes 1993); and dynamical feedback between convection and large-scale gravity waves (Grabowski and Moncrieff 2001).

In summary, the mesoscale organization of convection is integral to the following aspects (prioritized in terms of increasing physical complexity):

- **Numerical weather prediction:** Mesoscale organization is not truly consistent with the scale-separation principle of parametrization. State-of-the-art global numerical prediction models operate with grid lengths of about 30km. Regional prediction models have long operated at this resolution, and nowadays grid-lengths <10km are common. Regional numerical weather prediction models and the highest resolution global models can generate organized systems, but sometimes for the wrong reasons: at coarse resolution the mesoscale organization of convection gets distorted. Understanding this surrogate behavior is important in mesoscale models and also in high-resolution global numerical weather prediction models (section 4).
- **Climate research:** Mesoscale organization is not represented in convective parametrizations used in climate models, whose resolution is typically 200-300 km. Indeed, mesoscale organization is totally absent from climate models. This omission has unknown consequences. Note that convectively generated cirrus, which has important effects on radiative transfer, is a product of upper-tropospheric outflow from organized convection. This evinces the relevance of convective organization in a climate context.

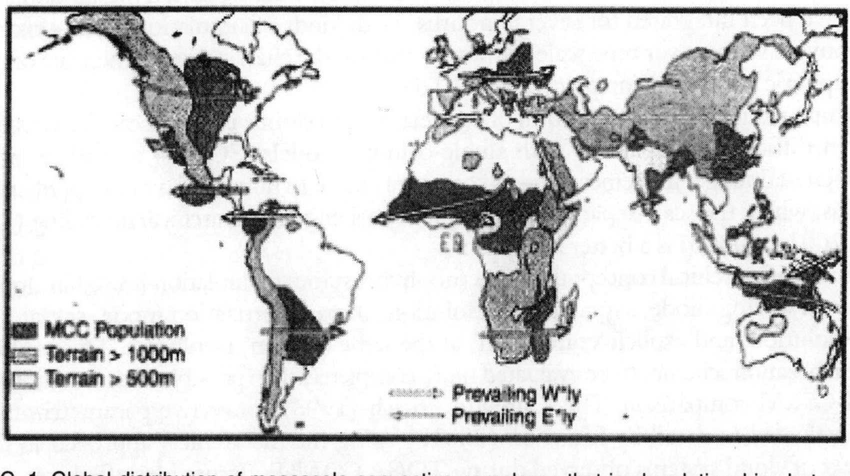

FIG. 1. Global distribution of mesoscale convective complexes that imply relationships between mountain ranges, mean flow and rainfall distribution. [Laing and Fritsch (1997)]

- **Water and energy science:** The atmospheric branch of the water and energy cycles has a time scale comparable to the time scale of convective-radiative-dynamical equilibrium (about two weeks) that CSRM can handle with considerable fidelity over a wide dynamic range. Quantifying the atmospheric water is an important problem that CSRMs are beginning to comprehensively address. Integration of space-based measurements with the CSRM simulation approach is a next step.

A complete understanding of physically complex systems requires unifying principles: an ultimate goal is to represent the collective effects of cloud systems, conceptually, as a field theory for a moist continuum. It is conceivable that this lofty goal could be met for tropical cloud systems because they are approximately in a state of convective-radiative-dynamical equilibrium. Explicit numerical simulation is a pathway to such an objective. The remainder of this paper illustrates progress toward this goal.

2. CLOUD SYSTEM SIMULATION

Cloud-system-resolving models are powerful tools for quantifying interactions between convection and large scales, the effects of clouds on atmospheric radiation, and the atmospheric component of the water and energy cycles. This led to an entirely new approach to parametrization wherein cloud-resolving models are embedded in general circulation models as *super-parametrizations* (section 6).

2.1 Cloud-system-resolving Models

Recall that the mesoscale dynamics attendant to cloud systems is fully resolved with about a 1-km grid-length, hence the terminology *cloud-system-resolving models*. No geophysical model is capable of simulating cloud systems from first principles — cloud-microphysics, radiative transfer and turbulence must be parameterized. The minimal cloud-microphysical parametrization has yet to be defined and will be different in weather and climate applications.

The dynamic range of CSRMs enables explicit week-long simulations in three-dimensional 1000-km domains. Two-dimensional CSRMs have planetary-scale domains and have been integrated for several months. Both kinds of simulation greatly exceed the convective turnover time scale (~ hours) and are the characteristic time scale of the atmospheric water and energy cycles (~10 days).

Improving the parametrization of convection and clouds in large-scale models has proven difficult. Comparison with single-column models (SCM) – essentially how parametrizations get implemented — is commonly used. In numerical weather prediction models, where the scale-separation principle is challenged, hierarchical modeling (Liu *et al.* 2001a, 2001b) is a better approach.

In the hierarchical concept the same (nonhydrostatic) formulation is used in cloud-system-resolving mode, say, at 1-km resolution; in parametrization mode, say, at 15-km resolution; and explicit 'convection' at the same (15-km) resolution. This enables parametrization schemes to be evaluated more completely than possible with a traditional SCM-CSRM comparison. The Kain and Fritsch (1993) convective parametrization was evaluated in detail by Liu *et al.* (2001b) using the hierarchical approach in the context of cloud systems observed during TOGA COARE. Issues identified are:

- Overly deep convection and extent of convectively generated cirrus, a result of excessive detrainment of condensate and water vapor.

- Simulations sensitive to the magnitude of moisture grid-scale feedback from the convective parametrization.
- Parameterized overshoot-generated cooling around cloud tops too strong, resulting in cold bias.
- Over-prediction of lower-level moisture attributed to the parameterized downdrafts being too moist.

These deficiencies are thought to stem from the single-plume (bulk) model used in the parametrization scheme, which does not treat the trimodal vertical distribution of convection (i.e., cumulonimbus, cumulus congestus, and shallow convection). A trimodal structure is realized in the CSRM and is also observed over the tropical oceans as shown by Johnston *et al.* (1999).

2.2 Super-parametrization

The aforementioned cloud-resolving convection parametrization (CRCP) or *super-parametrization* is a spin-off of cloud-resolving modeling. It is a precursor of global cloud-resolving capability that may be practicable within a decade in numerical weather prediction. The CRCP approach was pioneered in an aqua-planet study of tropical intraseasonal variability by Grabowski (2001) who used the global nonhydrostatic model of Smolarkiewicz *et al.* (2001). This concept was applied as a super-parametrization by Khairoutdinov and Randall (2001) in a climate model. In both studies, two-dimensional CSRMs were used instead of conventional convective parametrization. Super-parametrization alleviates key uncertainties because the following are treated explicitly:

- precipitating convection and its mesoscale organization;
- cloud fractional area and cloud overlap (in radiative transfer codes);
- cloud-microphysics are coupled to cloud-scale and mesoscale dynamics;
- convective enhancement of surface fluxes; and
- convectively generated gravity waves.

3. LARGE-SCALE CONTROL AND FEEDBACK

As Figure 2 illustrates, parametrization involves the concept of *control and feedback*, whether or not scale-separation is satisfied. In other words, the large-scale forcing (advection of temperature and moisture generates convective available potential energy,

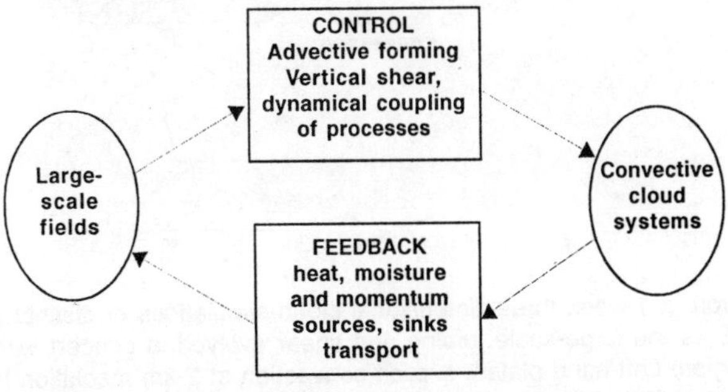

FIG. 2. Control and feedback in atmospheric convection

CAPE) and vertical shear control the amplitude and structure of convection. The convective transports feedback to modify the large-scale fields.

The Global Atmospheric Research Programme (GARP) Atlantic Tropical Experiment (GATE) conducted in tropical eastern Atlantic in 1974, was the first multiscale field experiment whose primary objective was to improve convective parametrization in numerical weather prediction models. The scale-separation aspect strongly influenced the design of the observing network.

Datasets of unprecedented dynamic range were produced from GATE but only now, a generation later, have computers advanced to a point where convection can be simulated in three spatial dimensions over long periods of time. This procedure realized synthetic data sets of unprecedented fidelity. A key finding was the ubiquitous nature of mesoscale organization (Houze and Betts 1981), which actually contradicts the scale-separation principle. The GATE cloud-system simulation used large-scale advective forcing provided by the objective analysis of sounding data, whose accuracy and representativeness is vital. Note that the ECMWF reanalysis (ERA 40) at T159 resolution includes all GATE soundings. This reanalysis will provide forcing for future CSRM simulations in larger domains than possible at present.

The snapshots on the left-hand-side of Figure 3 show three distinct regimes of convection realized by a CSRM. Convective intensity is a function of the large-scale forcing and the dymamical organization is controlled by the vertical shear of the large-scale flow, both of which evolve throughout the week-long simulation in response to the passage of an easterly wave passed over the observing network. Non-squall cloud clusters occur during large forcing and weak shear, highly organized squall clusters during strong shear and medium forcing, and scattered convection during weak shear

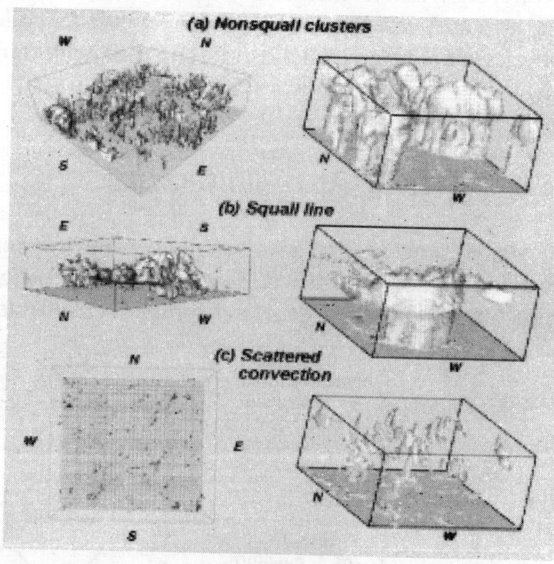

FIG. 3. Snapshots from a 1-week three-dimensional cloud-simulations of distinct regimes of convection during GATE as the large-scale forcing and shear evolved in concert with easterly wave passage over the region. Left-hand plates, explicit convection at 2-km resolution [Grabowski et al 1998]. Right-hand plates, Kain-Fritsch parameterization at 15-km resolution. [Liu et al. (2001a)]

and weak forcing. These regimes evolve spontaneously, proving the control/feedback principle in Figure 2. Comparison of the right-hand and left-hand plates in Figure 3 shows that the regime transitions are similar in the resolved and parameterized convection, but detailed analyses reveal significant errors (Liu *et al.* 2001a, 2001b).

Numerically generated (synthetic) datasets have detail, consistency and dynamic range beyond what is possible to obtain from observations alone. They are therefore useful for the development of parametrizations of convection and cloud-radiation-interaction (Wu and Moncrieff 1996, 2001).

The only truly multiscale field campaign, other than GATE, is the Tropical Ocean-Global Atmosphere (TOGA) Coupled Ocean Atmosphere Response Experiment (COARE) conducted November 1992 - February 1993. Resolved convection simulations of COARE cloud systems have been evaluated and compared to parametrizations (Wu and Moncrieff 2001). For example, it was shown that cloud parametrization under-predicted condensate and cloud fraction. As a result, the radiative flux at top and bottom of the atmosphere did not match observations, and the individual components of the energy budget were incorrect. The representation of cloud geometry and inhomogeneity in the radiation scheme are now being improved in order to produce more accurate radiative heating profiles for climate models.

4. ORGANIZED CONVECTION AND NUMERICAL WEATHER PREDICTION

Difficulties occur in the treatment of organized convection in weather prediction models as a result of the (artificial) partition between parameterized convection and 'resolved' convection that commonly occurs in the higher-resolution prediction models. This is a poorly understood aspect in regional-scale models at resolutions of 10–30km and also in high-resolution global NWP models at resolutions of 40-90 km. It follows that this problem is of considerable importance, although it gets surprisingly little attention.

Anticipating this problem and assuming that that the scale of organization relative to horizontal resolution is a key aspect, Moncrieff and Klinker (1997) examined tropical super-clusters associated with westerly wind burst in the tropical western Pacific. Note that westerly bursts are the precursors of the tropical intraseasonal oscillation, also referred to the Madden-Julian oscillation (Madden and Julian 1972). Super-clusters have a typical horizontal scale of ~1000km and consist of families of cloud clusters (i.e., mesoscale convective systems). Moncrieff and Klinker analyzed a supercluster within a westerly wind burst in the ECMWF model at a resolution of about 90 km. The westerly burst resembled an eastward-propagating twin-cyclone with squall-like convection at its leading edge (Figure 4).

Parameterized convection dominated during the early stages and resolved-scale circulation during the mature stage of this particular supercluster. But the amplitude of the resolved-scale heating was much larger than typically seen in the stratiform region of real mesoscale convective systems. This behavior stems from the 'resolved' circulation being coupled to the large-scale flow in the horizontal momentum equation at too large a scale, causing unrealistic feedback through the thermodynamic and moisture equations. It is concluded that this surrogate behavior associated with organized convection is an

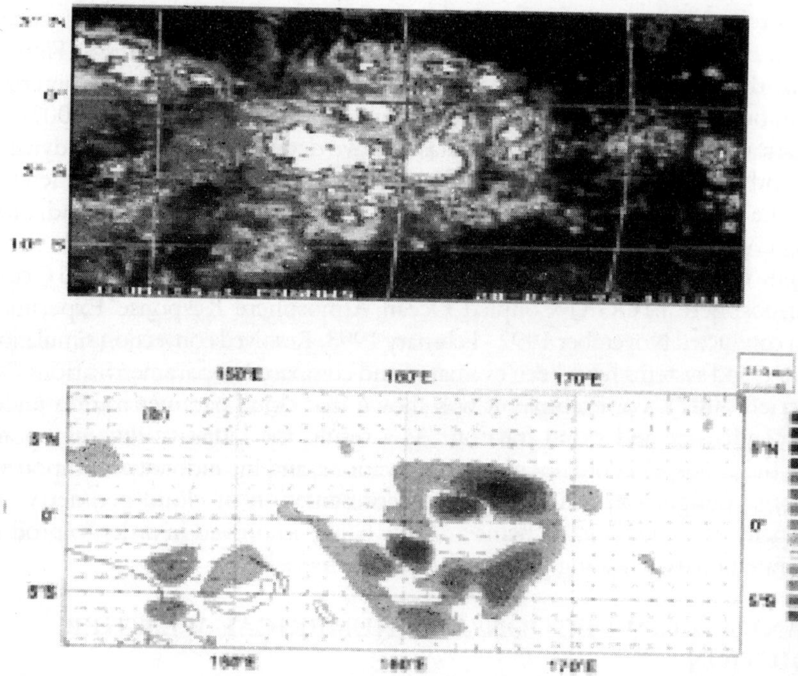

FIG. 4. a) Satellite image of a supercluster occurring within a westerly wind burst during TOGA COARE, suggesting it consists of a family of cloud clusters. b) The same supercluster in the ECMWF operational model at T213 resolution, which assumes the form of a large-scale coherent squall-like system. [Moncrieff and Klinker (1997)]

issue in both regional and global models. Furthermore, its effects are manifested not only in terms of thermodynamics (e.g., humidity and temperature structure, and precipitation distribution) but also as dynamics (e.g., convective momentum transport/ mean-flow interaction).

The explanation of this surrogate behavior can be cast in terms of the dynamics of convection in a sheared environment. The propagation of organized convection hinges on the system-scale airflow organization that is, in turn, a function of the wind shear as quantified by dynamical models (Moncrieff 1981). Specifically, the organization, persistence and propagation of a convective system is a response shear flow to latent heating and evaporation. Only in certain conditions of CAPE and shear can propagation occur and it depends on bulk properties such as the convective Froude number. This gives prospect for understanding the basic mechanisms responsible for the interaction of convective organization with large scales. Finally, these ideas define simple models for use in a future parametrization of organized convection (Moncrieff 1992).

5. CONVECTIVELY COUPLED GRAVITY WAVES

As aforementioned, the control and forcing of convection by the large scales and the feedback of convection is basic to parametrization, whether or not the scale separation principle is satisfied. Until recently, the large-scale forcing had to be specified and the convective response calculated in CSRMs (because the computational domains were not large enough). Consequently, the control-feedback concept in Figure 2 was compromised because spontaneous scale-selection could not occur. Modern computers have removed this artificial restriction, at least in two spatial dimensions. The results of such a calculation are now summarized.

Grabowski and Moncrieff (2001) examined the large-scale organization of tropical deep convection in their idealized two-dimensional 40-day cloud-resolving simulations. The initial state had a constant 10 ms^{-1} easterly wind and a uniform sea surface temperature. A prescribed temperature tendency mimics the mean radiative cooling of the tropical troposphere. A 20,000-km computational domain allows interactions among moist convection, mesoscale organization and surface exchange to occur over a realistic range of scales.

The simulated multiscale organization of convection and its eastward propagation resembles the multiscale (supercluster/cloud cluster) organization evident in satellite observations. The large-scale organization was explained in terms of convectively coupled large-scale gravity waves (non-rotating analogues of equatorial Kelvin waves). Referring to Figure 5, mesoscale convective systems organized several hundred kilometers in scale move westward within the envelope of convection that propagates eastward. This dual-scale behavior is reminiscent of that seen in satellite images (e.g., Nakazawa 1988). Interestingly, convective momentum transport was shown to be intimately

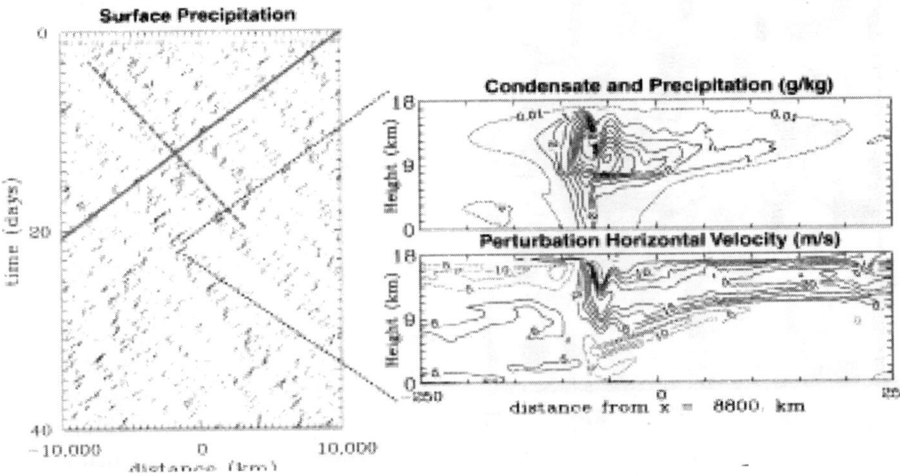

FIG. 5. Left-hand diagram, space-time depiction of the simulated rainfall pattern from a CSRM simulation of convectively coupled waves over an ocean having a constant-sea-surface-temperature and 10 m/s mean-flow to which the domain-averaged flow is relaxed. The large-scale envelops of convection (dotted) propagate eastward, heavily precipitating convective systems embedded within this envelope travel westward (full). Right-hand diagrams, squall-like systems that strongly perturb the wind field and effect momentum transport that accelerates the mean flow. [Grabowski and Moncrieff (2001)]

involved in the feedback process: momentum transport by the highly organized squall-line-like convective systems maintained the type of shear-flow conducive to their existence. This simulation is an explicit realization of the conceptual relationship between the squall systems and the environment expressed in Moncrieff (1992; Figure 11). The dynamical model therein explains the momentum transport properties of the numerically simulated convective systems.

6. THE TROPICAL INTRASEASONAL OSCILLATION (TIO)

The TIO is a fundamental mode of tropical variability on timescales of 30-60 days but has proven difficult to properly realize in either numerical weather prediction models or climate models. While the basic reasons are not understood, it is supposed that deficiencies in the parametrization of convection are responsible, so it helps to treat convection explicitly. But the intraseasonal oscillation is a global-scale phenomenon. Unlike the Kelvin-wave-like dynamics of the convectively coupled waves in the preceding section, the MJO also displays Rossby-wave behavior. As a result, the TIO is intrinsically three dimensional.. Along with its planetary scale, this means that cloud-resolving simulation of the TIO is impracticable at present.

Grabowski (2001) applied CRCP to couple cloud-scale processes to the large-scale dynamics using the anelastic nonhydrostatic global model developed by Smolarkiewicz and Margolin (1997). The experiment featured radiative-convective-dynamical equilibrium on a rotating constant SST aquaplanet. The global CRCP simulations feature a pronounced large-scale organization of convection within the equatorial wave guide.

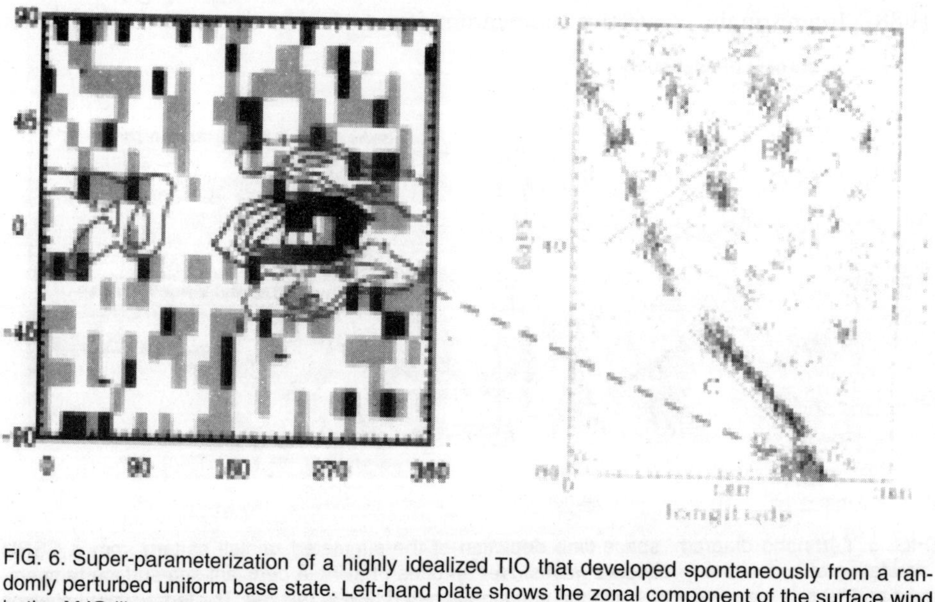

FIG. 6. Super-parameterization of a highly idealized TIO that developed spontaneously from a randomly perturbed uniform base state. Left-hand plate shows the zonal component of the surface wind in the MJO-like system, with easterlies in red and westerlies in blue. Right-hand plate is a time-distance plot of the surface precipitation that shows a coherent eastward-propagating MJO-like disturbance. This develops after about 50 days of simulation, and a period of higher-wavenumber variability. [Grabowski (2001)]

Prominent equatorial superclusters develop in these idealized simulations which bear a strong resemblance to the TIO. Momentum transport by mesoscale convective systems have an important effect on the large-scale organization of convection.

The super-parametrization concept is now being set onto a dynamical basis using nonlinear mathematical models and mapping principles that unify organized convection dynamics with Rossby gyres on an equatorial beta-plane.

7. INDO-US COLLABORATION

Unlike the eastern Atlantic and the western Pacific, the Indian Ocean region has not been the venue for a field campaign of the scope of GATE and TOGA COARE. Small focused field campaigns have been conducted such as the Bay of Bengal Monsoon Experiment (BOBMEX) in 2000 and the Arabian Sea Monsoon Experiment (ARMEX) in 2002. These campaigns are opportunities for quantying precipitating convection and its life cycle during the monsoon regime.

7.1 Mesoscale Prediction and Dynamics

Coastal weather associated with the summer monsoon are manifested as synoptic disturbances in the Western Ghats region, an aspect to be examined in ARMEX. The combination of regional weather prediction models and CSRM enables this problem to be examined comprehensively. Because the large-scale forcing can be provided by MM5 analysis, a CSRM approach can be adopted in the spirit of the GATE and TOGA COARE simulations. The CSRM can quantify the dynamics of convection in the formation of the low-level monsoon flow and its interaction with the Western Ghat mountains. Because CSRMs generate cloud-scale and mesoscale quantities not available from conventional observations alone, they are a benchmark for the evaluation and improvement of convection and cloud parametrization schemes used in the NCMRWF NWP models.

NCMRWF conducts real-time mesoscale forecasting over the Indian region using the MM5 and ETA models. The initial and boundary conditions for the models are prepared from the upper-air and surface fields obtained from the NCMRWF global data assimilation and forecast system. Mesoscale data assimilation as used in MM5 would be applied to observations obtained during BOBMEX and ARMEX.

Evaluation of regional forecasts in terms of precipitation and heavy rainfall events associated with pre-monsoonal thunderstorms, severe storms in the Bengal locale (Nor'westers) and monsoon depressions are important for understanding weather systems and improving prediction skill. Nor'westers, which are accompanied by local flooding and destructive winds, evidently resemble West African squall lines. This resemblance should be quantified.

Sensitivity experiments testing different physical parametrization schemes need to be carried out, and this is where cloud-system-resolving models can play a key role.

7.2 Tropical Intraseasonal Oscillation

The Indian Ocean is a breeding ground of the TIO, which is the largest organized convective phenomenon on Earth, and whose effects extend to planetary scales through the Rossby-wave teleconnection mechanisms. This remote response has important implications for midlatitude weather, notably toward the deterministic limit of numerical prediction (about two weeks). The mechanisms responsible for the generation and maintenance of large-scale convective heating can be quantified using CSRMs and super-parametrization. A poorly understood aspect of tropical variability is the interaction between the TIO and the Indian monsoon.

Global models with parameterized convection have varied success in realizing tropical intraseasonal oscillation. The multi-scale organization of convection is a key player. In the summer season, the TIO is complicated by the presence of monsoon winds in the Indian Ocean sector and large convective heating caused by monsoon rains north of the equator. Studies have shown that a branch of the eastward-propagating convection-related anomalies moves northward over the Indian Ocean and strongly affect the weather of the region.

8. CONCLUSION

Progress in modeling of tropical convection using explicit simulation and super-parametrization was reported herein. This has implications for the study of convection-related issues in the India/Indian Ocean locale. A three-pronged (regional, cloud-system-resolving and dynamical) modeling approach is viable. Recognizing that an adequate sounding network is probably not available, a practicable way to proceed with the observationally related part of the CSRM work is to obtain large-scale forcing and lateral boundary conditions from MM5 that, in turn, is driven by global model analysis. The purpose of the theoretical-dynamical modeling is to explore basic aspects, several of which have been pointed out herein.

REFERENCES

Arakawa, A., and W. H. Schubert, 1974: Interaction of a cumulus cloud ensemble with the large-scale environment: Part I: *J. Atmos. Sci.,* **31**, 674-701.

Carbone, R.E., J.D. Tuttle, D. Ahijevych, S.B. Trier, 2002: Inferences of Predictability Associated with Warm Season Precipitation Episodes, *J. Atmos. Sci.*, **59**, 2033-2056.

GCSS 2000: GEWEX Cloud System Study (GCSS) Second Science and Implementation Plan, *IGPO Publication Series*, **34**, 43 pp.

Grabowski, W.W., 2001: Coupling cloud processes with the large-scale dynamics using the Cloud-resolving Convection parametrization (CRCP). *J. Atmos. Sci*, **58**, 978-997.

Grabowski, W.W., and M.W. Moncrieff 2001: Large-scale organization of tropical convection in two-dimensional explicit numerical simulations. *Quart. J. Roy. Meterol. Soc.*, **127**, 445-468.

Grabowski, W. W., X. Wu and M.W. Moncrieff, 1998: Cloud-resolving Modeling of Cloud Systems during Phase III of GATE. Part II: Effects of Resolution and the Third Spatial Dimension. *J. Atmos. Sci.,* **55**, 3264-3282.

Houze, R.A., Jr., and A.K. Betts, 1981: Convection in GATE. *Rev. Geophys. Space Phys.*, **19**, 541-576.

Johnston, R.H., T.M. Rickenbach, S.A. Rutledge, P.E. Ciesieleski and W.H. Schubert, 1999: Trimodal characteristics of tropical convection. *J. Climate*, **12**, 2397-2418.

Kain, J.S., and J. M. Fritsch, 1993: Convective parametrization for mesoscale models: the Kain-Fritsch scheme. *Representation of Cumulus Convection in Numerical Models. Meteorol. Monog.*, **46**. 165-170.

Kairoutdinov, M. F., and D.A. Randall, 2001: A cloud resolving model as a parametrization in the NCAR Community Climate System Model: Preliminary results. *Geophys. Res. Lett.*, **28**, 3617-3620.

Laing, A.G., and J.M. Fritsch, 1997: The global population of mesoscale convective complexes. *Quart, J. Roy. Meteor. Soc.*, **123**, 2756-2776.

Liu, C., M.W. Moncrieff, and W.W. Grabowski, 2001: Hierarchical modeling of tropical convective systems using explicit and parameterized approaches *Quart. J. Roy. Meteorol. Soc.*, **127**, 493-515.

Liu, C., M.W. Moncrieff, and W.W. Grabowski, 2001: Explicit and parameterized realizations of convective cloud systems in TOGA COARE. *Mon. Wea. Rev.*, **129**, 1689-1703.

Madden, R., and P. Julian, 1972: Description of global-scale circulation cells in the tropics with a 40-50 day period. *J. Atmos. Sci.*, **29**, 1109-1123.

Mapes, B. E., 1993: Gregarious Tropical Convection. *J. Atmos.Sci.*, 2026–2037.

Moncrieff, M.W., 1981: A theory of organized steady convection and its transport properties. *Quart. J. Roy. Meteorol. Soc.*, **118**, 819-950.

Moncrieff, M.W., 1992: Organized convective systems: archetypal dynamical models, mass and momentum flux theory, and parametrization. *Quart. J. Roy. Meteorol. Soc.*, **118**, 819-950.

Moncrieff, M.W., 1995: Mesoscale convection from a large-scale perspective. *Atmos. Res.* **35**, 97-112.

Moncrieff, and E. Klinker, 1997: Mesoscale cloud systems in the Tropical Western Pacific as a process in general circulation models. *Quart. J. Roy. Meteorol. Soc., 123*, 805-827.

Moncrieff, M.W., S.K. Krueger, D. Gregory, J.-L. Redelsperger, and W.-K. Tao, 1997: Objectives of the GCSS Working Group on Precipitating Convective Cloud Systems. *Bull. Amer. Met. Soc.*, **78**, 831-845.

Moncrieff, M.W., 1998: Legacy of TOGA CORE for GEWEX. Proceedings of a Conference on the TOGA Coupled Ocean-Atmosphere Response Experiment (COARE), Boulder, CO, 7-14 July 1998, WCRP-**107**, WMO/TD-No. 940, 80-94.

Smolarkiewicz, P. K., and L.G. Margolin, 1997: On forward-in-time differencing for fluids: an Eulerian/semi-Lagrangian nonhydrostatic model for fluids. *Atmos. Ocean Special Ed.*, **35**, 127-152.

Tao, W.-K., S. Lang, J. Simpson, and R. Adler, 1993: Retrieval algorithms for estimating the vertical profiles of latent heat release. *J. Meteor. Soc. Japan*, **71**, 685-7000.

Tripoli, G. J. and W.R. Cotton, 1989: Numerical simulation of an observedorogenic mesoscale convective system Part 2: Analysis of governing dynamics. *Mon. Wea. Rev.*, **117**, 305-328.

Wu, X., and M.W. Moncrieff, 1996: Recent progress in cloud-resolving modeling of TOGA COARE and GATE cloud systems. *Proceedings, Workshop on New Insights and Approaches to Convective Parametrization*, ECMWF, Reading, United Kingdom, November 4-7, 1996, 128-156.

Wu, X., and M.W. Moncrieff, 2001: Long-term behavior of cloud systems in TOGA COARE and their interactions with radiative and surface processes. Part III: Effects on the energy budget and SST. *J. Atmos. Sci.*, **58**, 1155-1168.

Mesoscale and Cloud-Resolving Scale Simulation of a Heavy Precipitation Episode and Associated Cloud System Using MM5 Model

SOMESHWAR DAS

National Center for Medium Range Weather Forecasting, New Delhi – 110003, Email: *somesh@ncmrwf.gov.in*

ABSTRACT

The MM5 mesoscale model has been configured to run at cloud-resolving scale to study a heavy rainfall episode over Delhi that occurred during 13-15 August, 2001. Results have been compared with observations and those obtained from the global T80 forecast model of NCMRWF. Study shows that the T80 model severely underestimated the rainfall. The prediction was relatively better by the MM5 at 30 km resolution, though the values were still underestimated. The rainfall prediction improved considerably by the MM5 model at 1 km resolution when the cloud microphysics parameterization based on NASA/ Goddard Cumulus Ensemble Model (GCEM) was used.

1. INTRODUCTION

The MM5 model is a 5th generation PSU/NCAR Mesoscale Model (limited area), non-hydrostatic, terrain following sigma coordinate, designed to simulate or predict mesoscale & regional scale atmospheric circulation (NCAR, 2002). It can be configured to run from global scale to cloud scale in one model. It has multiple nesting capability. The model can be run in both 2-way and 1-way nesting mode. It has options for a wide variety of advance physical parameterization schemes. The model can be run using routine observations and, has provision for 4-dimensional data assimilation. Several options are available for the choice of physical processes in the MM5.

The model has been running at NCMRWF on Real-time basis for making mesoscale weather forecasts over the Indian region (Das, 2002a). The model is run on a triple nested domain at 90, 30 and 10 km resolutions. The model has been studied for several applications including mountain weather forecasting over the Indian region and the Orissa supercyclone over the Bay of Bengal by Das (2002b), Das and Singh (2002) Das and Dasgupta (2002).

Since MM5 is a regional model, it requires initial and lateral boundary conditions to run. The initial and lateral boundary conditions are obtained from the operational global model of NCMRWF. Initial and boundary values of upper air and surface fields are obtained from the outputs of the global T80 model through an interface program

(Das, 2002a). The boundary values are updated every 12 hours. The Four-Dimensional Data Assimilation (FDDA) is not carried out with MM5 at present. The data assimilation is carried out by the global model and its outputs are used for running the MM5.

The MM5 model has been used for different applications at cloud resolving scale by several investigators. Su *et al.* (1999) used the MM5 model at 2 km resolution to study the TOGA-COARE convective systems. Schultz and Albers (2001) presented a technique for diabatic initialization of cloud-resolving forecast models such as MM5, RAMS, ARPS and WRF that allows these models to prognose the precipitating cloud systems such as thunderstorms and snowstorms. Powers and Davis (2002) used the MM5 model at cloud resolving scale to study the tropical cyclogenesis. However, no study has been carried out by the MM5 model to study heavy precipitation events at cloud resolving scale during the summer monsoon over the Indian region.

The present study describes the application of the MM5 model at cloud resolving scale (1 km) to study the case of a heavy rainfall episode over northern India surrounding Delhi during monsoon-2001. Section 2 presents the synoptic conditions associated with the heavy rainfall event. Numerical experiments have been conducted to study the impact of horizontal resolution and different cloud microphysics parameterization schemes on the simulation of the heavy rainfall episode. Section 3 presents the results of the experiments. Finally, a summary of the results is presented in section 4.

2. SYNOPTIC CONDITIONS AND NUMERICAL EXPERIMENTS

Heavy rainfall occurred over most parts of north India during 13-15 August, 2001 associated with an active monsoon trough over the Indo-Gangetic plain and well marked low pressure circulation. Axis of the monsoon trough passed through Amritsar, Delhi, Kanpur and Sagar Islands during the period. Rainfall amounts ranging from 1 to 19 cm were recorded in 24 hours over Delhi and neighborhoods. The global T80 model of NCMRWF (summarized in Table 1) was unable to predict the intensity of rainfall observed over the region.

In order to investigate the predictability of the rainfall event, a nested mesoscale MM5 model (version 3.4 of PSU/NCAR) has been used to investigate the heavy rainfall episode over the Indian region. Table 2 shows a summary of the model as implemented at NCMRWF.

Numerical experiments have been conducted with the MM5 model to study the impact of horizontal resolutions and different cloud microphysics parameterizations, on the simulation of the heavy rainfall episode, which are summarized in the next section.

2.1 Numerical Experiments

Three experiments have been conducted to investigate the rainfall episode by configuring the model at different resolutions and by using different physics options. Table 3 presents the summary of experiments.

The model has been configured for non-hydrostatic, two-way nesting and 23 vertical levels. It has been run using the Grell (1993) scheme for cumulus parameterization and, the non-local closure scheme for boundary layer parameterization based on the Medium Range Forecast (MRF) model Planetary Boundary Layer (PBL) scheme

TABLE I: Description of the NCMRWF Global Spectral Model

Model Elements	Components	Specifications
GRID	Horizontal	Global Spectral- T80 (256 x128)
	Vertical	18 Sigma Layers (.995, .981, .96, .92, .856, .777, .688, .594, .497, .425, .375, .325, .275, .225, .175, .124, .074, .021)
	Topography	Mean
	Prognostic Variables	Relative Vorticity, Divergence, Virtual Temperature, Log of surface pressure, Water vapor mixing ratio
DYNAMICS	Horizontal Transform	Orszag's Technique
	Vertical Differencing	Arakawa' s Energy conserving scheme
	Time Differencing	Semi-Implicit, 900 sec
	Time Filtering	Robert's Method
	Horizontal Diffusion	Second order over quasi-press. Surface, scale selective
PHYSICS	Surface Fluxes	Monin-Obukhov similarity theory
	Turbulent Diffusion	Non-local closure
	Radiation	SW- Lacis & Hansen,Harshvardhan et al. (NASA/Goddard) L W -Fels & Schwarzkopf
	Deep convection	Kuo scheme modified
	Shallow Convection	Tiedtke's Scheme
	Largescale condensation	Manabe' s Scheme
	Clouds	Slingo's Scheme
	Rain Evaporation	Kessler's Scheme
	Land surface Process	Pan's (3-Layer Soil Temperature, Bucket Hydrology for Soil Moisture)
	Air-Sea Interaction	Roughness length (Chamock), SST, SH & LH (Bulk Form.)

of NCEP (Hong and Pan, 1996) for the boundary layer parameterization. A 5-layer soil model has been used for the land surface parameterization. The cloud process is based on simple ice scheme of Dudhia (1993) in EXPT1 and EXPT2, while the explicit cloud scheme of NASA (Tao and Simpson, 1993) has been used in the EXPT3. High resolution topography and vegetation data from USGS have been used to prepare the lower boundary conditions over the region. Initial and lateral boundary conditions have been prepared using the surface and upper air analysis fields of the NCMRWF global model and the boundary conditions have been updated every 12 hours as mentioned earlier.

TABLE 2: Summary of the MMS model as implemented at NCMRWF

Domain:

- Horizontal (Triple Nested)
  - Outer Domain-I (39 E- 121 E, 11 S -51 N) at 90 kln resolution
  - Middle Domain-2 (64 E -100 E, 4 N -45 N) at 30 kln resolution
  - Inner Domain-3 (73 E- 83 E, 32 N -40 N) at 10 kln resolution
  - Staggered Arakawa B-Grid
- Vertical: 23 Levels (Sigma-Hybrid), Being increased to 42
- Time Steps: Domain-I: 270 S, Domain-2: 90 S, Domain-3: 30 S
- Topography: USGS (Interpolated depending on resolution)
- Vegetation/ Land use 25 Categories (USGS)

Dynamics:

- Non-Hydrostatic
- Two-way Nesting
- Prognostic Equations: U, V, T, q, Ps, (Optional: TKE, cloud water, rain, ice, snow, graupel/ hail)
- Time Integration: Semi-implicit
- Boundary Conditions
  - Time-dependent / Nest
  - Relaxation/ inflow-outflow
  - Upper Radiative Condition
- **Horizontal Diffusion**
  - Fourth order for inner domains
  - Second order for the coarser domain
- **Four Dimensional Data Assimilation**
  (To be implemented shortly)

Physics:

- Cumulus Parameterization: Grell (Simplified Arakawa-Schubert)
- PBL Parameterization: MRF PBL
- Explicit Moisture Schemes: Simple Ice (Dudhia)
- Radiation Scheme: Simple cooling
- Land Surface Processes: Five layer soil model

TABLE 3 Summary of Numerical Experiments

Experiments	Scale	Description
EXPT1	Mesoscale (90 and 30 km)	Double nesting with Physics options: Grell-convection, MRF PBL, Simple ice and Radiation (Dudhia) and 5-layer soil model.
EXPT2	Mesoscale + Cloud scale (27,9,3,1 km)	Four nests with Physics options as in EXPT1
EXPT3	Mesoscale + Cloud scale (27,9,3, 1 km)	Four nests with Physics options as in EXPT1, but with explicit clouds of NASA and cloud radiation interaction in third and fourth nests

3. RESULTS

Figure 1a shows the observed rainfall based upon the estimates of satellite and rain gauge. It may be noted here that the observed rainfall are averaged over 150 km resolution and hence, it may not be able to represent the mesoscale features. The diagram shows wide spread rainfall over most parts of India between 13-14 August, 2001. Rainfall values of the order of 19 cm were recorded over the Delhi region (Figure 1b). This diagram also shows that there is a large gradient in the observed rainfall within Delhi itself. For instance, the North and East Delhi including the Hindon

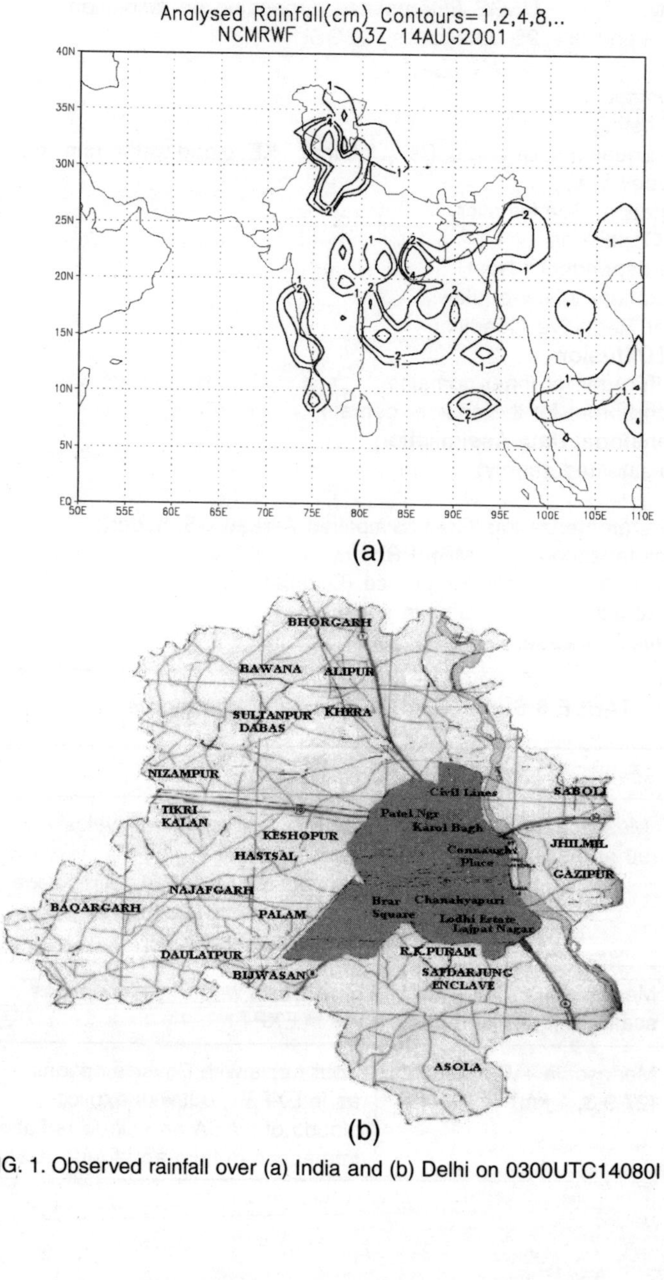

FIG. 1. Observed rainfall over (a) India and (b) Delhi on 0300UTC140801 in cm.

Air Force station reported rainfall values of about 16-19 cm, while the West and Southwest Delhi including Palam airport reported only 2 cm of rainfall as seen in the Figure 1b. The 24 hours forecast based on the global T80 model severely underestimated this rainfall and predicted less than 1 cm of rain around Delhi (Figure 2b). It is obvious that we need a high resolution mesoscale model in order to predict such variations in the rainfall over a small area. The MM5 model (EXPT1) predicted slightly better rainfall (Figure 2a) compared to T80, but the amounts around Delhi were only 2-3 cm. The model predicted rainfall value of 18.3 cm, but that was about 100 km north of Delhi.

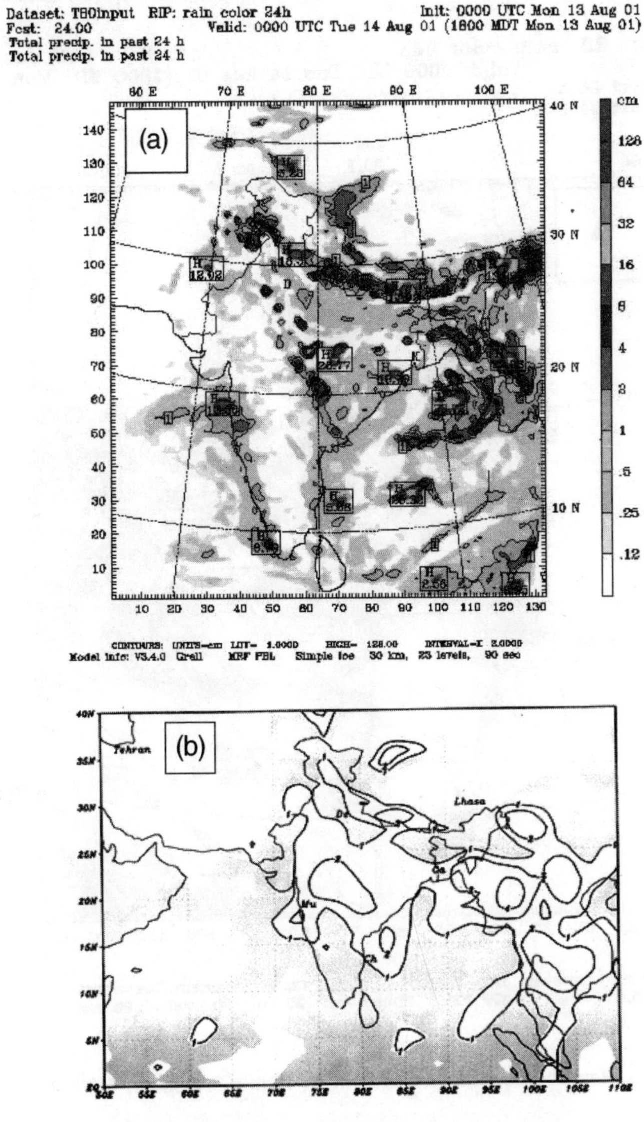

FIG. 2. Rainfall forecast (24 hrs) valid on 0003UTC140801 using (a) MM5 at 30 km grid and (b) Global T80 model.

The heavy rainfall band predicted by the operational MM5 were displaced by 100-300 km north and east of Delhi. It may be noted that the T80 model uses Kuo-Anthes scheme for cumulus parameterization and the non-local closure scheme for boundary layer parameterization (MRF-PBL).

Experiments showed that the rainfall forecast improved significantly when the resolution was increased to 1 km (EXPT2) in the MM5. However, the maximum value of rainfall obtained from this experiment around Delhi region was only about 4 cm (Figure 3a), which is still less than the observed value of 19 cm over the region. The rainfall forecast was improved further in the EXPT3 when the explicit cloud scheme of

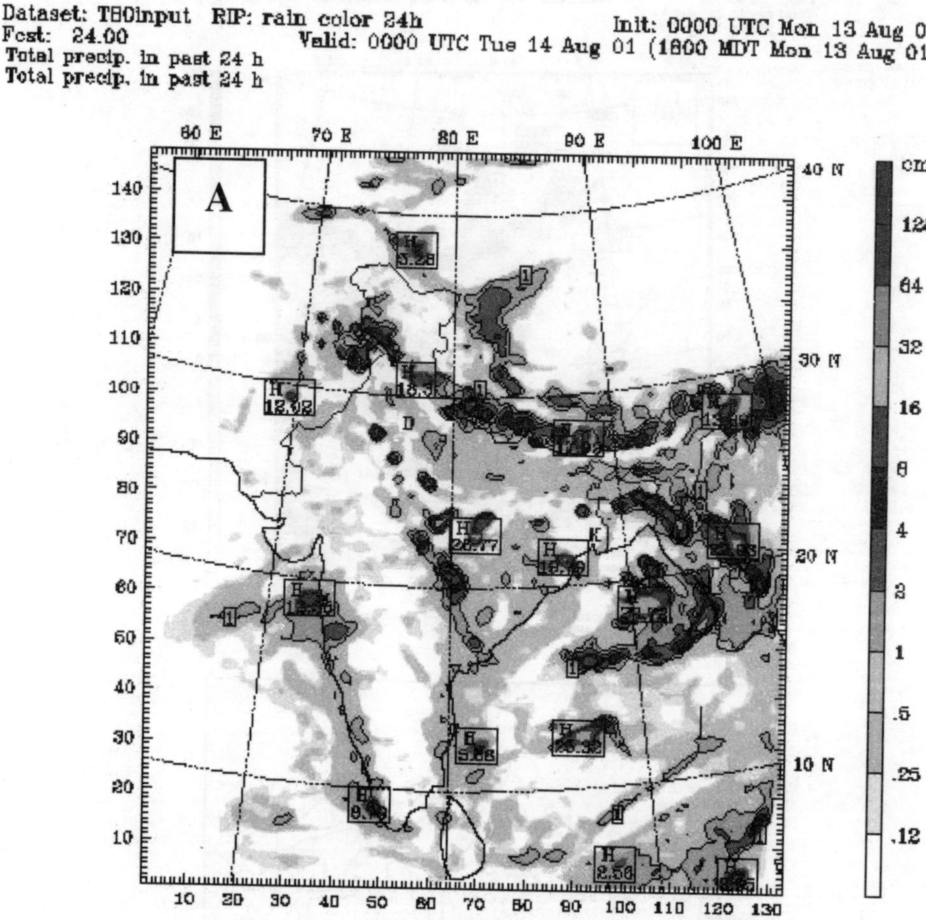

FIG. 3a. Rainfall forecasts around Delhi at 1 km resolution by EXPT2

NASA was used in the inner domains at 3 and 1 km resolutions. This experiment produced rainfall values about 4-8 cm over the region, which was better compared to the EXPT2. These results indicate that when the resolution alone is increased, there is a marginal increase in the rainfall as seen from EXPT1 (Figure 2a) and EXPT2 (Figure 3a). However, when the cloud microphysics is explicitly resolved treating the cloud liquid water, ice, snow and graupel separately by prognostic equations as in the EXPT3 (Figure 3b), the rainfall is improved significantly. Further experiments are required to examine the impact of other physical parameterizations on the rainfall forecasts.

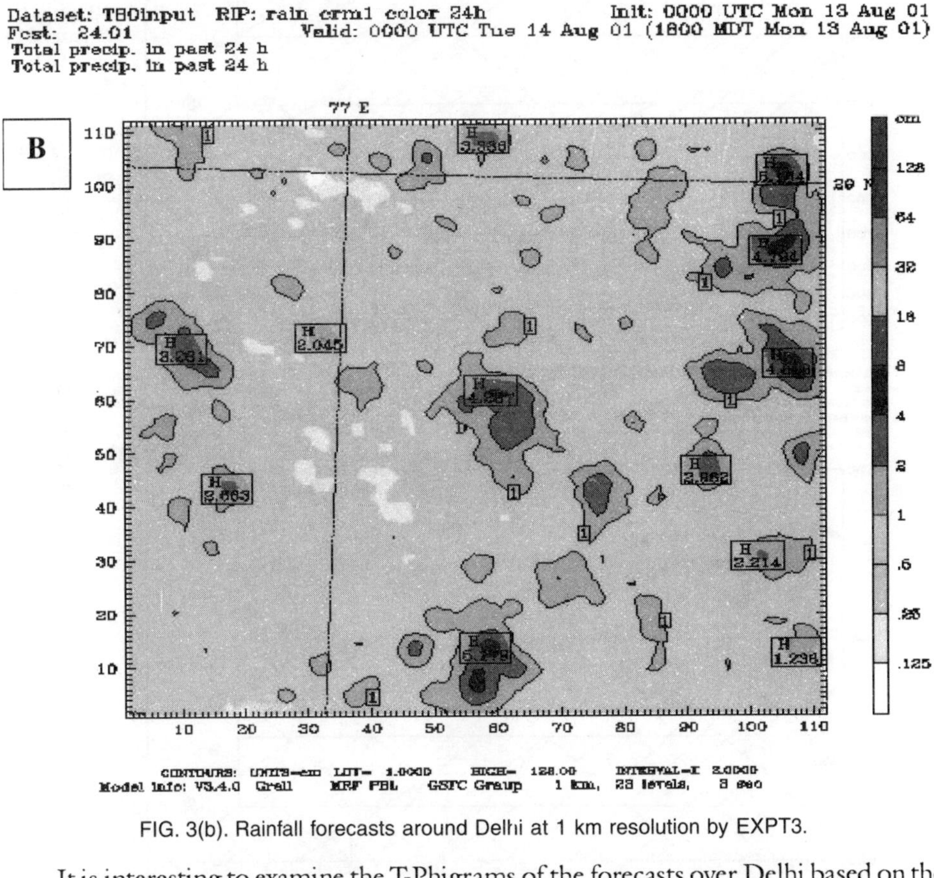

FIG. 3(b). Rainfall forecasts around Delhi at 1 km resolution by EXPT3.

It is interesting to examine the T-Phigrams of the forecasts over Delhi based on the 30 and 1 km resolution runs as shown in the Figure 4. The diagram indicates that vigorous mixing by clouds and resulting precipitation led to decrease in the convective Available Potential Energy (CAPE) at higher resolution. The magnitudes of CAPE at the two resolutions were 1527 and 203 respectively. Larger precipitation at 1 km resolution also led to cooling of land (26.5 K) compared to the 30 km resolution in which the land was warmer (28.5 K). The diagram also shows stronger easterly winds particularly at the upper levels over Delhi in the high resolution forecast indicating the strengthening of the monsoon trough.

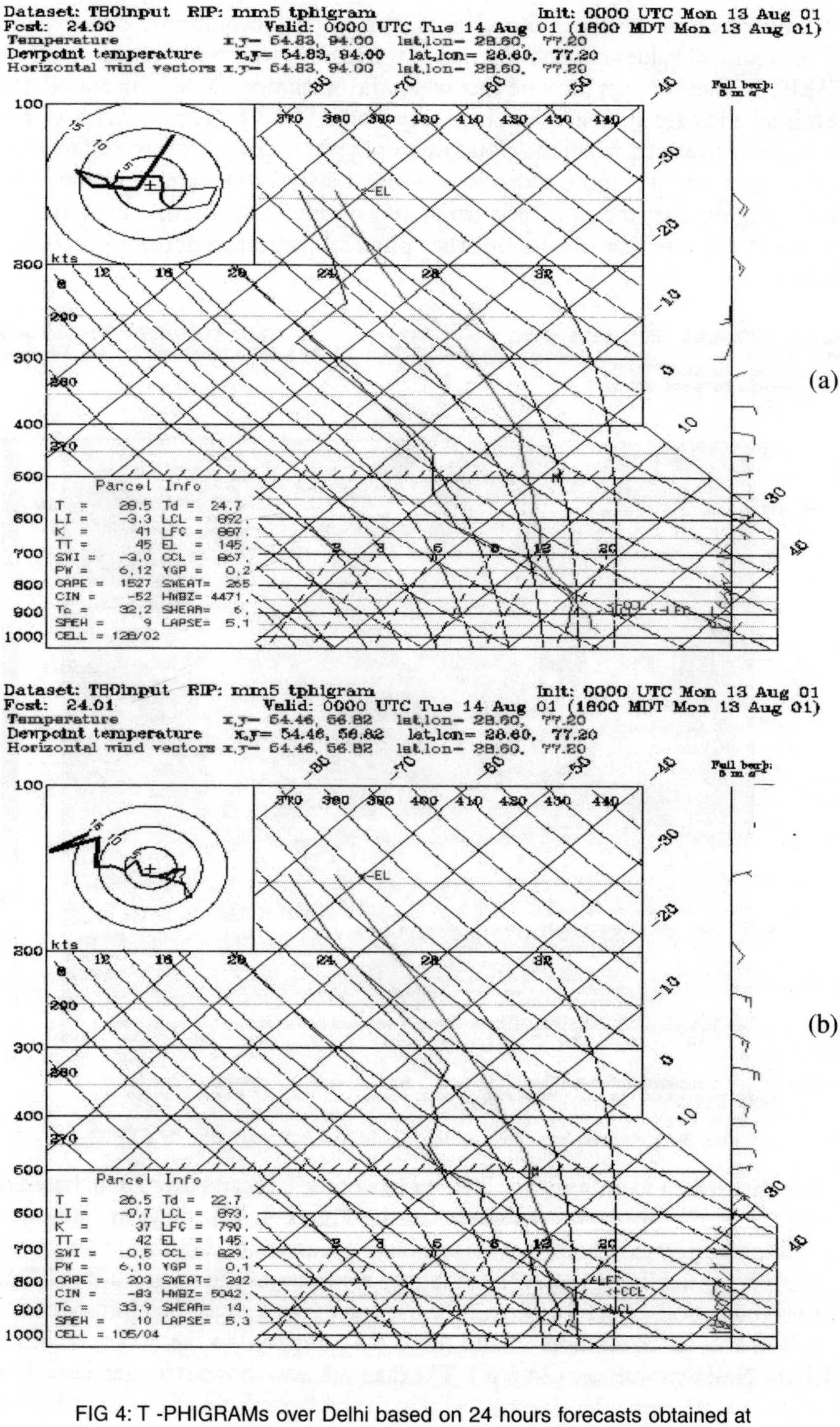

FIG 4: T -PHIGRAMs over Delhi based on 24 hours forecasts obtained at (a) 30 km and (b) 1 km resolution.

Figure 5 depicts the vertical cross section of simulated total cloud water (liquid + ice) mixing ratio and the total precipitation mixing ratio (snow + graupel + rain water) from northeast to southwest direction across Delhi. The diagrams clearly indicate that taller clouds were present towards northeast of Delhi producing intense precipitation. These values need to be compared with observations from radar and satellite.

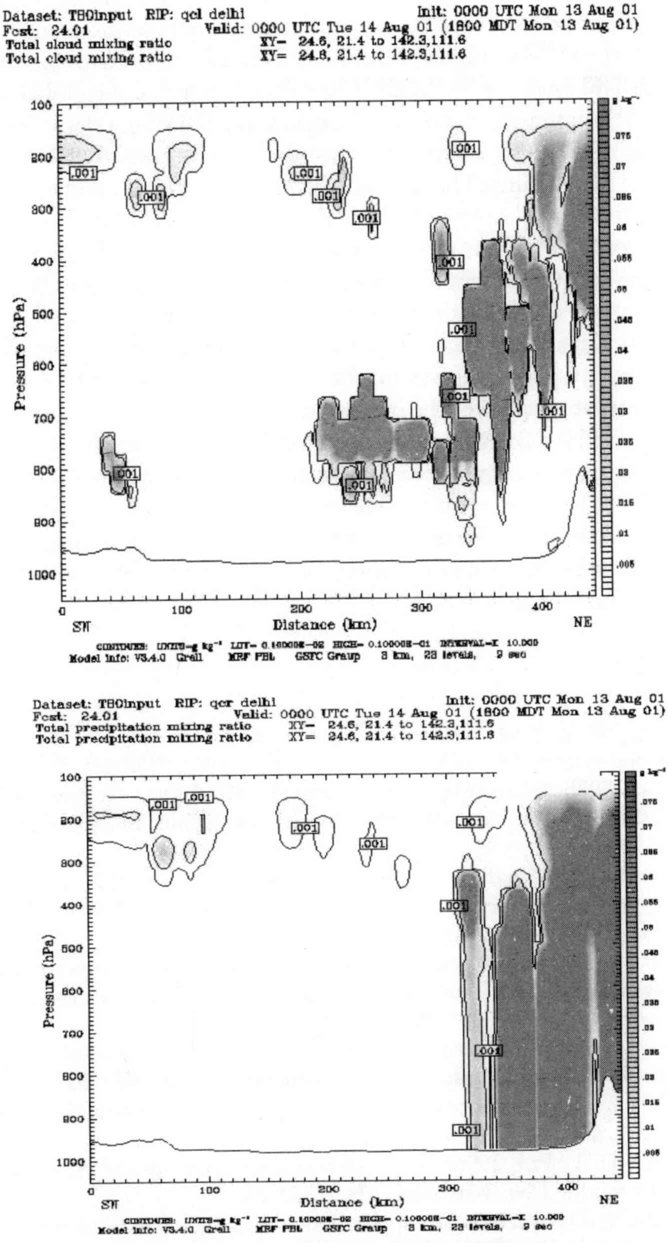

FIG 5: Forecasts (24 Hr) of (a) total cloud water {liquid + ice} and (b) snow+graupel+rain water mixing ratio in 9 kg^{-2} around Delhi.

4. SUMMARY

Forecasting intense precipitation and variation of rainfall over a small area is still a challenging problem for the numerical modelers. An investigation is carried out to study such a heavy precipitation episode including large gradients in the rainfall over Delhi. Experiments have been conducted at mesoscale (30 km) and cloud-resolving scale (1 km) using the MM5 model to simulate a heavy precipitation episode around Delhi region during 13-14 August, 2001. The forecast values have also been compared with the global T80 model of NCMRWF. Results indicate that the global model severely underestimated the intensity of rainfall. The operational MM5 model at 30 km resolution over the Indian region also failed to forecast the intense rainfall observed over Delhi, but it predicted heavy rainfall bands about 100-300 km north and east of Delhi. The simulation of rainfall was improved at cloud resolving scale (1 km) when explicit cloud scheme based on NASA was used in the model. Results indicate that increase in resolution alone is not sufficient to produce the observed high intensity of rainfall. The rainfall is improved significantly when the cloud microphysics is explicitly resolved treating the cloud liquid water, ice, snow and graupel separately by prognostic equations. Impact of other physical processes on the variation of rainfall in a mesoscale model remains to be investigated over the monsoon region.

Analysis of T-PHIGRAMS of the predicted fields indicate that vigorous mixing by clouds and resulting precipitation leads to decrease in the convective Available Potential Energy (CAPE) at higher resolution. Larger precipitation at high resolution also leads to cooling of land compared to the coarser resolution. Results showed strong easterly winds at upper levels over Delhi indicating the strengthening of the monsoon trough in the high resolution forecasts.

REFERENCES

Das Someshwar, 2002a: Real Time mesoscale weather forecasting over Indian region using MM5 modeling system. *NCMRWF Research Report No.NMRF/RR/3/2002.* 77pp.

Das Someshwar, 2002b: Evaluation and verification of MM5 forecasts over the Indian region. *Twelfth PSU/NCAR Mesoscale Model User's Workshop, 24-25 June, 2002, Boulder, CO, USA.* 77-81.

Das Someshwar and S.V. Singh, 2002: Mitigation of meteorological hazards over mountains using high resolution weather prediction model. *Conf. on sustainable mountain development in South Asia,* 19-21 Sept. 2002, Darjeeling, India.

Das Someshwar and Munmun Das Gupta, 2002: Mountain Weather Forecasts Over the Indian Region Using MM5 Model. *Workshop on Mesoscale Modeling with special emphasis on Mountain Weather Forecasting,* 29-30 July 2002, NCMRWF, New Delhi.

Dudhia, J, 1993: A non-hydrostatic version of the PSU/ NCAR mesoscale model: Validation tests and simulation of an Atlantic cyclone and cold front. *Mon. Wea. Rev.,* **121**, 1493-1513.

Grell, G. A., 1993: Prognostic evaluation of assumptions used by cumulus parameterization. *Mon. Wea. Rev.*, **121**, 764-787.

Hong, Y.-H. and H.-L. Pan, 1996: Nonlocal boundary layer vertical diffusion in a medium range forecast model. *Mon. Wea. Rev.,* **124**, 2322-2339.

Powers J.G. and C.A. Davis, 2002: The application of the MM5 to tropical cyclogenesis at cloud resolving resolution. *12th PSU/NCAR Mesoscale Model User's Workshop, 24-25 June, 2002, Boulder, CO, USA. pp 141-144.*

Schultz, P. and S. Albers, 2001: The use of three dimensional analysis of cloud attributes for diabatic initialization of mesoscale models. *Preprints, 14th Conf. On Numerical Weather Prediction, Fort Lauderdale, Amer. Meteor. Soc., J122-J124.*

Su, H., S. S. Chen, and C.S. Bretherton, 1999: Three-dimensional week-long simulations of TOGA-COARE convective systems using the MM5 mesoscale model. *J. Atmos. Sci.,* **56,** 2326-2344

Tao, W.-K, and J. Simpson, 1993: The Goddard Cumulus Ensemble model. Part -I: Model description. *Terres., Atmos. and Oceanic Sci.*, **4**, 35-72.

Performance of Mesoscale Eta Model over Indian Region

E.N. RAJAGOPAL & G.R. IYENGAR

National Centre for Medium Range Weather Forecasting
Department of Science and Technology
New Delhi – 110 003, INDIA
E-mail: rajagopal@ncmrwf.gov.in

ABSTRACT

A workstation version of the Eta model has been installed at NCMRWF. The Eta model forecasts for select cases of tropical cyclone, western disturbance and monsoon depression are compared with the global model and actual observations. In general, Eta model forecasts are found to be better than operational global model over Indian region. Eta model produces better orographic precipitation namely, the rainfall along the Himalayan mountain ranges, Western and Eastern Ghats.

1. INTRODUCTION

The Indian subcontinent is dominated by a wide variety of weather phenomena including monsoons, cyclones, western disturbances, norwesters, squall lines and severe thunderstorms. These phenomena operate at different scales ranging from mesoscale to large-scales, which gives rise to wide-spread rainfall and, at times very heavy snow fall.

NCMRWF has been actively involved in the establishment of mountain weather forecasting system over northwest India in association with DRDO and IMD. The National Centers for Environmental Prediction (NCEP), USA developed a step-mountain eta (η) coordinate model generally known as the Eta model and was made operational in 1993 (Black, 1994). With the availability of more computer power at NCMRWF, it is now possible to run mesoscale models like Eta and MM5 in real time in an operational environment.

Recently a workstation version of Eta model (Version 0.2) which was released in March 2000 has been installed on a SGI-Origin 200 workstation at NCMRWF (Rajagopal and Iyengar, 2001, Rajagopal *et al.* 2001). The main features of the Eta model as implemented at NCMRWF is given Table 1.

2. MODEL DESCRIPTION

The mesoscale Eta model is a hydrostatic model that we have run with a horizontal grid spacing of 32 km and 38 vertical levels, with layer depths that range from 20 m in the planetary boundary layer to 2 km at 50 mb. The eta coordinate, defined by Mesinger

TABLE 1 Description of Mesoscale Eta Model (Workstation Version 0.2 March 2000 Release, NCEP)

Grid: Semi-staggered Arakawa E-grid
Horizontal Resolution: 32 km
Vertical levels: 38
Model Domain: 60 – 100 E & 0 – 40 N (Variable)
Model Time step (Dt): 90 seconds
Topography: Silhouette step topography
Model Dynamics: (Mesinger et al., 1988; Black, 1994)·

- Fundamental prognostic variables – T, u, v, q, Ps, TKE, cloud water/ice·
- Inertial gravity wave adjustment – forward-backward scheme (Dt=90s)·
- Vertical advection
    - Euler-backward scheme
    - centred in space
    - piecewise linear for moisture variables
- Horizontal advection
    - modified Euler-backward scheme
    - Janjic advection in space
    - Conservative, (nearly) shape-preserving scheme for moisture variables
    - Upstream advection near boundaries

Model Physics: (Janjic, 1994; Mon. Wea. Rev.)

- Betts-Miller-Janjic convection
- Mellor-Yamada level 2.5 turbulent exchange
- GFDL Radiation
- Explicit cloud water/ice prediction
- 4 layer soil scheme with vegetation
- D^2 horizontal diffusion

Initial and Lateral boundary conditions: One way interaction

- NCMRWF global analysis and 6 hourly global model forecasts

Time-dependent Surface Fields:

- SST, Snowdepth,
- Surface analysis from NCMRWF global model

(1984), was used in order to remove the large errors which are known to occur when computing the horizontal pressure gradient force, as well as the advection and horizontal diffusion, along a steeply sloped coordinate surface, such as the sigma surfaces. This coordinate system makes the eta surfaces quasi-horizontal everywhere as opposed to sigma surfaces which can be steeply sloped. Thus, this model should show marked improvements in areas with widely varying topography. Because the eta coordinate is pressure based and normalized (i.e. quasi-horizontal), it leads to a much simpler solution of the equations of motion in areas such as the pressure gradient force, horizontal advection, and diffusion. The eta coordinate is defined by the relationship:

$$\eta = \left(\frac{\rho - \rho_T}{\rho_{sfc} - \rho_T}\right)\left[\frac{\rho_{ref}(Z_{sfc}) - \rho_T}{\rho_{ref}(0) - \rho_T}\right]$$

where T refers to the top of the domain (25 mb), sfc is at the model's lower boundary, and ref refers to a reference pressure level that is a function of distance above sea level (Black 1994).

2. 1 Horizontal Grid

The semi-staggered Arakawa E grid (Arakawa and Lamb 1977) is the basis of the model's horizontal structure. This grid staggers the mass variables and the wind variables and is designed to minimize errors associated with geostrophic adjustment and topographic forcing that occur in other horizontal grid domains. The E grid lies upon a rotated latitude-longitude framework. This coordinate system is created by simply rotating the earth's entire geographic latitude-longitude grid so as to place the intersection of the equator and the prime meridian (or 0 degrees north and east) over the center of the forecast area. In doing this, the convergence of the meridians, which occurs as one moves away from the equator towards the poles, is minimized over the forecast area, thus leading to a more uniform and evenly spaced grid. Each grid box consists of a mass point surrounded by four velocity points, all of which lie along parallels and meridians of the rotated latitude-longitude coordinate system.

2.2 Model Topography

The model topography is represented as discrete steps whose tops coincide exactly with one of the model's 38 vertical layer interfaces (Black 1994). In determining their elevations, each 32 km horizontal grid box is first divided into 16 subboxes. Mean elevations for each of these 16 subboxes are calculated from official United States Geological Survey (USGS) topographical data. Using these values, the maximum mean value from each of the four rows and four columns are determined, resulting in eight intermediate terrain values. The mean of these eight values are taken to yield an intermediate value for the step height. Having already determined the height of each model layer interface based on the standard atmosphere and the specified distribution of vertical resolution, the final step elevation is found by moving the mean either up or down to match the closest layer interface in the model domain.

Eta model topography over Indian region was calculated from USGS topographical data. The Palghat Gap in the Western Ghats and 2 peaks (Nilgiri and Anai Mudi) on either sides of the gap are clearly resolved in model topography (Figure 1), however, the peak heights in model topography are lesser than the actual.

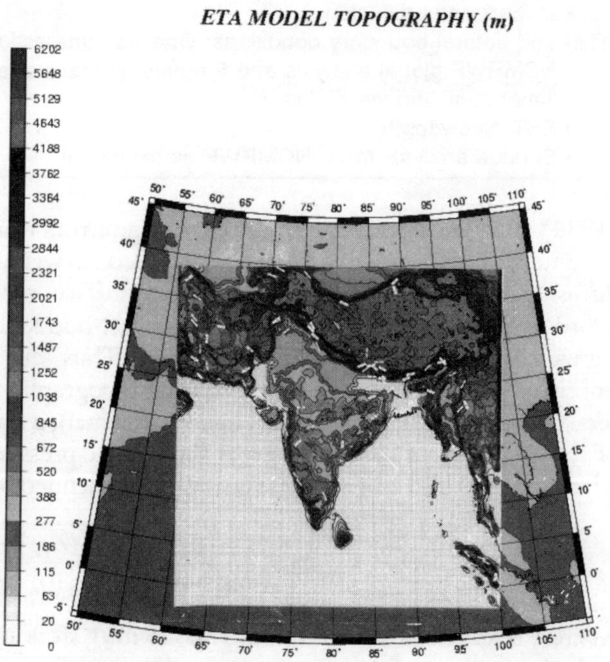

FIG. 1: Eta model (32km/L38) topography over Indian region (in metres)

2.3 Lateral Boundary Conditions

The model's boundary data on its single outermost row of points are obtained by direct interpolation from the Global Spectral Model (GSM). At inflow boundary points, all of the prognostic variables are determined by the GSM data, while at outflow points, the velocity components tangential to the boundary are extrapolated from the interior of the integration domain. The values of the second outermost row are a blend of those along the boundary and those in the third row which are part of the true integration domain.

2.4 Prognostic Variables

The primary prognostic variables in the Eta model are temperature, specific humidity, horizontal wind components, surface pressure, and turbulent kinetic energy. More recently, cloud water and ice have also become prognostic variables. A split-explicit integration approach is used in producing forecasts based on these quantities. This means that during each time step, after each process is computed, each of the primary variables is updated and the integration continues to the next time step when both the primary variables and any advective processes are updated. The fundamental time step in the Eta model used here is 90 seconds, with twice that time step required for advective processes.

3. PHYSICAL PROCESSES

3.1 Convective Precipitation

Convective precipitation, based on the Betts-Miller cumulus parameterization (Betts 1986; Betts and Miller 1986) with some of the modifications based on work by Janjic (1994), is computed every eight adjustment time steps, or about every ten minutes. In this parameterization scheme, non-precipitating shallow convection (i.e. fair weather cumulus) serves to carry moisture upward and maintain low-level temperature inversions. Deep convection (i.e. thunderstorms) transports heat and moisture upward and produces precipitation. For both types of convection, model profiles of temperature and specific humidity are constructed at each grid point using the values that are present in the model. These model profiles are compared to profiles derived from numerous observations based on actual convection. The model profiles are then relaxed toward these field-generated reference profiles. The precipitation is deduced from the net negative change of specific humidity in the model's deep convective cloud; if the net change is positive (i.e. net evaporation occurred rather than condensation), no adjustment of the variables is made at that grid point.

3.2 Large-scale Precipitation

The model calculates large-scale precipitation by using an explicit cloud water parameterization scheme (Zhao *et al.* 1997). The mixing ratio of cloud water and ice has recently been added to the temperature, specific humidity, wind components, surface pressure, and turbulent kinetic energy as a fundamental prognostic variable. This explicit cloud water parameterization takes into account the physical processes of evaporation, condensation, melting, freezing, sublimation, and deposition which occur in the atmosphere. The amount of condensed cloud water is predicted throughout the domain and a three-dimensional array indicates whether that condensate is liquid or solid if supersaturation occurs in a model grid box. The cloud phase is strictly a function of temperature: (a) if T> 0° C then the cloud is liquid; (b) if T<–15°C then the cloud is ice; (c) for intermediate temperatures, the cloud will be ice only if cloud ice existed in the same grid box or any box directly above during the previous time step. Thus, ice is not allowed to develop in the cloud process until the temperature reaches –15° C. However, if ice exists in the layer above a grid box which is between –15°C and 0°C, then ice will also form in this grid box. This simulates a process called "seeding" which happens often in nature as ice crystals fall into a warmer cloud below it producing larger ice crystals in that cloud and ultimately precipitation. Cloud water is allowed to evaporate or sublimate when the relative humidity drops below a critical value. This critical value differs for locations in the model which are located above water or above land. The model does not allow for more than one phase of cloud water within any given grid box, although the grid-scale precipitation within a box can be both liquid and solid. It also does not consider supercooled water below –15°C. There is currently no direct interaction between the cloud water scheme and the Betts-Miller convection parameterization, thus the convective scheme produces no cloud water.

3.3 Radiation

The current version of the radiation package used in the model is one developed at GFDL, USA. It requires the specification of only three levels of cloud (low, middle, and high). These are determined from the cloud amounts in all of the model layers which in turn were calculated by a simple relation based on the relative humidity. The model's carbon dioxide and ozone distributions are taken from climatology and held constant. The initial surface albedo is also taken from climatology, but is allowed to evolve during the forecast. However, snow cover is calculated from analyzed data allowing the albedo to reflect some reality. The atmospheric temperature tendencies arising from the radiative effects are applied after every adjustment time step.

3.4 Land Surface Processes

A new soil package has been developed for the Eta model, which provides a more rigorous description of the ground processes (Chen *et al.* 1997). The model now employs four predictive soil layers of 10, 30, 60, 100 cm depth. The prognostic temperature for these layers is valid at their centers. Internal vertical heat and moisture fluxes are computed. The initial soil temperature and moisture are taken from the 6-hour forecast from the Global Data Assimilation System. Soil type, vegetation type, and green vegetation fraction are specified as functions of geographic location and the latter also as a function of time. Variability in these quantities permits a much improved handling of such factors as ground permeability, transpiration effects, and albedo. Heat exchange with the atmosphere takes place either at the bare ground interface or on the plant canopy surface. Moisture exchange occurs with the bare soil and with condensed water on plant surfaces as well as through transpiration via leaf pores. Currently, the soil and vegetation fields are derived from 1 × 1 degree data which are then extrapolated to the model grid points, but are expected to be taken from much higher resolution sources in the future.

3.5 Planetary Boundary Layer

The vertical turbulence exchange in the Eta model uses a modified Mellor-Yamada Level 2.5 scheme (Black 1994). In this approach, turbulent kinetic energy (TKE) is a fully prognostic variable which is recalculated every eight advective time steps. The predicted TKE is then used to compute the exchange coefficients for the transfer of heat, moisture, and momentum between adjacent model layers. The exchange coefficients are used to modify the prognostic variables in the grid box through which the transfer is occurring. Surface fluxes of moisture and heat between the model surface and the first model layer are computed using Monin-Obukov functions (Black 1994).

4. MODEL INPUTS

4.1 Initial and Lateral Boundary Conditions

NCMRWF global (T80/L18) analysis and 6 hourly global model forecasts are used for initial and lateral boundary conditions. Wind, Temperature, relative humidity and geopotential height interpolated to 26 pressure levels at 1° x 1° global resolution are used

4.2 Time-dependent Surfaces Fields

Daily NCEP 1° global SST analysis , 47-km daily USAF global snow depth analysis and 23-km daily NESDIS/NOAA snow cover (NH only) downloaded from *ftp.ncep.noaa.gov* are used. Soil moisture at 2 layers (surface & root zone), soil temperatures at 3 layers and global Sea-Land-Ice mask are extracted/generated from surface analysis of NCMRWF global analysis-forecast system.

5. RESULTS AND DISCUSSIONS

5.1 Test Runs

Initial and lateral boundary conditions for the model test runs were taken from the 6 hourly WAFS (World Area Forecast System) files downloaded from NCEP. Soil moisture , Soil temperatures and global Sea-Land-Ice mask were taken from surface analysis file of from NCEP/AVN model. Eta (32km/L38) model was run on-real time basis with WAFS data for issuing weather forecasts for the Indian Army Everest Expedition 2001 during April-May. The Eta model's temperature, wind, clouds and rainfall forecasts over Eastern Himalayan region were found to be very useful for issuing Special Weather Bulletins for the Expedition.

5.2 Select Cases

The forecast results of (1) active monsoon (13 August 2001), (2) tropical cyclone (15 October 2001), (3) northeast monsoon surge (21 December 2001) and (4) western disturbance (15 January 2002) cases are discussed in this study. The forecast results of both Eta and global (T80L18) models are compared with analysis/observations and the results are described below.

a) Active monsoon (13-16 August 2001)

During this period the circulation features was characterized by a strong low level westerly flow over the Arabian Sea and adjoining peninsula. A low pressure are formed over the Bay of Bengal and adjoining areas of West Bengal and Orissa on 13th , which persisted over the same are till the next day. It then moved north-westwards and became less marked on 16th. Monsoon was active over the northwest and central parts of India and along the west coast of India.

Figure 2(a, b, c) shows the analyzed rainfall (Mitra *et al.*, 1997), day 1 forecast of T80L18 model and ETA model valid for 14th August.

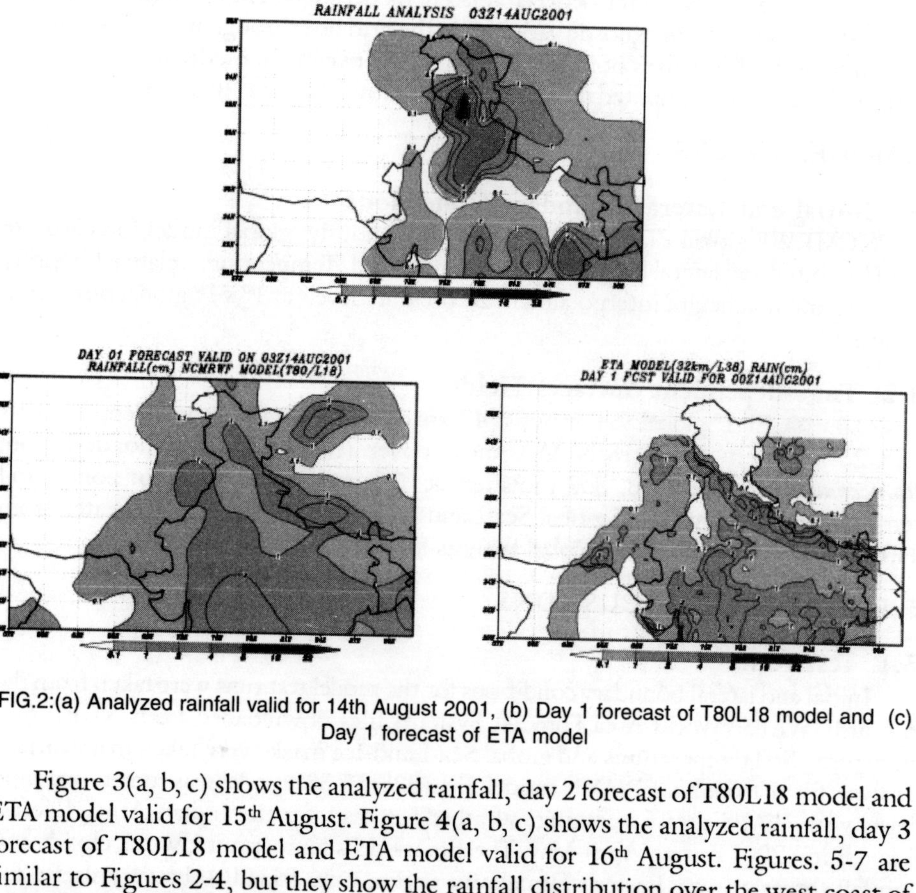

FIG.2:(a) Analyzed rainfall valid for 14th August 2001, (b) Day 1 forecast of T80L18 model and (c) Day 1 forecast of ETA model

Figure 3(a, b, c) shows the analyzed rainfall, day 2 forecast of T80L18 model and ETA model valid for 15th August. Figure 4(a, b, c) shows the analyzed rainfall, day 3 forecast of T80L18 model and ETA model valid for 16th August. Figures. 5-7 are similar to Figures 2-4, but they show the rainfall distribution over the west coast of India and the adjoining peninsula.

It can be seen from the figures that the ETA model forecasts show more amounts of rainfall as compared to the T80L18 model over the northwest and central parts of India as well as along the west coast of India.

b) Tropical Cyclone (15-17 October 2001)

During this period a depression formed in west central Bay of Bengal on 15th October and intensified into a cyclonic storm in the same evening. It moved WNW direction and crossed near AP coast near Nellore on 16th morning. Nellore and neighboring areas recorded rainfall amounts of 15-25 cm on 03 UTC 16th October. It remained practically stationary near Nellore till 17th Morning and weakened thereafter into a LPA. Coastal AP recorded rainfall amounts of 15-23 cm on 03 UTC 17 October.

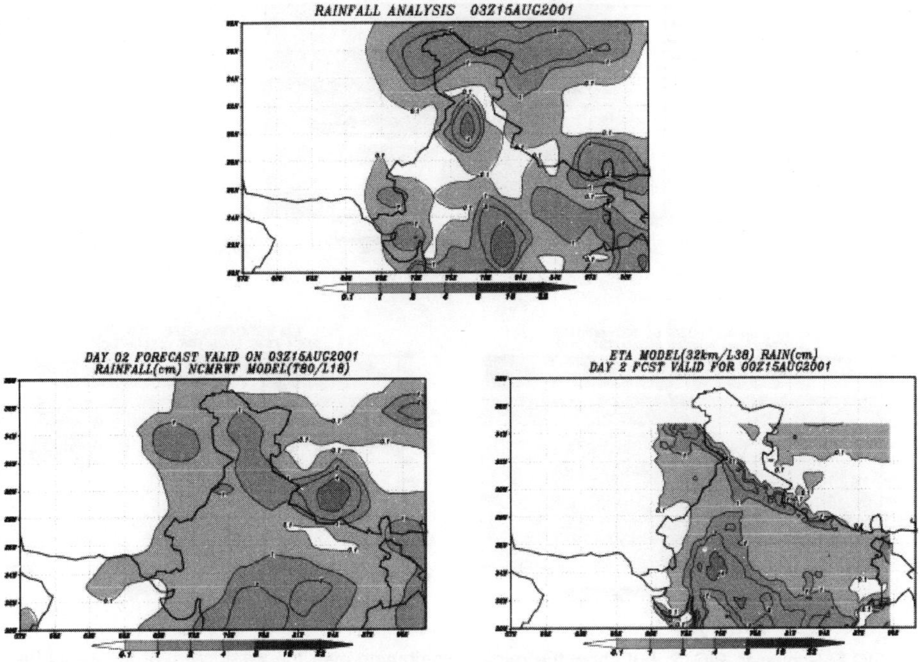

FIG.3:(a) Analyzed rainfall valid for 15th August 2001, (b) Day 2 forecast of T80L18 model and (c) Day 2 forecast of ETA model

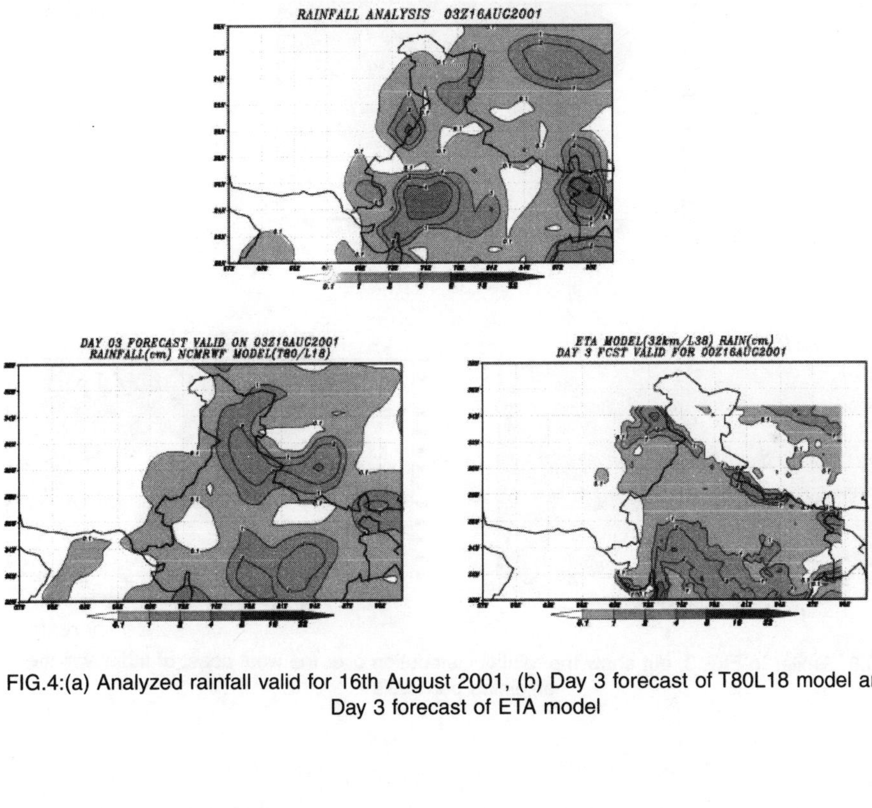

FIG.4:(a) Analyzed rainfall valid for 16th August 2001, (b) Day 3 forecast of T80L18 model and (c) Day 3 forecast of ETA model

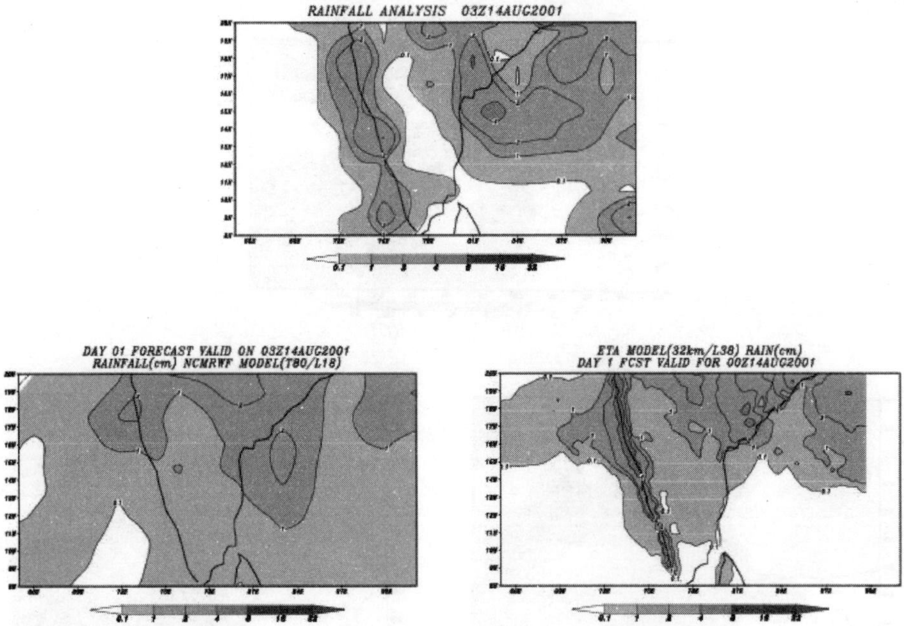

FIG.5: Similar to FIG. 2, but show the rainfall distribution over the west coast of India and the adjoining peninsula

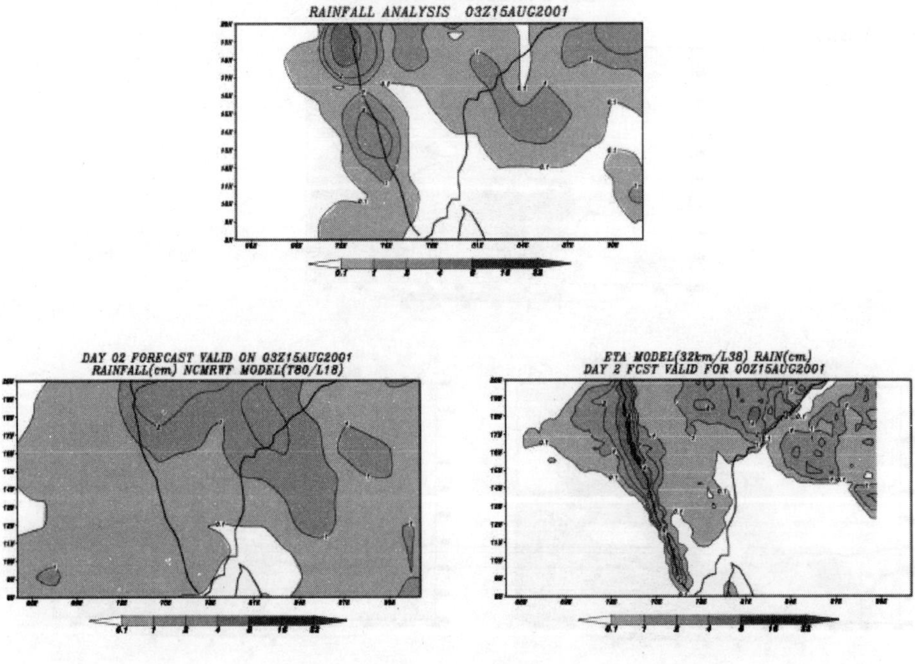

FIG.6: Similar to FIG. 3, but show the rainfall distribution over the west coast of India and the adjoining peninsula

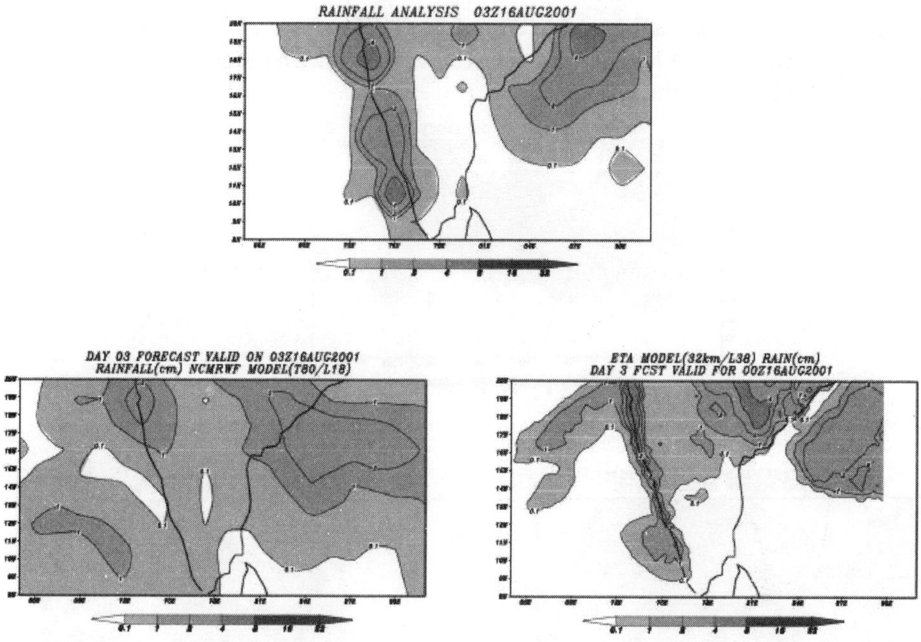

FIG.7: Similar to FIG. 4, but show the rainfall distribution over the west coast of India and the adjoining peninsula

Figure 8(a, b, c) shows the analyzed rainfall, day 1 forecast of T80L18 model and ETA model valid for 16th October.

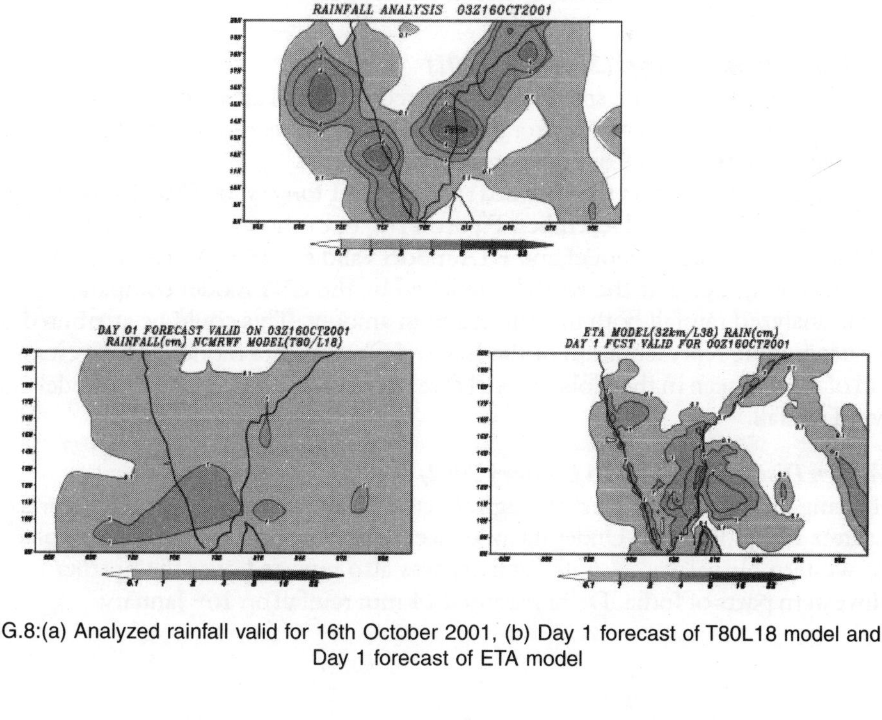

FIG.8:(a) Analyzed rainfall valid for 16th October 2001, (b) Day 1 forecast of T80L18 model and (c) Day 1 forecast of ETA model

Figure 9(a, b, c) shows the analyzed rainfall, day 2 forecast of T80L18 model and ETA model valid for 17th October. The rainfall associated with the tropical cyclone has been better predicted by the ETA model.

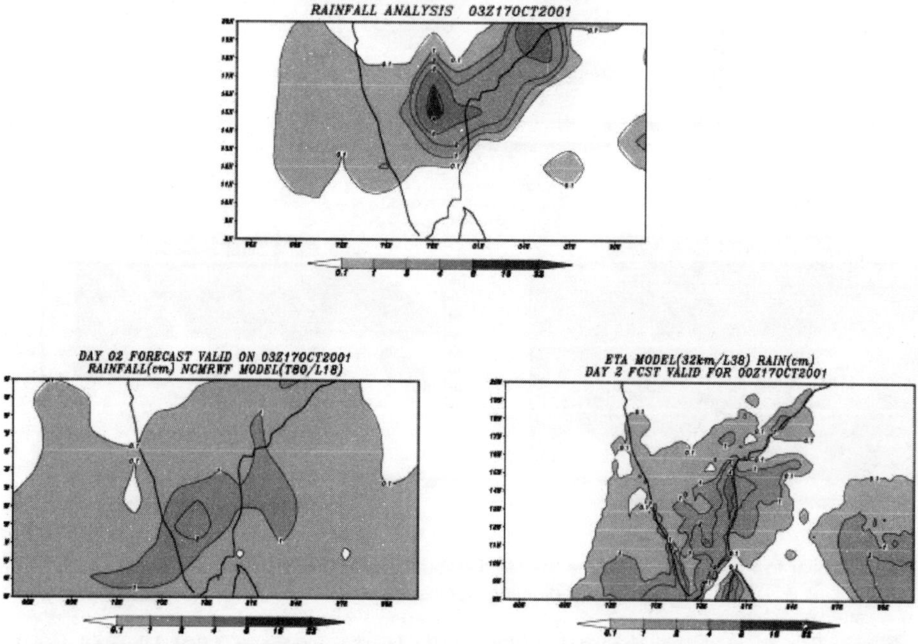

FIG.9:(a) Analyzed rainfall valid for 17th October 2001, (b) Day 2 forecast of T80L18 model and (c) Day 2 forecast of ETA model

c) Northeast Monsoon Surge (21-24 Dec 2001)

During this period wide spread rain occurred in coastal areas of Tamil Nadu and south coastal AP associated with a strong surge of northeast monsoon. Rainfall amounts of 5-10 cm were reported from the region.

Figure 10(a, b, c) shows the analyzed rainfall, day 1 forecast of T80L18 model and ETA model valid for 22nd December. Figure 11(a, b, c) shows the analyzed rainfall, day 2 forecast of T80L18 model and ETA model valid for 23rd December. It can be seen from the figures that the rainfall predicted by the ETA model compares better with the analyzed rainfall both in pattern and in amount. This could be attributed to better orographic representation of the Eastern Ghats in Eta model. The westward spread of rainfall seen in the global model forecasts is not seen in both Eta model and analyzed rainfall.

d) Western Disturbance (13-16 January 2002)

During this period a westerly trough affected the northern and northwest parts of India after a long dry spell. Under its influence, rain/snow occurred over many places in the western Himalayas. Rainfall activity was also reported over the northern and northwestern parts of India. Delhi recorded 14 mm rainfall on 16th January.

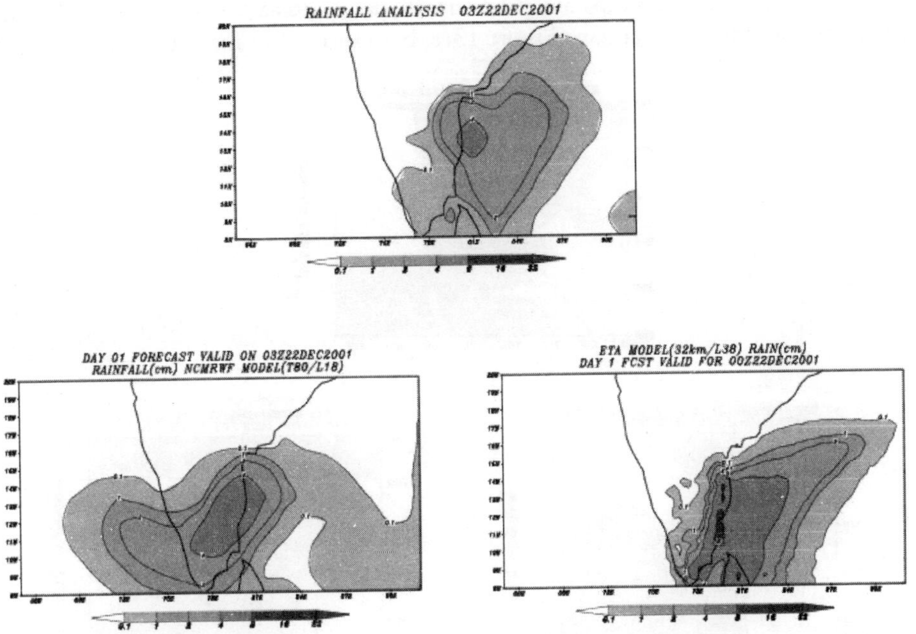

FIG.10:(a) Analyzed rainfall valid for 22nd December 2001, (b) Day 1 forecast of T80L18 model and (c) Day 1 forecast of ETA model

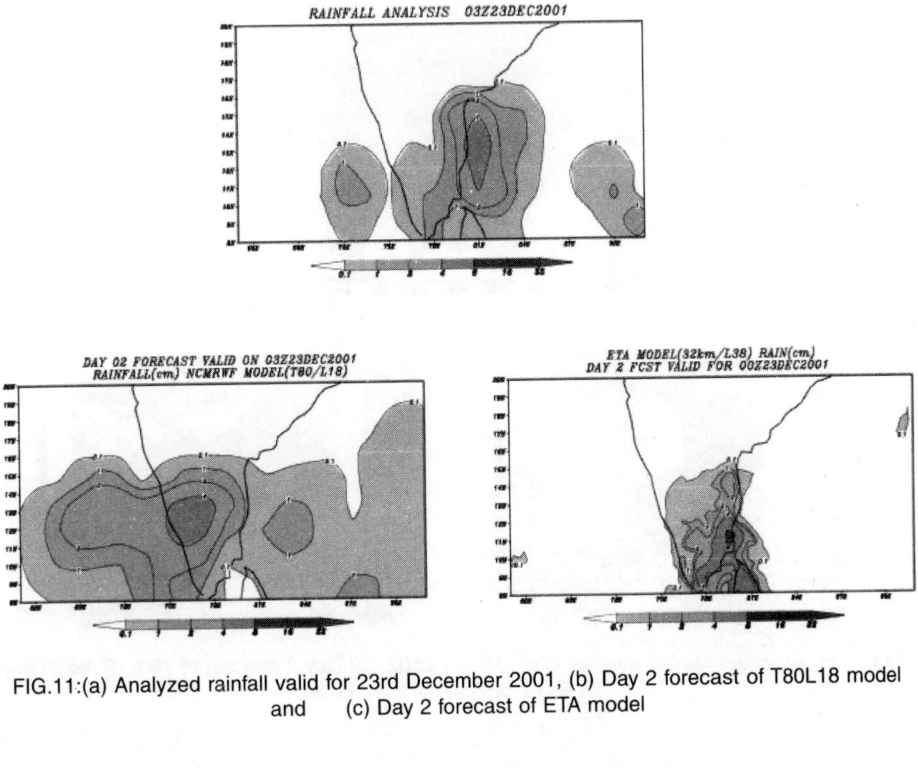

FIG.11:(a) Analyzed rainfall valid for 23rd December 2001, (b) Day 2 forecast of T80L18 model and (c) Day 2 forecast of ETA model

Figure 12(a, b, c) shows the analyzed rainfall, day 1 forecast of T80L18 model and ETA model valid for 16th January. Figure 13(a, b, c) shows the analyzed rainfall, day 2

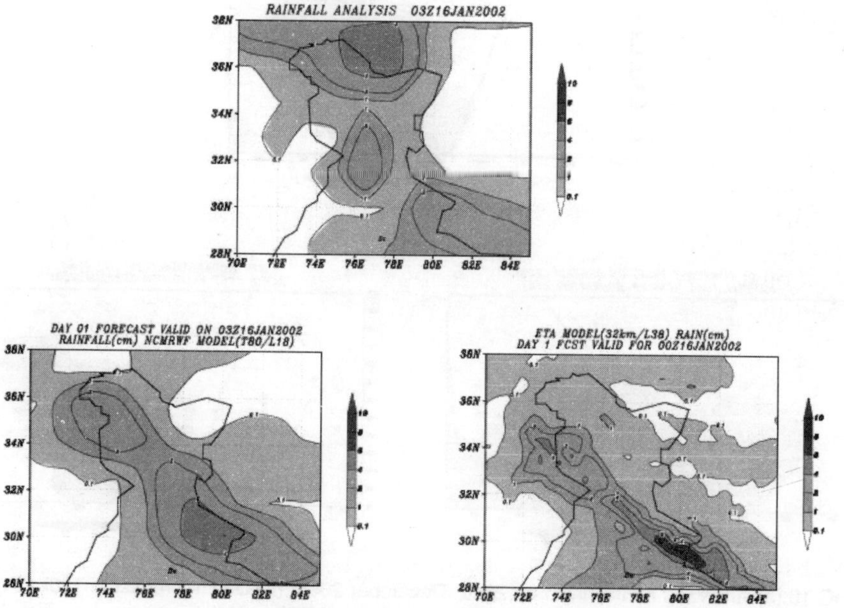

FIG.12 (a) Analyzed rainfall valid for 16th January 2002, (b) Day 1 forecast of T80L18 model and (c) Day 1 forecast of ETA model.

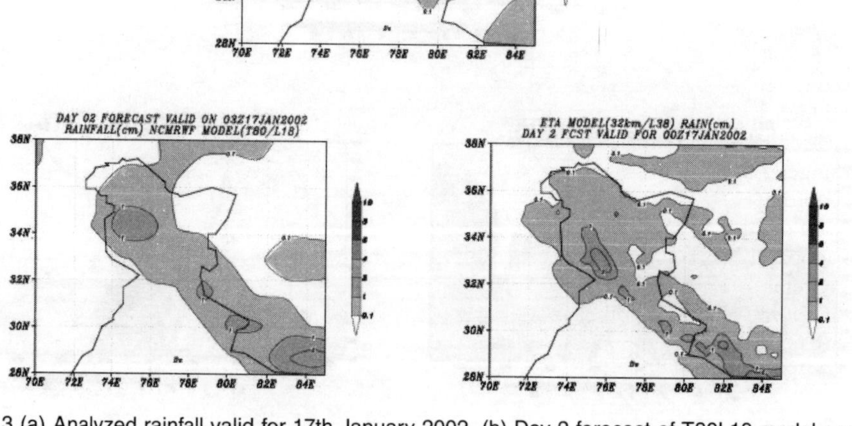

FIG.13 (a) Analyzed rainfall valid for 17th January 2002, (b) Day 2 forecast of T80L18 model and (c) Day 2 forecast of ETA model forecast of T80L18 model and ETA model valid for 17th January. It can be seen from the figures that the rainfall over Delhi and surrounding areas on 16th January (Figure 12) is better predicted by Eta model. The clear weather observed at Delhi region on 17th is predicted well by both models (Figure 13).

6. CONCLUSIONS

In general, Eta model forecasts are found to be better than operational global (T80L18) model over Indian region. Eta model produces better orographic precipitation namely, the rainfall along the Himalayan mountain ranges, Western and Eastern Ghats. Considering the better performance of Eta model over Indian region, it is being operationally run for 48 hrs daily since 1 January 2002. The model predicted rainfall, winds and geopotential heights are also displayed daily on NCMRWF website (*www.ncmrwf.gov.in*).

Acknowledgements. The authors gratefully acknowledge NCEP, USA for providing the model. The authors are thankful to Dr. Matthew Pyle, NCEP for providing the topography data over Indian region and for his support in implementing the model at NCMRWF. Drs. S.V. Singh and R.K. Paliwal are thanked for their keen interest in the work.

REFERENCES

Arakawa, A., and V.R. Lamb, 1977: Computational design of the basic dynamical processes of the UCLA general circulation model. *Methods Comput. Phys.*, 17, 173-265.

Betts, A.K., 1986: A new convective adjustment scheme. Part I: Observational and theoretical basis. *Quart. J. Roy. Meteor. Soc.*, 112, 1306-1335.

Betts, A.K., and M.J. Miller, 1986: A new convective adjustment scheme. Part II: Single column tests using GATE wave, BOMEX, and arctic air-mass data sets. *Quart. J. Roy. Meteor. Soc.*, 112, 693-709.

Black, T.L., 1994: The new NMC mesoscale eta model: Description and forecast examples. *Wea. Forecasting*, 9, 265-278.

Chen, F., Z. Janjic and K. Mitchell, 1997: Impact of atmospheric surface layer parameterization in the new land-surface scheme of the NCEP mesoscale Eta numerical model. *Bound. -Layer Meteor.*, 85, 391-421.

Janjic, Z.I., 1994: The step-mountain eta coordinate model: Further developments of the convection, viscous sublayer, and turbulence closure schemes. *Mon. Wea. Rev.*, 122, 927-945.

Mesinger, 1984: A blocking technique for representation of mountains in atmospheric models. *Riv. Meteor. Aeronaut.*, 44, 195-202.

Mesinger, F., Z.I. Janjic, S. Nickovic, D. Garvrilov and D.G. Deaven, 1988: The step-mountain coordinate: model description and performance for cases of Alpine lee cyclogenesis and for a case of an Appalachian redevelopment. *Mon. Wea. Rev.,* 116, 1493-1518.

Rajagopal E.N., G.R. Iyengar, R.K. Paliwal and S.V.Singh, 2001: Prediction of severe weather systems over Indian region. *Proceedings of International Workshop on Observation and Forecasting of Severe Weather Associated with KEOP, METRI/KMA, 13-14 Nov., 2001*, 9-12.

Zhao, Q., T.L. Black and M.E. Baldwin, 1997: Implementation of the cloud prediction scheme in the Eta model at NCEP, *Wea. Forecasting,* 12, 697-711

Challenges of Representing Land-Surface Processes in Weather and Climate Models over the Tropics: Examples over the Indian Subcontinent

DEV DUTTA S. NIYOGI[1], ROGER PIELKE SR.[2], KIRAN ALAPATY[3], JOSEPH EASTMAN[2], TEDDY HOLT[4], U.C. MOHANTY[5], SETHU RAMAN[1], T.K. ROY[7], Y.K. XUE[6]

[1]North Carolina State University, [2]Colorado State University, [3]MCNC - Environmental Modeling Center, [4]Naval Research Laboratory, Monterey, [5]Indian Institute of Technology, Delhi, [6]University of California at Los Angeles, [7]International Institute for Population Sciences, Mumbai

E-mail: dsniyogi@unity.ncsu.edu

ABSTRACT

Land-surface processes are an important driver for weather and climate systems over the tropics and particularly the Indian subcontinent. Realistic representation of land-surface processes over the Indian region will help accurate simulations of environmental processes at micro, meso, and regional climate scale. However, in order to achieve these potential benefits, it is necessary to develop a strategy through the Indo-US Forum on Science and Technology that will address and overcome the different challenges associated with the representation of the land-surface processes over the Indian region. In this review, seven focus areas are identified: (i) development of a high-resolution surface data set for soil and vegetation /biophysical characteristics; (ii) testing of different biophysical Parametrizations in the land-surface schemes including ecological/ photosynthesis based land surface models at different scales; (iii) developing techniques to assimilate surface meteorological and land surface data in to regional analyses; (iv) development of regional climate studies to identify the radiative and other biogeophysical forcings that are critical to this region so as to include them in GCM studies; (v) studying the recirculation of evaporation, transpiration, and precipitation as part of the Indian monsoon system as well as effects of land – atmosphere interactions on heavy precipitation and tropical storm track and intensity; (vi) assessing the effect of regional aerosols and soot on global climate through changes in land – atmosphere interactions, cloudiness, and water vapor recirculation; and (vii) initiation of studies related to the inclusion of population growth and corresponding energy usage and land use change as a forcing within regional climate models.

1. INTRODUCTION

Land-surface processes form a critical pathway of energy and mass transfer. The surface can be a source as well as a sink of energy and mass transfer. For instance, the surface is a sink for momentum transfer; and absorbs the incident radiation during the day. At night, the land surface releases this stored radiative energy in the form of longwave emission. In addition to the energy exchange, the land surface is also known to be a significant source–sink mechanism for the greenhouse gases. Similarly, it is well established that vegetated surfaces can also act as a source for trace gases and other pollutants, including pathogens. Various such examples can be cited which show the significance of land-surface processes in energy and mass exchanges for different environmental applications affecting the short–term weather as well as regional climate simulations.

The role of land-surface processes is particularly interesting in the tropics. Often this is because the lack of frontal and synoptic activities (particularly in the absence of the monsoons) make the local surface interactions a dominant forcing for the atmospheric boundary-layer processes. Further, our understanding of the interactions between tropical surface and atmosphere exchanges is still relatively limited compared to that in the midlatitudes. This may be due to the lack of detailed field studies, as well as the extreme nature of the hydrological cycle (drought and flooding) in the tropics. Additionally, tropics support a large global population, often with limited agricultural and water resources. This could lead to important societal implications when the regional weather and climate are altered, or if it could be better predicted.

In this review, we discuss the different aspects of these land-surface interactions, with specific reference to the Indian subcontinent. The Indian subcontinent is an important region meteorologically and an ideal study domain for several reasons. These include the availability of long observational data series; interactions between land–air–sea and topographical features (such as the Western Ghats); presence of a variety of landscapes and soil textures over different biogeographical regions; and intense landuse change in recent years due to agricultural and industrial advances and large population growth.

In the following section, we describe the basic components of the land-surface processes within a meteorological model. Section 3 presents discussion on some of the challenges in representing land-surface processes in weather and climate models over the Indian subcontinent. Concluding remarks are presented in Section 4.

2. COMPONENTS OF LAND-SURFACE PROCESSES AND LAND-ATMOSPHERE INTERACTIONS

Land-surface processes are interactive at all scales — micro, regional, and global. The impact land-surface processes have on weather and climate can be appreciated through the energy balance equation. The surface turbulent heat energy can be partitioned as:

$$Q_N + Q_H + Q_{LE} + Q_G = 0; \text{ and } Q_N = Q_S\,(1\text{-A}) + Q_{LWD} - Q_{LWU}$$

where Q_N is the net radiation, Q_H is the sensible turbulent heat flux, Q_{LE} is the turbulent latent heat flux, Q_G is the ground heat flux, Q_S is the solar irradiance, A is the surface albedo, and Q_{LWD} and Q_{LWU} are the downward and upward longwave component of the solar irradiance.

The changes in the surface energy balance could be due to a number of factors primarily linked to land-surface characteristics. For example, the Bowen ratio of the surface will depend on the soil moisture availability, as well as on the evaporation–transpiration feedback as related to surface vegetation (coverage, density, as well as biophysical and soil textural properties). The surface albedo change is also a function of vegetation as well as soil moisture availability. Thus the hydrological and thermal effects associated with soil moisture–transpiration–evaporation cause coupled feedbacks.

The surface energy and mass transfer at leaf and canopy scale leads to differential surface heat fluxes as a function of landuse and the soil state. The spatial heterogeneities in the lateral surface fluxes often form a sufficient condition to trigger local (meso-) scale circulations and low-level convergence. Such circulations associated with moist air-dry air interfaces can lead to a preferential zone for cumulus formation and enhanced deep convection and inland thunderstorms. Even at larger scales (hundreds of kilometers in space, and years in time scale) changes in the vegetation and surface characteristics significantly correlate with the precipitation and temperature anomalies. Thus, through numerous nonlinear processes, leaf and canopy scale systems interact with mesoscale and regional-scale events at a time scale of a few hours to days and beyond. Figure 1 summarizes this coupled scale interaction and feedbacks from the leaf to the regional climate. Thus issues at the core of land-surface process studies include process representation, initialization of the surface base state, prognostic evolution of the surface state, analysis and validation of the different interactive features, impact on the weather and climate model performance, and predictability of the seasonal to regional climate.

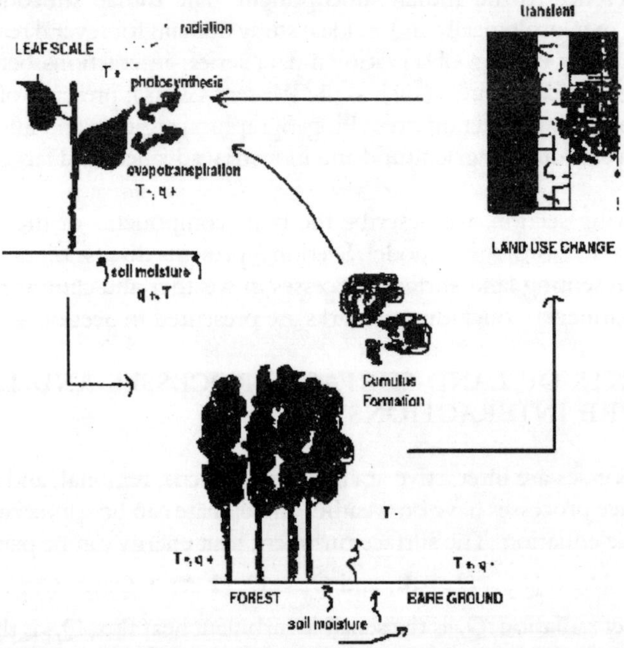

FIG. 1. Land-surface processes impact processes that can be characterized through the balance between energy and hydrological components at the earth's surface. Gradients in surface fluxes, circulation, convection and cumulus formation, precipitation and landuse change, and regional climate change are all factors assoicated with the surface energy and the hydrologic balance. In the figure, T and q represent temperature and humidity effects, while + and – sign indicate a potential increase or decrease in the temperature (T) or the humidity (q) effect.

3. CHALLENGES IN REPRESENTING LAND-ATMOSPHERE INTERACTIONS OVER THE INDIAN SUBCONTINENT

The multitude of interactions and the feedbacks associated with land-surface processes makes its accurate representation critical as well as challenging. In this section, we summarize some of the broad areas that need to be considered in developing a robust land-surface process representation for the tropical systems.

3.1 Resolution and Surface Data Representation

The earth's land surface is characterized by inhomogeneities due to variations in features such as vegetation types, soil types and moisture, canopy layers, urban effects, and topography. These inhomogeneities produce gradients in surface fluxes of momentum, heat, and moisture that drive three-dimensional circulations and significantly alter the low-level mesoscale thermodynamic and dynamic structure and aerosol distribution. For example, variations in soil moisture alone have been shown to enhance mesoscale features such as dryline formation, precipitation recycling, and convective initiation. To account for the effects of such land-surface heterogeneities, a numerical weather prediction model must be able to accurately represent the evolution and distribution of these inhomogeneities. Furthermore, a numerical model run at sufficient horizontal grid spacing that can resolve these inhomogeneities must have accurate databases that also resolve the relevant surface characteristics, such as soil texture, vegetation greenness, and landuse.

Current land-surface Parametrizations are based on several empirical relations and default characteristics associated with the surface as defined through different vegetation and soil texture look-up tables. The minimal input requirements for a typical land-surface model would include vegetation (or landuse) type, green vegetation fraction (temporally-varying, typically monthly or bi-weekly), and soil texture classes (at specified depths). Other parameters often provided by look-up tables include soil slope, background albedo (snow-free and snow-covered surface), and fractional and predominant vegetation types (if tiling techniques are used). From these basic parameters, a variety of other important land-surface parameters such as vegetative characteristics, and soil hydraulic and thermodynamic properties can be determined. The horizontal and temporal resolution of these basic parameters becomes more critical for accurate forecasts as the model grid spacing is reduced. Typically over the continental United States all of these parameters are currently available at both horizontal (~1 km or less) and temporal (~daily or weekly) grid increment commensurate with current weather prediction model resolutions. However, particularly outside the United States, the soil databases suffer from lack of sufficient horizontal resolution (~5-min to 1-degree). The impact of degradation in resolution outside the United States on model forecasts is not fully known.

Even with the existence of these high resolution databases, it is critically important in this regard to be able to validate the accuracy and representativeness of these parameters associated with the different landuse categories. Field studies and evidence from local experts have shown that soil texture (and even vegetation) classes used in land-surface models can be very different from ground truth observations, particularly in regions outside the continental United States, where observations are very limited.

With a heavy dependence on remotely-sensed observations, validation with ground truth exercises as well as remote-sensed data analysis and ingestion in the weather and regional climate models becomes even more crucial. This issue continues to be a significant challenge as models are being run with finer grid spacing, and the impact of heterogeneities and surface characteristics becomes more significant.

3.2 Biophysical Parametrizations in the Land-Surface Schemes

Improving the physics of the surface atmosphere exchange processes is another challenge in representing the tropical land-surface processes. Figure 2 shows the potential evolution of the land surface or evaporation and transpiration Parametrization schemes over the decades. As shown in the figure, initial models considered large-scale dynamical relations for assessing water vapor exchange between the different atmospheric layers. This was introduced in the climate models in early 1970s along with the so-called 'bucket' models. Parallel models used uncoupled Priestley–Taylor and Penman Monteith-type vegetation-based relations. This was generally followed by the advent of the coupled modeling studies involving an explicit feedback between the atmosphere onto the vegetation, followed by the change in the vegetation-induced environmental changes onto the model boundary conditions. One of the prominent features of this Parametrization continues to be the Jarvis–type approach. This involves modulating a predefined landuse/vegetation type based minimum stomatal resistance (Rsmin) value on the basis of environmental changes such as radiation, air temperature, humidity, and soil moisture. Some of the present-day models adopt the bucket and the Jarvis-type

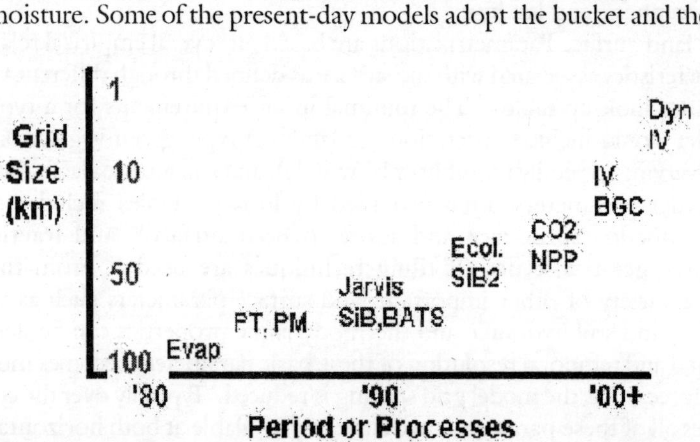

FIG. 2. Schematic representation of the development of vegetation schemes in land-surface models. Evap is large-scale evaporation schemes, PT, PM correspond to Priestley—Taylor alpha approach, and the Penman Monteith technique, Jarvis is the Rsmin-based Jarvis-approach, SiB and BATS are the Simple Biosphere model and the Biosphere Atmosphere Transfer Schemes. Ecol. is the ecological model such as in the SiB2, CO2 refers to the inclusion of explicit carbon effects using NPP or the net primary productivity consideration, IV EGC corresponds to interactive vegetation with biogeochemical models, and Dyn IV corresponds to the potential for dynamically-interactive vegetation models in which the vegetation state will change as a prognostic, coupled variable with the model meteorology. The period corresponds to the last 3 decades or so and is not precise. The grid size corresponds to the typical grid-scale spacing/model resolution adopted for model simulations. This is considered to provide the scale effect and is not accurate either. In general, the model grid spacing has decreased considerably and more detailed vegetation schemes are becoming increasingly popular.

Parametrizations in some of the simpler land-surface schemes both for coupled as well as uncoupled studies for understanding land–atmosphere interactions and biosphere feedbacks on the environment.

Presently, a number of advances in the plant physiology and the crop modeling as well as remote-sensing algorithms for retrieving radiation biospheric feedbacks have been reported, as discussed earlier. This has resulted in the application of stomatal/ transpiration models of varying complexities, to be applied for simulating the land-atmosphere interactions. These models can be categorized as: a diagnostic radiation-based photosynthesis approach that utilizes a radiation-use efficiency approach and vapor pressure deficit feedback for providing the transpiration (latent heat flux) or water vapor exchange pathways; and an iterative/interactive photosynthesis approach which involves representing stomatal conductance-carbon assimilation/photosynthesis relations for water vapor and carbon exchanges. With the advent of high-resolution terrestrial remote-sensed data and techniques for assimilating this information into terrestrial models, there have been newer opportunities for adopting and developing such interactive and dynamically-evolving vegetation and land surfaces (as against a static entity of the weather and climate modeling system). Recent efforts in coupling ecological and photosynthesis-based models mark a definitive step towards such dynamical models, however, there are still very limited studies concerning the interactive, two-way coupled vegetation and atmosphere models (e.g., see Lu *et al.* 2001).

An outstanding question that remains to be addressed is: are there significant advantages in adopting more complex photosynthesis-like schemes over the tropical/ Indian subcontinent region?

Indeed, over the continental U.S. the representation of the stomatal resistance/ transpiration processes yield significant differences in the model outcome. As an example, Niyogi *et al.* (1998) tested the different stomatal resistance representations for their ability to represent the land-surface processes active for transpiration and water vapor exchange. They found each of the transpiration schemes converged on different scenarios, with the more detailed photosynthesis- based scheme showing the closest agreement with the observations. In a subsequent study, Alapaty *et al.* (2002) performed a series of summertime runs over the U.S. using a coupled mesoscale model with and without a photosynthesis–transpiration model. They found the model performance was comparable with no one model performing better than the other. Hence it is important to review the performance of the different land-surface modeling approaches over the Indian domain for different weather and climate regimes. Therefore, although a number of studies have been reported in the literature, and some were indeed pioneering in nature, they often resort to the use of limited field observations, or have been episodic in nature, or not focused over the Indian region. Further, the study objectives tend to be focused on comparing one scheme over other to show its ability to reproduce land-atmosphere interactions leaving a number of important questions related to land – atmosphere interactions unanswered. Thus a concerted effort to study and understand the limitations and advantages for adopting one approach over another is needed.

3.3 Coupling Observations with Modeled Analysis

The application of the different schemes needs to be evaluated with local field studies and observations. This is particularly important for land-surface processes since the variability in the land–atmosphere response over different geographical domains continues to be an important feature for climate models. As an example, Niyogi *et al.* (2001) showed that for similar meteorological forcing in the midlatitudes and the semi-arid tropics, there are similarities in the direct responses while the interactions or the indirect feedback pathways could be quite different. The arid tropical regimes are dominated through vegetative pathways (via variables such leaf area index, stomatal resistance, and vegetal cover), while the mid-latitudes show soil wetness (moisture) related feedback. Additionally, for the mid-latitudinal case, the vegetation and the soil surface acted in unison, leading to more interactive exchanges between the vegetation and the soil surface. On the other hand, the water-stressed semi–arid tropical surface showed response either directly between vegetation and the atmosphere or the soil and the atmosphere with very little interaction between the vegetation and the soil variables. Thus, the semi–arid tropics would require explicit bare ground and vegetation fluxes consideration, while the effective (combined vegetation and soil fluxes) surface representation used in various models may be more valid for the midlatitudinal case. This also implied that with higher resource (water) availability, the surface invested more in the surrounding environment. On the other hand, with poor resource availability (such as water stress in the tropical site), the surface components retain individual resources without sharing. The uncertainty associated with the input conditions and the memory of the errors with the subsurface initial conditions for the land-surface models also become important. The uncertainty is largest for extreme conditions (very wet or very dry) and during landscape transitions during the monsoons (Niyogi *et al.* 1999).

To better understand the land–atmosphere interactions over the Indian subcontinent, a number of field experiments have been recently undertaken. The measurements, by their very nature are point representations and need to be interfaced with models for developing three–dimensional analyses over the study region. One such challenge in developing the land-surface process representation in the tropical perspective is to develop and test different surface assimilation techniques coupled with a land-surface model. One such approach was developed by Alapaty *et al.* (2002) for the MM5 modeling system. In this study, the errors in the model-simulated scalar or flux variables were converted into equivalent soil moisture and soil temperature corrections following an inverse technique, which are then used in a sequential assimilation approach for model integration. Figure 3 shows a typical time series with and without the nudging of the surface variables with the land-surface scheme. The ability of the model to simulate the observations better increases considerably due to such inversion/correction techniques. Such approaches can be easily adopted for developing realistic surface–atmosphere interaction assessments over the Indian region, where mesoscale monitoring is not available, and synoptic/frontal activity is often minimal.

3.4 Modeling the Regional Climate over the Indian Subcontinent

Regional climate models provide a powerful tool to study several of the outstanding questions related to the biophysical representation over the Indian subcontinent. Figure 4 shows the typical land use over the Indian subcontinent. The socioeconomic outlook in the Indian subcontinent is weather sensitive. With newer trade and agricultural opportunities, there is an increased need for regional climate information.

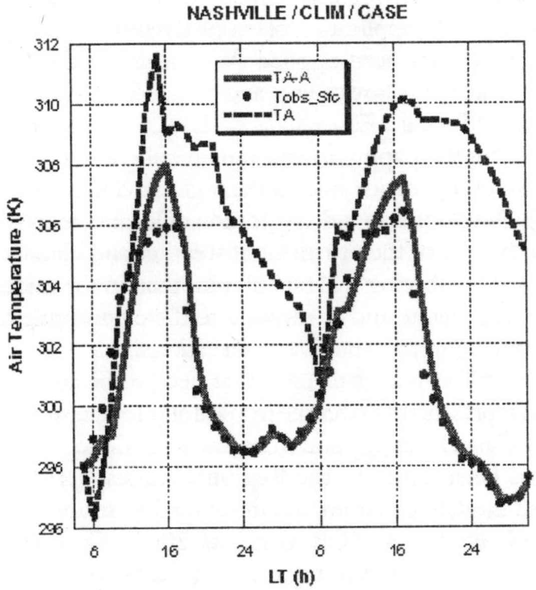

FIG. 3. Typical time series of model-simulated surface temperature using the climatological bucket scheme within MM5. The model was integrated over Nashville. TN using tower observations. The FASDAS/nudging approach of Alapaty *et al.* (2002) as a applied for connecting the initial estimates of soil moisture and/soil temperature using a nudging correction.

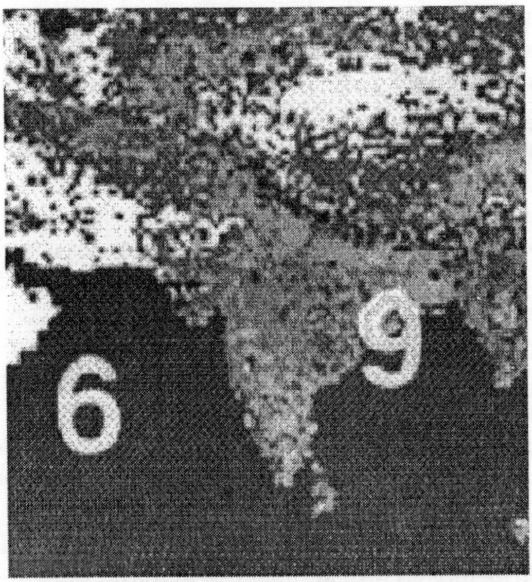

FIG. 4. Typical landuse over the Indian subcontinent. Using the global vegetation classification database, the agricultural landuse (Simple Biosphere Model ver-2-SiB2, vegetation type 9) and the tropical grassland and fallow (SiB2 vegetation type 6) appear to be the most predominant landuse category. Note that this will change for a given region, with seasonality, and there will be significant heterogeneity which can local scale weather and variability in the outcome.

This will require detailed, realistic regional climate models to be integrated over the study regions. As an example, the application of coupled regional atmospheric-ecological models is a new tool which has been applied successfully in the U.S. to assess the sensitivity of seasonal climate to landuse change and biogeochemical feedbacks associated with elevated carbon dioxide (e.g., Eastman *et al.* 2001; see also references in Pielke 2001). These are ideal tools to apply to the Indian subcontinent where vegetation processes are expected to exert a major role on the region's climate system, as discussed in Pielke *et al.* (2002a). Development and application of locally- representative regional climate models should be one of the priorities of weather and climate programs over the tropics. The nature of the forcing will be different than in the midlatitudes and the ability of the surface to regenerate and to respond to the environmental forcing needs to be assessed with regionally-representative observations.

However, there are still several challenges that need to be addressed to make the regional climate models representative over the tropical and Indian subcontinental region.

As an example, the General Energy and Mass Transfer Model (GEMTM, Chen and Coughenour 1994) has been linked to the Regional Atmospheric Modeling System (RAMS). This coupled modeling system was developed to study regional climate for the Great Plains of the United States (Eastman *et al.* 2001). Consequently, the biome types used in the model are biased to this region's vegetation. Subsequent to this initial development phase, the coupled modeling system has been adapted to accommodate other regions around the globe. There are ongoing applications directed towards Australia, South America, and the Caribbean. Presently, the coupled system is being applied to study the regional climate over the Indian subcontinent (Pielke *et al.* 2002a). In each of these studies it has been found that new vegetation classes, and vegetation and land characteristics which define the regional features, need to be added to the RAMS and GEMTM model parameters lists. As an example, over the Caribbean region, it was found that seasonal tropical forest, dry tropical forest, closed and open tropical rainforest, coniferous and broadleaf rainforest, and mangrove vegetation classes needed additional treatment in the models. An immediate solution to this problem is to develop a detailed wide literature survey in order to determine gross biophysical parameters specific to these vegetation types. Similar research was required in the other regional applications. In order to address global applications of the coupled modeling system, a team of collaborating researchers need to create a database specific to the Indian region similar to the Olson's Global Ecosystem classification system (Olson 1994), for example. Presently, there are 94 classes described in this classification system, and once the database is completed, it will enable more reliable application of the coupled modeling system to the Indian subcontinent. Thus, the Parametrization tests need to explicitly consider development of biophysical constants and land-surface parameters that are needed for adopting the different models over the Indian region.

3.5 Indian Monsoonal Circulation and Hydrological Recirculation

The regional climate models provide an efficient means of evaluating land-atmosphere interactions. However, the importance of land-atmosphere interactions is not limited to long-term regional climate simulations alone. Raman *et al.* (1998) investigated the influence of soil moisture and vegetation variation on the simulation of monsoon circulation and rainfall using a coupled land–atmosphere model. For this purpose, a simple land-surface Parametrization scheme was incorporated in a three-dimensional regional high resolution nested grid atmospheric model. Monsoon features such as the Somali jet, monsoon trough, and tropical easterly jet were found to be sensitive to land-surface process representation. Inclusion of realistic land surface resulted in not only a realistic short-range weather numerical prediction but also an enhanced short-range prediction of the hydrological cycle including precipitation over the Indian subcontinent. The gradients in the surface latent heat fluxes over land increased the low-level convergence in their study. The increased heat flux gradients, especially latent heat flux, further enhanced the convection and precipitation providing realistic diabatic forcings in the model. This further intensified the large-scale circulation as the strengths of both the low-level Somali jet and the upper-level tropical easterly jet increased because of the realistic land-surface conditions. Stronger meridional circulation caused by large-scale divergent flow and stronger vertical motion further enhanced evaporation. This provided better simulation of the monsoon circulation and its associated precipitation for summer monsoon circulation features and rainfall over India as a direct feedback of better land-surface representation.

Similarly, a general circulation model-based sensitivity study by Ferranti *et al.* (1999) indicates that though the ITCZ propagation is a dynamical feature, the temporal characteristics of the fluctuations between the two TCZ regimes, are influenced by the land-surface processes. Indeed, the low-frequency intraseasonal monsoon variability was also found to be enhanced by hydrological surface feedbacks. When the land-surface processes are not considered, the active and break regimes are significantly altered and the changes appear to depend on the land-surface conditions prior to the monsoon onset. Interestingly, the time-mean monsoon circulation, and the statistics of the intraseasonal oscillations (such as frequency of occurrence and mean amplitude), is also modified by the surface feedbacks, highlighting the importance of land-surface processes in the interannual predictability of the time-mean monsoon simulations.

Thus, improvements in the representation of land-surface processes will have a significant impact on the short- to medium-range forecasting as well as regional climate studies over India. The impact of changes in the surface landuse and hydrology on the subsequent precipitation patterns still needs to be addressed as a dynamic landuse change problem. The interactive vegetation growth and seasonal landuse change aspects need to be developed and tested with reference data over the Indian subcontinent. Specifically, there are several crop growth and ecosystem-based models that have been developed for different agricultural practices. Lu *et al.* (2001) developed one such model successfully over the central U.S. to study regional-scale two-way interactions between the atmosphere and biosphere. The atmospheric and ecosystem models exchanged information on a weekly time step, while the ecosystem model produced vegetation characteristics, which are an essential input for the land-surface schemes. Their results show that seasonal vegetation phenological variation strongly influences regional climate patterns through its control over land-surface water and energy exchange. The coupled model captures the key aspects of weekly, seasonal, and annual feedbacks between the atmospheric and ecological systems. In addition, it has demonstrated its usefulness as a research tool for studying complex interactions between the atmosphere, biosphere, and hydrosphere. The rapid greening and the modulation of the vegetation as a function of precipitation and monsoon activity over India is a complex land–

atmosphere issue, which requires such dynamic vegetation models to be explicitly developed and tested over the Indian region.

There are still unresolved issues pertaining to the use of such models for mixed farming and agricultural practices that are predominant over the Indian region. So there is a need to develop simple techniques for assimilating remote-sensed information over land in the weather and climate models, and for precipitation and temperature-based plant growth relations in the model. However, development of such relations will require detailed studies involving surface and remote-sensed observations, development of assimilation techniques, and ensemble analysis of the model meteorological fields, and plant growth algorithms to provide regionally- representative feedbacks due to short-term landuse land-cover changes (such as during the monsoon onset).

3.6 Regional Climate Feedbacks on Global Climate

Another factor that is pertinent over the Indian region, and is not represented well in the mesoscale and regional climate models, is the radiative feedbacks affecting the terrestrial water vapor exchange. The environmental conditions in the Indian subcontinent are conducive to particulate and aerosol transport. Using the Indian Ocean Experiment (INDOEX) observations, Raman *et al.* (2003) developed a dynamical scenario explaining the aerosol transport as a feedback between land–air–sea interactions of the land plume originating over the Indian subcontinent. These aerosol plumes can interact with the direct radiative forcing on the surface that can affect the Bowen ratio of the surface vegetation. This has implications for hydrological feedback as well as the carbon source/sink potential for the region and hence the plant growth, resulting landuse change and the regional climate. This effect of aerosol–induced radiative feedback of the terrestrial landscape is an important energy exchange pathway, which is currently poorly represented in the weather and climate models. The land-based aerosol plumes also significantly influence the microphysical processes associated with the cloud and precipitation, and this feature needs to be investigated further at micrometeorological to monsoonal scales.

In addition to the meteorological aspects of the weather and climate scenario developments, the Indian subcontinent has an important role for the global carbon cycle studies. This effect of the land-surface feedback will be addressed separately in studies such as Pielke *et al.* (2002a) and is not discussed here. Overall, one of the important aspects of the carbon management is to understand the ability of the terrestrial landscape to be a carbon sink. Studies are needed which will identify the conditions for carbon source/sink pathways over the Indian subcontinent. The interaction of the monsoonal cycle with the carbon cycle and the feedback on the regional productivity is still one of the unresolved issues in the global carbon cycle. Issues relating to the scaling and integration of ecological land-surface models in weather and climate models are also poorly represented and the magnitude of the impact of such errors is also not understood. This coupled representation of the water, energy, carbon and other trace gases and aerosol processes needs to be achieved in weather and climate models and tested over the Indian subcontinent.

3.7 Population Change as a Driver for Landuse Land-Cover Change in the Indian Subcontinent

An important component of the land-surface models in regional climate studies is to understand landuse change and factors affecting regional landuse changes. Present investigation of the impact of the landuse change is more in a post-analysis mode with little ability to predict such changes. One of the important stresses for the Indian subcontinent is the population increase and its impact on regional landuse.

The Indian population was about 350 million in 1947, and has since tripled to 1027 million per the 2001 census. The growth of population in the country peaked during 1961-71. Even though the average annual growth rate has registered a significant deadline, the rate in the recent decade was 1.9 percent. Population density of 77 km² in 1901 increased to 117 km² in 1951 and to 324 km² in 2001. These changes in the population densities and the migration patterns from rural to urban centers have caused significant changes in the regional landscapes.

Thus, a pertinent question to ask is what will be the size of Indian population in the foreseeable future, and what could be the resulting impact on the landscape modification. The recent national population policy contemplates that the country would achieve the replacement level fertility (an approximate level of total fertility rate of 2.1) by 2010. According to demographers, however, this seems overtly optimistic, and a more realistic assumption is that the replacement level fertility would be achieved by the year 2026. Irrespective of whether the goal is reached in 2010 or 2026, the population of the country is estimated to reach a level of 1.4 billion in 2026. In other words, the country will add about 400 million persons to its present population in another 25 years.

The impact of this population increase along with increasing consumption levels can be easily seen on our natural and man-made resources. For instance, availability of water per capita has shrunk from 5236 m³ at the time of independence to 2227 m³ by 1990-91. Water requirement is estimated to increase from 552 km³ in 1990 to 750 km³ in 2000 and further to 1050 km³ by 2025. Utilization of ground water is estimated to increase from 190 km³ in 1990 to 350 km³ by 2025.

Deforestation, in particular, is one of the major adverse consequences of this rapid population growth and development. With the enormous increase in human population and development works like roads, dams, mines, industries etc., massive biotic pressure is brought about on the country's forest resources. Encroachment is generally on marginal land, often on hill slopes, unsuitable for cultivation. Denudation of the slopes leads to erosion and siltation of rivers and reservoirs, resulting in inadequate recharge of ground water. The processes to meet the increasing demand for energy (both due to population growth and increasing standard of living) have serious environmental implications. The demand for total primary commercial energy increased from 25.5 metric tons of energy (MTOE) in 1953-54 to 212.9 MTOE in 1994-95 representing an increase of more than 8 times over a period of 40 years. It is likely to increase to 770 MTOE by 2011-2012. In particular demand for electricity is likely to increase by about 3.5 times by 2011-2012. The coal demand is estimated to increase 2.6 times by 2011-2012 (776 million tons) as compared to 1996-97 (296 million tons).

Thus population change can cause a stress on the socioeconomic growth and result in land use change, as well as anthropogenic particulates and greenhouse gas consumption. Therefore, incorporating the effects of population change in regional climate models stands out as an important challenge facing the regional climate community today. Studies are needed to link population models with the regional models as a biophysical feedback, and this remains a unique problem particularly for the Indian subcontinent.

4. CONCLUSIONS

Pielke *et al.* (2002b) propose a regional climate change potential as an intrinsic component of the climate system in which the surface processes provide the driving mechanism. The modulation of the Asian monsoons and the processes over the Indian subcontinent are largely surface driven, and hence models and policies which include the critical land–atmosphere interaction processes as well as socioeconomic couplings are needed. Towards this development, studies are initially needed to investigate the importance of different land–atmosphere components, the role of dynamic vegetation changes coupled with hydrology, aerosols, and other natural as well as anthropogenic modulations on the interannual variability of the Indian monsoon.

Specifically, this Indian region shows significant heterogeneity in the landuse land cover, which will directly affect simulation of weather and climate features such as convection initiation, localized heavy rainfall and variability in the precipitation patterns, precipitation and effects associated with the landfalling tropical systems, and even the monsoonal flow simulation. Hence a wide variety of weather processes will be directly affected depending on the land-surface model considered in the meteorological model. Further, landuse/land-cover changes and dynamical evolution need to be simulated along with soil and surface variables initial conditions.

Finally, studies involving validation and evaluation of the different land-surface schemes are necessary using observations available over Indian regions (such as the land-surface experiments in Anand-LSPX and Bangalore-VEBEX). These studies are needed using both site-specific field validations, as well as tests with episodic case studies for convection and storm events and from regional climate studies.

Questions such as what level of complexity is required (in the LSM) for simulating the tropical weather processes and what is their impact on model performance, are other immediate research priorities for land-surface models.

Another pertinent research area for the Indian and tropical regions is the effect of LSMs on landfalling tropical systems track, structure, and intensity. Thus, studies are needed which test different initialization techniques for the land surface particularly for summer conditions.

The feedback processes need to be explicitly coupled with the gradual but obvious changes in the population and regional landscape, which will form an important component of the regional climate studies.

Land-surface processes are an important driver for weather and climate systems over the tropics and particularly the Indian subcontinent. Realistic representation of land-surface processes over the Indian region will help accurate simulations of environmental processes at micro, meso, and regional climate scale. However, in order to achieve these potential benefits, it is necessary to develop a strategy which will address and overcome the different challenges associated with the representation of the land-surface processes over the Indian region. Some of the key challenges related to the issue of model database development, Parametrizations and biophysical scheme evaluations, model validations, and development of newer modeling approaches were identified and discussed in this brief review. A framework will be needed through the Indo-U.S. Forum on Science and Technology to develop a comprehensive, long-term priority to support research activities directly contributing to the challenges identified for the issue of land-surface models. Such a concerted, long-term effort will provide immense benefits in the form of improved simulations for better forecasting of natural disasters such as tropical cyclones, monsoon structure and their dynamics, and the regional climate. This in turn will provide synergistic feedback to the socioeconomic well-being and geopolitical stability of this region.

REFERENCES

Alapaty, K., N. Seaman, D.S. Niyogi, and A. Hanna, 2002: *J. Appl. Meteor*. Assimilating Surface Data to Improve the Accuracy of Atmospheric Boundary Layer Simulations, *J. Appl. Meteorol.*, **40**, 2068 – 2082.

Chen, D-X, and M.B. Coughenour, 1994: GEMTM: A general model for energy and mass transfer of land surfaces and its application at the FIFE sites. *Agric. Forest Meteor.*, **68**, 145-171.

Eastman, J.L., M.B. Coughenour, R.A. Pielke Sr., 2001: The regional effects of CO2 and landscape change using a coupled plant and meteorological model, *Global Change Biology*, **7**, 797- 815.

Ferranti, L., J.M. Slingo, T.N. Palmer and B.J. Hoskins, 1999: The effect of land surface feedbacks on the monsoon circulation. *Quart. J. Roy. Meteor. Soc.*, **125**, 1527-1550.

Lu, L., R.A. Pielke, G.E. Liston, W.J. Parton, D. Ojima, and M. Hartman, 2001: Implementation of a two-way interactive atmospheric and ecological model and its application to the central United States. *J. Climate*, **14,** 900-919.

Niyogi, D., S. Raman, and K. Alapaty, 1998: Comparison of four different stomatal resistance schemes using FIFE observations, Part 2: Analysis of terrestrial biosphere atmosphere interactions. *J. Appl. Meteor.*, **37**, 1301–1320.

Niyogi, D., S. Raman, and K. Alapaty, 1999: Uncertainty in specification of surface characteristics, Part 2: Hierarchy of interaction explicit statistical analysis. *Bound.-Layer Meteor.*, **91**, 341-366.

Niyogi, D., Y.-K. Xue, and S. Raman, 2001: Hydrological land surface response in a tropical regime and a midlatitudinal regime. *J. Hydrometeor.*, **3**, 39–56.

Olson, J., 1994: EROS Data Center Report 1 for the U.S. Geological Survey (USGS), September 28, 1994 (revised December 2001).

Pielke Sr., R.A., 2001: Influence of the spatial distribution of vegetation and soils on the prediction of cumulus convective rainfall. *Rev. Geophys.*, **39**, 151-177.

Pielke Sr., R.A., T. Chase, J. Eastman, D.S. Niyogi, and R.A. Pielke Jr., 2002a: A new perspective on climate change and variability: A focus on India. *Proc. Indian Natl. Acad. Science,* in preparation.

Pielke Sr., R.A., G. Marland, R.A. Betts, T.N. Chase, J.L. Eastman, J.O. Niles, D. Niyogi, and S. Running, 2002b: The influence of land-use change and landscape dynamics on the climate system- relevance to climate change policy beyond the radiative effect of greenhouse gases. *Phil. Trans. A*, Special Theme Issue, **360**, 1705-1719.

Raman, S., U.C. Mohanty, N.C. Reddy, K. Alapaty, and R.V. Madala, 1998: Numerical simulation of the sensitivity of summer monsoon circulation and rainfall over India to land surface processes. *Pure Appl. Geophys.*, **152,** 781-809.

Raman S., D.S. Niyogi, M. Simpson, and J. Pelon, 2003: Dynamics of the elevated land plume during the northeastern Indian monsoon and during the Indian Ocean Experiment (INDOEX). *Geophys. Res. Lett.* (*In Press*).

A New Approach to the Cumulus Parametrization Issue

T. N. KRISHNAMURTI & J. SANJAY*

Department of Meteorology, Florida State University, Tallahassee, FL 32306-4520
** Permanent Affiliation: Indian Institute of Tropical Meteorology, Pune 411 008, INDIA*
E-mail: tnk@io.met.fsu.edu

ABSTRACT

Can the superensemble methodology provide insights on physical parametrization? This question was recently addressed by us in the context of Numerical Weather Prediction (NWP). We feel, however, that the information provided here may be useful for seasonal climate modeling as well. In the NWP context, we have developed multi model forecasts from 6 versions of the Florida State University global spectral model (at a horizontal resolution of 170 waves, triangular truncation). These different versions deployed 6 different cumulus parametrization schemes; these models were identical in all other aspects including the initial states. Making the assumption that differences in short range (one day) forecasts of precipitation arise largely from differences in the cumulus parametrization, a superensemble methodology, following a recent study, was deployed to assign geographically distributed weights to convective heating for the different cumulus parametrization schemes. This was done after completion of some 85 experiments for each model for training phase of the superensemble. A new single spectral model was next designed that included the weighted sum of the 6 cumulus parametrization schemes strung out within this model. This model was next shown to outperform in NWP forecasts of precipitation compared to any of those models that used a single cumulus parametrization scheme. This merely suggests that no single, present scheme is superior to all other schemes over the entire tropical belt; they all seem to have some virtues over different geographical regions. This Unified collective scheme is physically based since it does carry mechanisms such as mass flux, moisture convergence, cloud detrainment, downdrafts, effects of sea surface temperature etc. that are explicitly carried within one or the other schemes. This collective scheme is however based on optimized weights for these processes that vary geographically. It is our premise that even if a new breakthrough in cumulus parametrization were to occur from the development of a new scheme, that scheme, at best, may only achieve a skill ranking of number three for precipitation forecasts. The first place, we noted, still belongs to a multi model superensemble, based on the optimal combination of 6 separate models. The second place belongs to the single model that utilizes a strung out weighted sum of many cumulus parametrization schemes within it. The individual member models have larger precipitation forecast errors compared to the two above. The skills, here, are evaluated using standard metrics such as correlations, root mean square errors and the equitable threat scores; finally we also present the vertical profiles of the apparent heat source and the apparent moisture sink that also confirm these above findings.

1. INTRODUCTION

This present paper addresses precipitation forecasts from six versions of the FSU (a list of acronyms is presented in Table 1) global spectral model (Krishnamurti *et al.* 1991). The six versions deploy six different cumulus parametrization algorithms, these include three different versions of the Arakawa-Schubert scheme, a modified Kuo scheme, Kerry Emanuel scheme and Zhang and McFarlane scheme. This paper next addresses the precipitation forecasts for the FSU multimodel superensemble, Krishnamurti *et al.* (2001) that is developed from this array of models. It is followed by the precipitation forecasts from a single Unified FSU global spectral model where a string of cumulus parametrization schemes are deployed within a single model. The relative weights for different cumulus parametrization schemes, within that single model, are based on multimodel superensemble forecasts of precipitation. These weights are determined from the performance of the individual models, each of which carries one cumulus

TABLE 1. LIST OF ACRONYMS

AMIP	Atmospheric Model Intercomparison Project
A-S	Arakawa-Schubert Cumulus Scheme
BMRC	Bureau of Meteorology Research Centre
CAPE	Convective Available Potential Energy
DMSP-SSM/I	Defense Meteorology Space Program-Special Sensor Microwave/ Imager
ECMWF	European Centre for Medium range Weather Forecasts
EMA	Emanuel Cumulus Scheme
FSU	Florida State University
GPROF	Goddard Profiling (algorithm)
GSFC	Goddard Space Flight Center
JMA	Japan Meteorological Agency
NCAR	National Center for Atmospheric Research
NCEP	National Center for Environmental Prediction
NOGAPS	Navy Operational Global Atmospheric Prediction System
NRL	Naval Research Laboratory
NWP	Numerical Weather Prediction
RAS	Relaxed Arakawa-Schubert Cumulus Scheme
RPN	Recherche en Prévision Numérique
SAS	Simplified Arakawa-Schubert Cumulus Scheme
T170	170 wave Triangular truncation
TRMM	Tropical Rainfall Measuring Mission
TSDIS	TRMM Science Data and Information Service
UTC	Coordinated Universal Time
ZM	Zhang and McFarlane Cumulus Scheme

parametrization scheme. Finally, this paper presents a comparison of the following precipitation forecasts:

a) Performance of individual cumulus parametrization algorithms.
b) Performance of Unified cumulus parametrization algorithms.
c) Performance of the FSU multimodel superensemble.

The multimodels used in this study are derived from the FSU global spectral model at the resolution T170. In all 85 separate forecast experiments were first carried out using each of the six different versions of the cumulus parametrization. Thus, a total of 510 experiments were carried to define what is called a training phase for a multimodel superensemble. The start dates for these experiments covered each of the days during the second week of April through June 2000 at 1200 UTC. Details on these experiments are presented in Shin and Krishnamurti (2002a,b).

The superensemble approach is a recent contribution to the general area of weather and climate forecasting, developed at the FSU, and is discussed in a series of publications by Krishnamurti *et al.* (1999, 2000a, 2000b and 2001). The novelty of this approach lies in the methodology, which differs from ensemble analysis techniques used elsewhere. This approach yields forecasts with considerable reduction in forecast error compared to the error in the member models, the 'bias-removed' ensemble averaged forecast, and the ensemble mean. In brief, the technique entails the division of a time line into two parts. One part is a 'training' phase, where forecasts by a set of member models are compared to observed fields with the objective of developing a least squares fit of the forecasts to the observations. Specifically, the observed anomalies are fit to the member model forecasts according to the classical prescription

$$O' = \sum_{i=1}^{N} a_i (F_i - \overline{F}_i) + \varepsilon_i , \qquad (1)$$

where F_i is the ith model forecast (out of N total models), $\overline{F}_i$ is the mean of the ith forecast over the training period, O' is the observed anomaly relative to the observed mean over the training period, the a_i values represent the regression coefficients and $_i$ is an error term. The a_i's are determined by requiring the summed squared error integrated over the training period $E = \sum_{i=1}^{N} \varepsilon_i^2$ to be as small as possible. A fit of this sort is performed for all model variables and at all model grid points for which reanalysis observations are available and typically yields some 10 million regression parameters. These may be thought of as bias correction weights. The second time line part is composed of genuine model predictions, *i.e.* the forecasts of the member models. The superensemble approach combines each of these forecasts according to the weights determined during the training period through the formulation

$$S = \overline{O} + \sum_{i=1}^{N} a_i (F_i - \overline{F}_i) , \qquad (2)$$

where the notation is defined above except the a_i values, which are the regression coefficients. The prediction 'S' is referred to as the "superensemble" forecast. This forecast should be contrasted with the more standard anomaly forecasts known as the biased - removed ensemble mean (E) or ensemble mean $(\hat{E})$ forecasts

$$E = \frac{1}{N}\sum_{i=1}^{N}(F_i - \overline{F}_i) \text{ or } \hat{E} = \frac{1}{N}\sum_{i=1}^{N}(F_i - \overline{O}). \quad (3)$$

The distinction between them comes in the weighting. Assigning all models a weight of unity in (2) not only illustrates the connection between the forecasts, but also illustrates how the training period attempts to identify and highlight good model performance. We have developed a real time NWP capability for the forecast of all basic variables such as winds, temperature, surface pressure, geopotential heights and precipitation. These are multianalysis-multimodel superensemble forecasts where eleven models are used on a daily basis. These include the daily operational forecasts from the NCEP, Canadian Weather Service RPN, Australian model for the BMRC, U.S. Navy's NOGAPS, the Japanese model for JMA and different versions of the in-house FSU global spectral model physically initialized using different rain rate algorithms (see section 1a.). In some sense, the construction of the superensemble is a post-processing of multimodel forecasts. This is still a viable forecast product that is being prepared experimentally in real time at FSU and is currently available on a real-time basis on the website http://lexxy.met.fsu.edu/rtnwp. Real-time hurricane track and intensity forecasts are another major component of superensemble modeling. This approach of carrying out a training phase followed by real-time forecasts has shown improved track and intensity predictions (up to 5 days) for the Atlantic hurricanes during 1999, 2000 and 2001. Improvements in track forecasts were 25% to 35% better than those of the participating member models. The intensity forecasts for hurricanes have been only marginally better than the best models. In some recent real-time tests during 1999, marked skill improvement in the forecasts of difficult storms such as Floyd and Lenny were noted where the performance of the superensemble was considerably better than that of the member models.

The area of seasonal climate simulations has only been addressed recently in the context of atmospheric climate models where the sea surface temperatures and sea ice are prescribed, such as the AMIP data sets. In this context, given a training period of some 8 years and a training data base from the ECMWF, the results exhibit improved skill compared to the member models and the ensemble mean. Preliminary work in this area (Krishnamurti *et al.* 2003) examines the difficulties involved with prediction of seasonal precipitation anomalies. Most individual member models perform poorly compared to climatology, whereas the superensemble appears to demonstrate precipitation skills slightly above those of climatology. The effectiveness of weather and seasonal climate forecasts from the superensemble methodology has also been assessed from measures of standard skill scores such as correlation against observed fields, root mean square (RMS) errors, anomaly correlations and the so-called Brier skill scores for climate forecasts (assessing skills above those of a climatology).

Why does the superensemble generally have higher skill compared to all participating multimodels and the ensemble mean? At each location and for all variables, the ensemble mean assigns a weight of 1/N to all N member models, including several poorer-performing models. As a result, assigning the same weight of 1/N to some poorer models was noted to degrade the skill of the ensemble mean. It is possible to remove the bias of models individually (at all locations and for all variables) and to perform an ensemble mean of the bias removed models. That, too, has somewhat lower skill compared to the superensemble, which carries selective weights distributed in space among all multimodels and for all variables. A poorer model does not reach the levels of the best models after its bias removal.

1.1 Summary of Past Work on Precipitation Superensemble

A major advance in precipitation forecasts has emerged from the use of a multimodel/multianalysis superensemble (Krishnamurti *et al.* 2001). "Multimodel" refers to different models whose forecasts are being assimilated for the construction of the superensemble. "Multianalysis" refers to different initial analyses contributing to forecasts from the same model. In this study, the multianalysis component is based on the FSU global spectral model initialized with TRMM and SSM/I data sets via a number of rain rate algorithms. Five different initial analyses for each day are deployed that define the multianalysis component. Those are based on different versions of rain rate estimates derived from TRMM and the DMSP-SSM/I satellites. The differences in the analyses arise from the use of these different rain rates within the physical initialization procedure outlined in Krishnamurti *et al.* (1991). The resulting initialized fields have distinct differences among their initial divergence, heating, moisture, and rain rate descriptions. The five multimodel forecasts with diverse horizontal and vertical resolutions are received from a number of global operational centers. The model output is then interpolated to a common grid of the lowest resolution member model. Generally, the resulting horizontal resolution is around 125 km. The global models include several different parametrizations of physical processes; effects of ocean, snow, and ice cover; and treatment of orography. The training data set is produced from the daily TSDIS operational files of the TRMM microwave radiometer-based rainfall estimates. In order to extend the global coverage, the observed data sets are augmented with the use of the SSM/I derived rain rates from the US Air Force polar orbiting DMSP satellites (F11, F13, F14, and F15).

2. VARIOUS CUMULUS PARAMETRIZATION SCHEMES USED IN THE PRESENT STUDY

A rather large number of papers have appeared in recent literature on the comparisons of the performance of various cumulus parametrization schemes; Kuo *et al.* (1996), Wang *et al.* (1997), Belair *et al.* (2000), Ferretti *et al.* (2000), Rajendran *et al.* (2002), Das *et al.* (2001) and several others. The impression one gets from these comparative studies, that are generally based on integrations starting from a few initial states, is that not one single cumulus parametrization scheme is superior to another, and that is often a function of the initial state. When we carried out 85 experiments with six different cumulus parametrization schemes (Shin and Krishnamurti 2002a,b) we did see an intersection of their skill score curves (as a function of time). This is depicted in Figure 1 where we show the precipitation RMS errors and correlations for the six member models (thin lines), the ensemble mean (blue, thick line) and the super ensemble (red, thick line) for day 1 and day 2 forecasts during the training phase over the global belt. This implies that the short-range precipitation forecast skills are often initial value dependent. Thus it appeared worthwhile to examine a Unified scheme, which has been determined from a larger experimentation program. The various cumulus parametrization schemes used in this study are; the FSU modified Kuo scheme

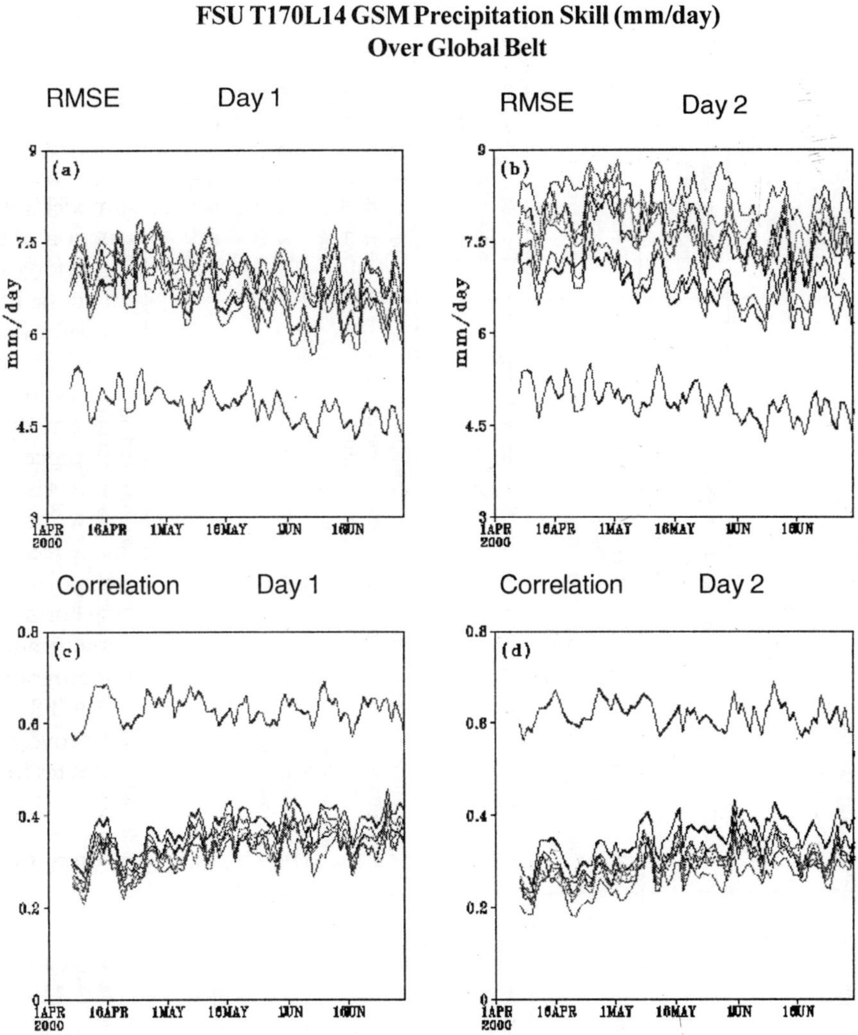

FIG. 1: Precipitation RMS errors (mm day^{-1}) [(a) and (b)], and correlations [(c) and (d)] of member models (thin lines), the ensemble mean (blue thick line) and the superensemble forecasts (red thick line) for day 1 and day 2 over the global belt. These results pertain to the training phase .The results of member models are clustered together in these illustrations while that for the superensemble with much higher skill stands out.

(Krishnamurti et al 1980, 1988), three versions of a simplified form of the Arakawa and Schubert (1974) cumulus parametrization known as GSFC RAS (Moorthi and Suarez 1992), NRL-NOGAPS RAS (Moorthi and Suarez 1999) and NCEP SAS (Pan and Wu 1995), the NCAR Zhang and McFarlane (1995) scheme and the NRL-NOGAPS Emanuel scheme based on Emanuel (1991) and Emanuel and Zivkovic-Rothman (1999). Table 2 lists the different features among these different cumulus parametrization schemes.

TABLE 2. Features among cumulus parameterization schemes

Sr. No:	Cumulus Scheme	Environmental Trigger (*Modulation of convection by large-scale*)	Cloud Model (*Treatment of cloud thermodynamic properties*)	Final State of Atmosphere (*Quantitative effects of convection on environment*)
1.	FSU Modified Kuo	Integrated vertical advection of moisture.	Moist adiabatic lapse rate.No downdrafts.	Tends towards local moist adiabat
2.	GSFC Relaxed Arakawa-Schubert	Rate of destabilization by advective changes.	Entraining plume model.Invokes single members of a cloud ensemble one after the other. Normalized cloud updraft mass flux linear function of height.No downdrafts.	Relaxes to steady state clouds in prescribed time.
3.	NRL Relaxed Arakawa-Schubert	Rate of destabilization by advective changes.	Includes evaporation of falling rain.With downdrafts.	Relaxes to steady state clouds in prescribed time.
4.	NCEP Simplified Arakawa-Schubert	Rate of change of stability.Upward vertical velocity at cloud base.	Only the deepest cloud considered.Moisture detrainment from convective clouds.Warming from environmental subsidence.Downdrafts and evaporation of falling rain included.	Adjusts toward an equilibrium cloud work function within a specified time.
5.	NCAR Zhang and McFarlene	Plumes of updraft ensemble need to be sufficiently buoyant to penetrate through locally conditional unstable lower troposphere.	Ensemble of entraining updrafts with associated saturated downdrafts.	Removes CAPE at an exponential rate with a specified time. Neutrally buoyant for undiluted reversible ascent of a parcel.
6.	NRL Emanuel	First level of neutral buoyancy for the undiluted, reversible ascent of near surface air is at a higher altitude than level of cloud base.	Sub cloud scale drafts using buoyancy sorting technique.Determines mass flux prognostically. Stochastic coalescence and Bergeron-Findeisen mechanism.Cloud water in excess of a threshold amount converted to precipitation.Saturated and unsaturated downdrafts.	Adjusts to a local quasi-equilibrium situation, does not depend on the relaxation of cloud work function.

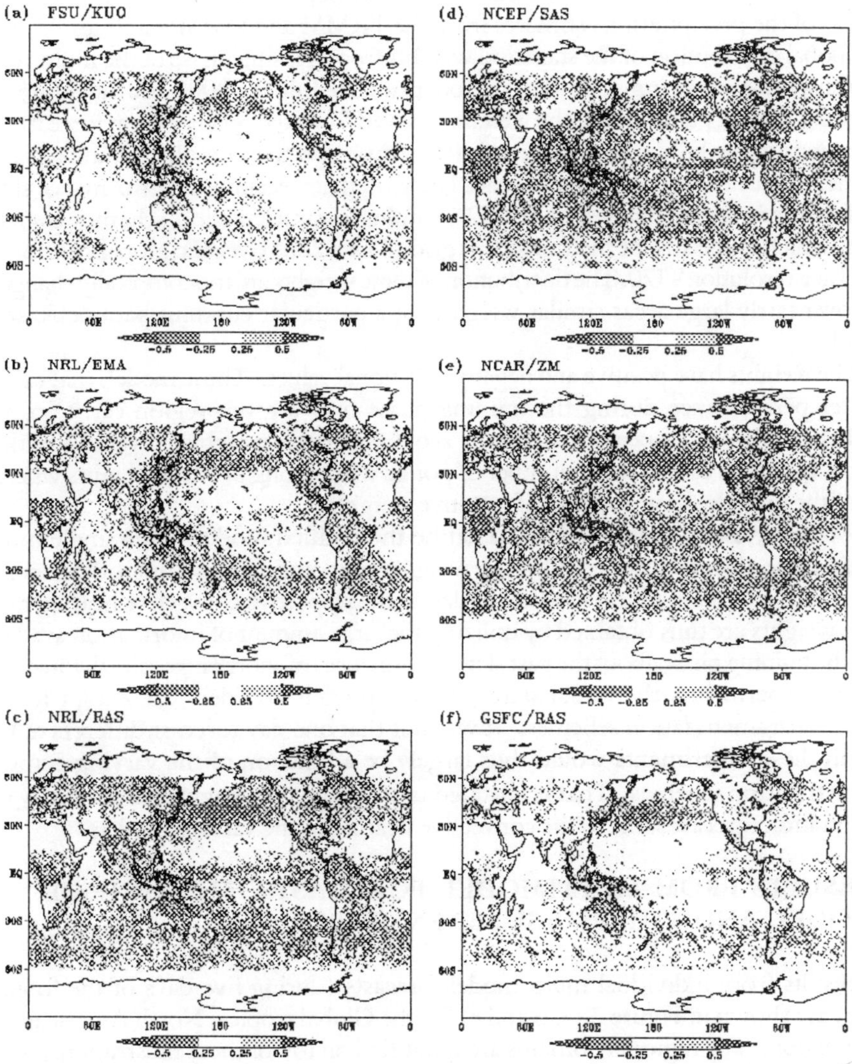

FIG. 2.The global distribution of weights for the six different cumulus parameterization schemes. These are derived from using one same model each time with a separate cumulus parameterization scheme. These weights are determined from the precipitation forecast skills during the training phase of the multimodel superensemble at T170 resolution.

3. DISTRIBUTION OF WEIGHTS AND DESIGN OF THE UNIFIED MODEL

We used 85 days of multimodel day one precipitation forecasts to construct the weights from the training phase. These weights are optimized by selecting those recent 85 days that exhibited highest skills for the precipitation superensemble.For determination of the weights during the training phase and for the forecast verification (in the forecast phase), observed precipitation estimates are obtained from a continual, once daily, analysis of the precipitation estimates from the TRMM and DMSP satellites.

Global distribution of the statistical weights from training phase of the multimodel superensemble covering a 85-day period prior to July 1, 2000 are illustrated in Figure 2. The six panels show the weights derived from the precipitation forecasts for six versions of the FSU model that made use of six respective cumulus parametrization schemes. Since the weights vary in space based on their recent past performance skills, in effect there are as many as 131,072 cumulus parametrization schemes over the globe, which is the size of the transform grid (512 x 256) over the global belt for the model at a resolution TI70. The distribution of these weights are monotonically changing, thus over fairly large areas similar weights for a particular cumulus parametrization prevails.

The weights have positive and negative fractional values. Those models that exhibit a better performance during the training phase over a given region carry positive weights and those that performed poorly acquire negative weights. Overall 4 of the 6 models appear to carry a large number of points with strong positive weights whereas two of the models show a spread of negative weights.

We make use of these weights to define the Unified model. This single model utilizes these aforementioned differing weights to incorporate within it the string of six cumulus parametrization schemes to define the respective convective heating rates. These weights are thus obtained by utilizing the large amount of information available from the training phase about the past skill performance (of short range rainfall forecasts). In these experiments the initial states and the models are identical, except for the cumulus parametrization schemes. It was felt that one day forecast differences that arise in these experiments would thus largely be a function of the varying cumulus parametrization schemes. That is what led us to assign the weights based on day one precipitation forecast skills to the convective heating of the various models.

4. RESULTS FROM MULTIMODEL FORECASTS, THE UNIFIED MODEL AND THE FSU SUPERENSEMBLE

The results from individual multi-model forecasts (during five days of the forecast phase) are shown in Figure 3 (a, b and c) for the Global tropics, North American and Asian region. The regional domains are identified on top of these illustrations. Here the results valid on days 1 and 2 of forecasts are shown. The top panels show the RMS errors and the bottom panels show the correlation of the predicted daily rainfall totals with the observed estimates. The vertical barb at the far right (black color) shows results from the FSU multi-model superensemble, the vertical barb next to it (red color) shows results from the FSU Unified model that incorporates 6 cumulus parametrization schemes with varying weights over the global belt. These are the results from the use of 131,072 combinations of the six cumulus parametrization schemes at the model transform grids. The remaining vertical barbs show the skill scores from individual FSU model members that use one of the same 6 cumulus parametrization schemes. Overall it is clear from this figure that the FSU superensemble has the least errors. The Unified model outperforms all of the single models. The improvements in

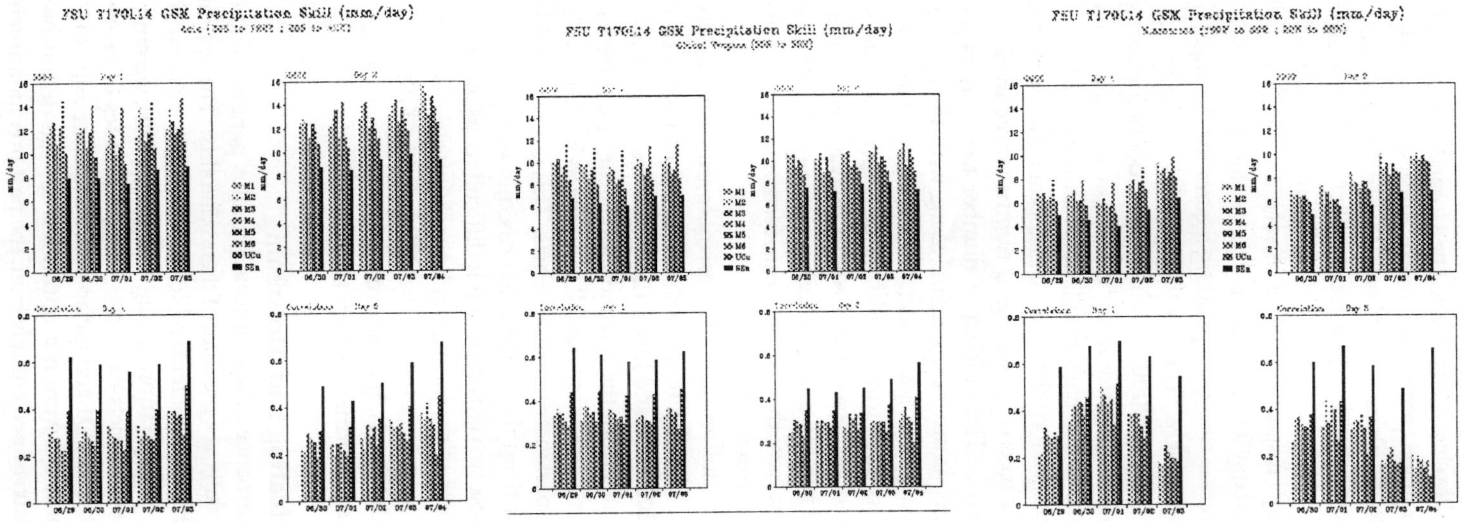

FIG. 3. The results of precipitation forecast skills for the forecast phase of the superensemble for days one and two of forecasts over the (a) Global tropics, (b) North American and (c) Asian regions. The forecasts dates are indicated at the bottom of each panel. The left 6 vertical barbs in each panel show the performance of the member models, the red color barb pertains to the unified model and the black barb that stands out belongs the multimodel superensemble. Units mm day^{-1}.

skill from the FSU superensemble clearly stand out. The Unified scheme appears to be a better way to go than to use a single cumulus parametrization scheme within a single model. The results pertaining to the global tropical belt (Figure 3a) show that the range of RMS errors for day 1 and day 2 of forecast for the member models are around 10.0 and 12.0 mm day^{-1} respectively, whereas corresponding figures for the superensemble are around 7.0 and 7.5 mm day^{-1} respectively. The correlations of the model rain to observed estimates for day 1 and day 2 of forecasts for the member models are around 0.30 and 0.25 respectively, whereas the corresponding figures for the superensemble are around 0.60 and 0.45 respectively. These results pertain to the forecast phase covering the period June 29 to July 4, 2000. The member model's forecast skills for day 1 and day 2 of precipitation are slightly lower than those of the Unified model, whereas superensemble has significantly higher skill. It is tempting to say that if a very much improved cumulus parametrization scheme were to appear on the scene tomorrow, its use in a single model might receive a ranking of number three in terms of skill scores for precipitation forecasts. The ranking of number two may be held by a Unified model that includes this future 'best scheme' along with the other available schemes in a weighted sense within a single model. The ranking of number one for rainfall forecasts may still be assigned to the multi-model superensemble, which is constructed from a training, and a forecast phase where all available versions of the cumulus parametrization were run separately within single models. Only some 85 separate multimodel experiments have thus far been run, further confirmation of these results could have a significant impact to NWP.

Precipitation forecasts from multimodel superensemble are shown for the global tropical belt in Figure 4(a, b). This figure includes observed estimates of rain (24 hourly total ending on July 1 1200 UTC and July 2 1200 UTC respectively for the year 2000), the predicted rainfall totals from the Unified model, the ensemble mean, the multimodel superensemble and the predicted rainfall from the individual models using the six different single cumulus parametrization schemes. This is a typical example of a tropical precipitation forecast from the models described in this paper. When we look at the different panels of these two illustrations it is difficult to assess a comparison of one panel against another. Some of the prominent heavy rains over southern Bay of Bengal and the western Pacific are noted in the NRL/RAS, GSFC/RAS, the superensemble and the FSU/KUO panels. It is the overall statistics over a longer period that seems to convey the relative performances better.

4.1 Vertical Structure of Heating and Moistening

For the evaluation of performance of the various cumulus parametrizations we have used precipitation forecast skills as a measure of their reliability. The optimization of the Unified scheme and the FSU superensemble forecast skills addressed above made use of the weights of the precipitation during the pre-forecast training phase. Thus the issue of how well these models carry the vertical profiles of heating and moistening (the apparent heat source, Q1 and the apparent moisture sink, Q2, Yanai, 1973) was indirect since Q1 and Q2 were not optimized for the definition of the proposed weights, those were determined entirely from the past precipitation forecast skills. It is difficult to bring in an optimization of three different parameters R (the rain

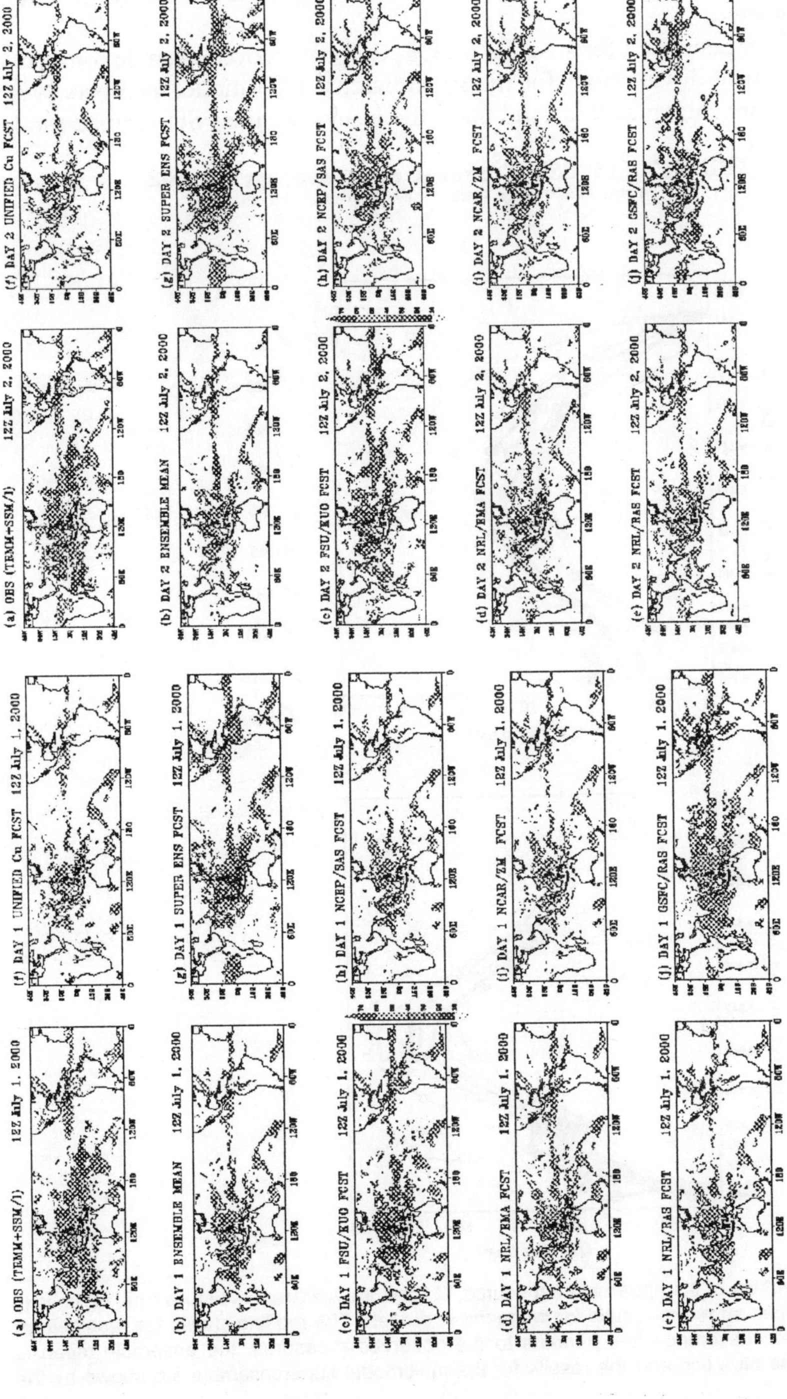

FIG. 4.A sample of observed rainfall estimates and those produced from model forecasts for day 1 and day 2 of forecasts from the following: Unified model, ensemble mean of model forecasts, multimodel superensemble and from six member models. Units mm day^{-1}.

rates), Q1, and Q2. Q1 and Q2 were calculated, after the fact, in order to assure that the Unified scheme was not running away towards unrealistic values after the forecasts had been completed.

Figure 5 illustrates day 1 forecasts for the Q1, Q2 profiles over a sample convective area. Here vertical distributions for the six individual FSU model members (with single scheme) are shown as thin solid lines, the results for the FSU Unified model

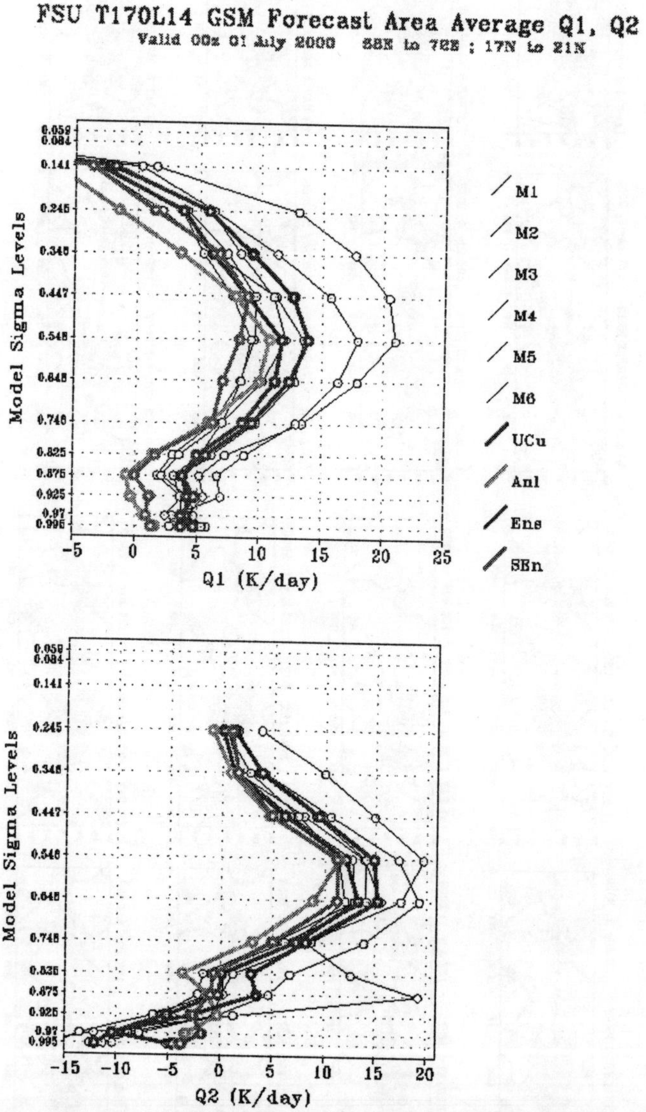

FIG. 5 Vertical profiles of the apparent heat source, Q and the apparent moisture sink, Q, units K day^{-1}. Thin solid lines show the results for the member models. The red line shows the results from the Unified model, the orange line pertains to the observed measures, the ensemble mean is represented by the blue line and the results for the multimodel superensemble are shown by the green line.

(single model containing weighted sum of six cumulus parametrization schemes) is shown as a red line, the ensemble mean as a blue line, the FSU multimodel superensemble results as a green line and the observed estimates are shown as a orange line. The observed estimates came from the analysis data sets of ECMWF. Here 24 hourly analyses were available, hence the time tendencies of the static energy and specific humidity were based on those time intervals. The advective terms, horizontal and vertical, were determined from the use of spectral transform method and finite differences respectively. Construction of vertical profiles for the multimodel superensemble involved optimization of the Q1 and Q2 weights at each model sigma levels from 85 experiments with the six individual model members during the training phase. It is clear from this diagram that the best results arise from the multimodel superensemble, the member models show a considerable spread and the Unified model appears to carry these profiles reasonably well compared to the observations.

5. SUMMARY AND CONCLUSIONS

Table 2 lists various physically based cumulus parametrization schemes that invoke a number of processes such as detrainment, massflux, moisture convergence, plume ensemble model, restoration of moist static instability, saturated and unsaturated downdrafts, buoyancy sorting, conversion of cloud and ice water, etc. The proposed Unified scheme is physically based since it does include all of these processes. There are in fact 131,072 cumulus parametrization schemes that were deployed over the global domain in this study. The superensemble methodology, which was needed to determine these globally distributed weights, is an important component of this study. Although, that many weighted combination of the 6 basic schemes may not have been necessary, it was felt that coding that degree of regional variability was a relatively simple matter. The rationale here is that a parameter such as moisture convergence provides a good measure of rain over some regions, such as the Eastern Atlantic, Reed *et al.* (1977), it is not the best measure of deep convection at some other places. Likewise, we perceive that no single mechanism, illustrated in Table 2, is the best measure for deep convective processes globally. Just as data uncertainties call for a host of initial perturbations in ensemble forecasts, Toth and Kalnay (1993) and Palmer and Tibaldi(1988), likewise we feel that the current parametrizations of physical processes carry a high degree of uncertainty. A Unified ensemble approach can provide benefits especially if they are cast in the spirit of the superensemble, where the weights determined for their individual past performance could be extended to develop a Unified physical parametrization. While carrying out these forecasts with individual models, we have deliberately limited the forecasts to one or at most two days, towards the design of this Unified scheme. The models used here are all identical except for the cumulus parametrization schemes and furthermore all of these models used the same set of initial states. Thus, we have tacitly assumed that very short-term differences in rainfall forecasts arise from the differences in their respective cumulus parametrizations.

Further improvements from the Unified model are possible if we construct the superensemble by optimizing Q_1 and Q_2 at each vertical level. Currently we have only used the past precipitation forecasts to construct and assign statistical weights for convective heating. If the weights were directly determined for Q_1 and Q_2 in addition to using the rain rates, the performance of the Unified model could have been improved even further.

Although the superensemble presently does not seem to provide any major physical insights, it does identify systematic errors and does reduce them drastically. It is this feature that makes performance of the superensemble considerably higher than the individual models or even the Unified model. The Unified model appears somewhat appealing, and one can in principle construct such Unified models for specific physical parametrizations such as the planetary boundary layer and radiative transfer. Thus collectively it may be possible to improve the performance considerably further from the design of a grand Unified model. In spite of those possible improvements we recognize that if the Unified model is improved within one single model it may still not be able to compete with a multimodel superensemble. The latter can provide a diversity from the global modeling community, which always seems to produce superior forecasts.

Acknowledgements. This work could not have been completed without the datasets provided by the following global weather centers, ECMWF, NCEP, NRL, BMRC, RPN and JMA. The precipitation estimates were derived from the microwave radiances provided by TRMM and DMSP satellites. This work was supported by the following grants NSF ATM-0108741, NASA TRMM NAG5-9662, NOAA NA06GP0512 and NOAA NA16GP1365.

REFERENCES

Arakawa, A. and W. H. Schubert, 1974: Interaction of a cumulus cloud ensemble with the large-scale environment, Part I, *J. Atmos. Sci.,* **31**, 674-701.

Bélair, S., M. A. Méthot, J. Mailhot, B. Bilodeau, A. Patoine, G. Pellerin, and J. Côté, 2000: Operational Implementation of the Fritsch–Chappell Convective Scheme in the 24-km Canadian Regional Model. *Wea. Forecasting,* **15**, 257–275.

Das, S., A. K. Mitra, G. R. Iyengar, and S. Mohandas, 2001: Comprehensive test of different cumulus parametrization schemes for the simulation of the Indian summer monsoon. *Meteor. Atmos. Phys.,* **78**, 227 –244.

Emanuel, K. A., 1991: A scheme for representing cumulus convection in large-scale models. *J. Atmos. Sci.*, **48**, 2313-2335.

Emanuel, K. A., and M. Zivkovic-Rothman, 1999: Development and evaluation of a convection scheme for use in climate models. *J. Atmos. Sci.*, **56**, 1766-1782.

Ferretti, R., T. Paolucci, W. Zheng, Guido Visconti, and P. Bonelli, 2000: Analyses of the Precipitation Pattern on the Alpine Region Using Different Cumulus Convection Parametrizations. *J. Appl. Meteor.,* **39**, 182–200.

Krishnamurti, T. N., Y. Ramanathan, H. L. Pan, R. J. Pasch, and J. Molinari, 1980: Cumulus parametrization and rainfall rates I. *Mon. Wea. Rev.,* **108**, 465-472.

Krishnamurti, T. N., S. Low-Nam, and R. Pasch, 1983: Cumulus parametrization and rainfall rates II. *Mon. Wea. Rev.*, **111**, 815-828.

Krishnamurti, T. N., and H. S. Bedi, 1988: Cumulus parametrization and rainfall rates III. *Mon. Wea. Rev.*, **116**, 583-599.

Krishnamurti, T. N., J. Xue, H. S. Bedi, K. Ingles, and D. Oosterhof, 1991: Physical initialization for numerical weather prediction over the tropics. *Tellus,* **43AB**, 53-81.

Krishnamurti, T. N., C. M. Kishtawal, T. LaRow, D. Bachiochi, Z. Zhang, C. E. Williford, S. Gadgil and S. Surendran, 1999: Improved weather and seasonal climate forecasts from multimodel superensemble. *Science*, **285**, 1548-1550.

Krishnamurti, T. N., C.M. Kishtawal, Z. Zhang, T. LaRow, D. Bachiochi, C. E. Williford, S. Gadgil and S. Surendran, 2000a: Multimodel ensemble forecasts for weather and seasonal climate. *J. Climate*, **13**, 4196-4216.

Krishnamurti, T. N., C. M. Kishtawal, D. W. Shin and C. E. Williford, 2000b: Improving Tropical Precipitation Forecasts from a Multianalysis Superensemble. *J. Climate*, **13**, 4217-4227.

Krishnamurti, T. N., S. Surendran, D. W. Shin, R. J. Correa-Torres, T. S. V. Vijayakumar, C. E. Williford, C. Kummerow, R. F. Adler, J. Simpson, R. Kakar, W. S. Olson, and F. J. Turk, 2001: Real-Time Multianalysis-Multimodel Superensemble Forecasts of Precipitation Using TRMM and SSM/I Products. *Mon. Wea. Rev.,* **129**, 2861-2883.

Krishnamurti, T. N., L. Stefanova, A. Chakraborty, T.S.V. Vijaya Kumar, S. Cocke, D. Bachiochi and B. Mackey, 2003: Seasonal Forecasts of precipitation anomalies for North American and Asian Monsoons. *J. Met. Soc. Jap.* Accepted.

Kuo, Y.-H., R. J. Reed, and Y.-B. Liu, 1996: The ERICA IOP 5 storm. Part III: Mesoscale cyclogenesis and precipitation parametrization. *Mon. Wea. Rev*.,**124**, 1409-1434.

Moorthi, S. and M. J. Suarez, 1992: Relaxed Arakawa-Schubert: A parametrization of moist convection for general circulation models. *Mon. Wea. Rev*., **120**, 978-1002.

Moorthi, S. and M. J. Suarez, 1999: Documentation of version 2 of relaxed Arakawa-Schubert cumulus parametrization with convective downdrafts. *NOAA Technical Report* # NWS/ NCEP **99-01**, 44 pp.

Palmer, T. N. and S. Tibaldi, 1988: On the prediction of forecast skill. *Mon. Wea. Rev.,* **116**, 2453-2480.

Pan, H. -L. and W. -S. Wu, 1995: Implementing a mass flux convection parametrization package for the NMC medium-range forecast model. *NMC Office Note*, No. **409**, 40pp. [Available from NCEP, 5200 Auth Road, Washington, DC 20233]

Rajendran, K., R. S. Nanjundiah, and J. Srinivasan, 2002: Comparison of seasonal and intraseasonal variation of tropical climate in NCAR CCM2 GCM with two different cumulus schemes. *Meteor. Atmos. Phys.,* **79**, 57-86.

Raymond, D. J., and A. M. Blyth, 1986: A stochastic mixing model for nonprecipitating cumulus clouds. *J. Atmos. Sci.,* **43**, 2708-2718.

Shin, D. W., and T. N. Krishnamurti, 2002a: Short-to-medium range superensemble Precipitation forecasts using satellite products. Part I: Deterministic forecasting. Submitted to AGU special issue, dedicated to *7th International Precipitation Conference*, 2001.

Shin, D. W., and T. N. Krishnamurti, 2002b: Short-to-medium range superensemble Precipitation forecasts using satellite products. Part II: Probabilistic forecasting. Submitted to AGU special issue, dedicated to *7th International Precipitation Conference*, 2001. Toth, Z. and E. Kalnay, 1993: Ensemble forecasting at NMC: The generation of perturbations. *Bull. Amer. Met. Soc.,* **74**, 2317-2330.

Wang, W. and N. L. Seaman, 1997: A comparison study of convective parametrization schemes in a mesoscale model. *Mon. Wea. Rev*., **125**, 252-278.

Yanai, M., S. Esbensen and J.H. Chu, 1973: Determination of bulk properties of tropical cloud clusters from large-scale heat and moisture budgets. *J. Atmos. Sci.*, **30**, 611-627.

Zhang, G. J., and N. A. McFarlane, 1995: Sensitivity of climate simulations to the parametrization of cumulus convection in the Canadian Climate Centre general circulation model. *Atmos.-Ocean*, **33**, 407-446.

Dynamical Simulation of Indian Summer Monsoon

K. KRISHNA KUMAR & M. MUJUMDAR
Indian Institute of Tropical Meteorology, Pune 411 008, India
E-mail: krishna@tropmet.res.in

ABSTRACT

India has a long history of monsoon rainfall prediction going back to late eighteenth century. While these early efforts were primarily empirical and statistical in nature, global climate community has made great strides during the last 5 decades or so in advancing our understanding of simulating global climate using general circulation models (GCMs) following well known physical and dynamical principles. Though considerable progress has been made in the recent decades in simulating mean monsoon rainfall patterns and associated variability over the Indian region to a considerable extent, a lot is still desired in improving the physics and resolution of the GCMs to achieve respectable skills of these models. Here we try to provide a general idea of the work that is being carried out on the dynamical simulation of Indian summer monsoon rainfall using different GCMs at some of the climate centers in India and not intended to be an exhaustive review of the status of the skills of different GCMs in general.

1. INTRODUCTION

The region of South Asia, particularly the Indian sub-continent, has a geographical location with unique characteristics, physically, culturally and economically dominated by the most important monsoon systems of the world. The most important climatological feature of the region is the occurrence of monsoons, the southwest or summer monsoon being the principal source of water and the northeast or winter monsoon meeting the water needs of the southern peninsular regions of India. The whole year's supply of water over a major part of the area is realized in just 3 to 4 months in the summer monsoon season, which makes the people critically dependent on the monsoon activity. The year-to-year or other wise known as the inter-annual variability of monsoon rainfall occasionally leads to large scale floods and droughts over different parts of India resulting in serious reduction in agricultural output affecting the regional and national economies. In view of the critical influence of such variability on agricultural and industrial production, forecasting of monsoon rainfall, at least a season in advance, assumes profound importance for policy making and planning mitigatory efforts. While the early efforts in forecasting monsoon rainfall have been primarily using various statistical methods (see Krishna Kumar *et al.* 1995 for a detailed review on monsoon forecasting), General Circulation Model (GCM) simulations were initiated in research mode by Climate Modeling group at IITM, Pune during 1997

(Mujumdar et. al. 2001) as shown in Table 1. Later other modeling groups in India also attempted GCM seasonal forecast.

TABLE 1: Description of Models used in Experimental Seasonal Forecasting at IITM, Pune

Aspects of GCM	*COLA Model*	*UKMO Unified Model*
Domain	Global (Spectral)	Global
Prognostic variables	ξ; D; T; q; ln(p_s)	P^{*}' u; v; T; q
Horizontal Resolution	3.75°lat * 3.75°long (T30)	2.5°lat. * 3.75°long. Arakawa-'B' grid
Vertical Resolution	18σ -levels	19 level hybrid coordinate
Time Integration	Semi-implicit (time step 15 min.)	Split explicit scheme, (time step 30 min.)
Convection	Relaxed Arakawa-Schubert Shallow convection (Tiedke 1984)	Mass flux convection scheme by Gregory
Radiation	Shortwave (Lacis and Hansen (1974)) (CALL 1hr) Long wave (Harshvardhan et al. (1987)) (CALL 3 hr)	Spectral bands for solar radiation and six bands for long wave radiation (Ingram et al., 1996)
Gravity wave drag	∇^4; Roberts time filter, Gravity wave drag	Based on sub-grid scale orographic variance and vertical velocity profile
PBL	Mellor - Yamada (1982) Simple biosphere (SIB) (Xue et al. (1991))	Multi layer soil temperature model and soil moisture prediction scheme and a 5 level boundary layer turbulant transport depending on Richardson number

The mechanisms responsible for inter-annual variability of seasonal mean climate can be thought of as a consequence of the internal dynamics (Palmer and Anderson, 1994) and the boundary forcing (Charney and Shukla 1981; Lindzen and Nigam, 1987). There is considerable evidence that the variance due to internal dynamics is relatively weak in the tropics as compared to extratropics (Mujumdar and Krishnan, 2001). It is also generally known that the inter-annual variability in the summer monsoon rainfall is largely determined by the slowly varying boundary forcing such as sea surface temperature, sea ice cover, land surface temperature, albedo, soil moisture and snow cover, though not all of them are equally important (Shukla, 1987). Variation in lower boundary conditions results from the coupling of the dynamics of the atmosphere to the dynamics of the oceans and to the hydrology of the land masses. The most dominant mode of variability of the tropical ocean-atmosphere system associated with El Nino/Southern Oscillation (ENSO), though has its origin in the tropical Pacific, extends to influence half the globe, and is also found to account for a significant part of variability in the Indian summer monsoon rainfall (Sikka, 1980; Pant and Parthasarathy, 1981; Rasmusson and Carpenter, 1983 and Krishna Kumar *et al.* 1995). This can be appreciated from the coincidence of a majority of large rainfall anomalies in India to be in association with the occurrence of ENSO events in the Pacific. Also, the profound influence of the ENSO on the monsoon rainfall can be understood from the striking similarity between the spatial correlations of monsoon rainfall and the SSTs in the Pacific and the El Nino related SST pattern in the Pacific (Figure 1). Apart from ENSO, the areal extent and thickness of Eurasian/Himalayan snow in preceding winter and spring seasons is also known to have a profound influence on the strength of

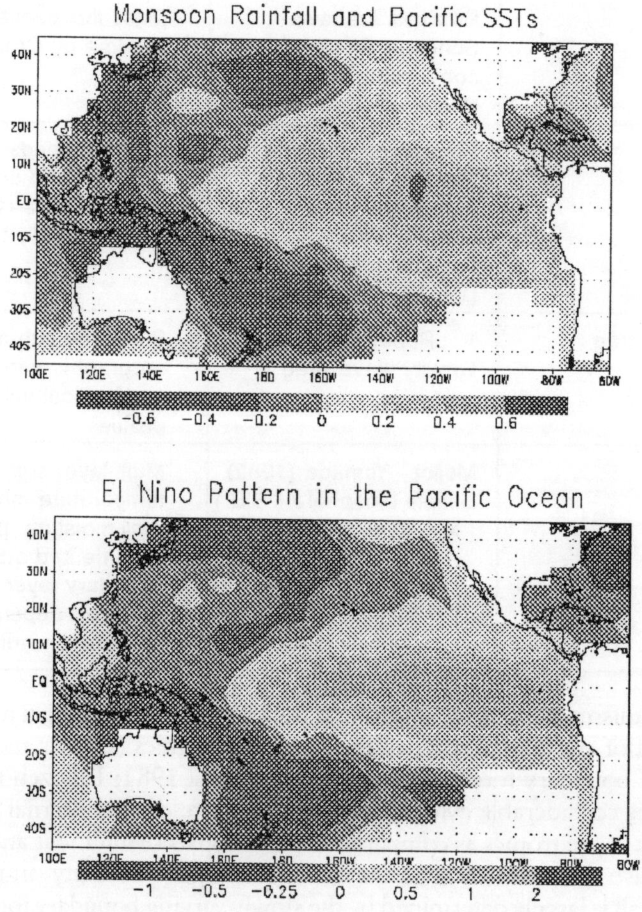

FIG . 1 Correlation between Indian monsoon rainfall and the Pacific SST (upper panel); SST anomalies in a typical El Nino event (lower panel).

ensuing monsoon, vis-a-vis the enhanced/reduced land-sea thermal gradient (see Kripalani and Kulkarni, 1999 for details on monsoon rainfall-snow connections). This suggests considerable predictability of the inter-annual variability of the monsoon system on seasonal time scales. That said, it must also be taken in to cognizance that significant inter-decadal changes occur in the strength of relationship between monsoon rainfall and various ENSO related precursors/indices (Krishna Kumar *et al.* 1999a,b) having implications on the skill of monsoon prediction.

An attempt has been made here to present a brief account of the work being carried out on the simulation/prediction of monsoon rainfall using GCMs by some of the groups and not intended to be a comprehensive review of all the work being done in India.

2. SENSITIVITY EXPERIMENTS

It has been observed that the impact of different El Nino events on the monsoon rainfall is not quite proportionate to their strength. For example, the strong El Nino events of 1982 and 1997 have produced monsoon seasonal rainfall anomalies, on all-India scale, of -13% and +2% respectively, while a relatively weak event of 1987 has produced -18% anomaly. Also, Krishna Kumar *et al.*, (1999b) have shown that the strength of the relationship between the monsoon rainfall and the ENSO has been considerably weakening in the recent decades. The reasons for this are generally attributed to (1) a southeast ward shift in the ENSO induced anomalous Walker circulation leading to reduced suppression of convection over the Indian sub-continent, thus favoring normal monsoon conditions and (2) increased surface temperatures over Eurasia in winter and spring as a consequence of recent mid-latitude continental warming trend, resulting in enhanced land-ocean thermal gradient conducive for a strong monsoon. In order to address these issues, a set of atmospheric general circulation model (AGCM) experiments have been conducted by forcing the ECHAM3 (T21) model of Max-Planck Institute for Meteorology (Hamburg) with different sea surface temperature (SST) patterns in the Pacific Ocean. The possible impact of recent continental warming on the connection between the Indian monsoon and ENSO is also addressed with an additional sensitivity experiment by coupling the atmospheric GCM with a simple mixed layer ocean model and doubling the CO_2 concentration with fixing ENSO related SST pattern in the ocean. The various experiments conducted are listed below:

1. Control Run with Climatological SSTs (CTL) - 20 years
2. Full ENSO Pattern (ENSO) - 7 years
3. East Pacific El Nino Pattern (NINO) -10 years
4. Dateline El Nino Pattern (NINODL) - 12 years
5. ENSO with Global Warming (ENSOGW) - 10 years

The pattern of SSTs used to force the GCM corresponding to the experiments 2-4 listed above are shown in Figure 2. While in experiment 5, the full ENSO pattern is imposed, the SST pattern in experiment 3 is similar to the observed SST patterns of 1982 and 1997 El Nino events, and the SST pattern imposed in experiment 4 is very close to that observed during 1987 event. Examination of the simulations from the control run showed that the model has a fairly good seasonal rainfall and other

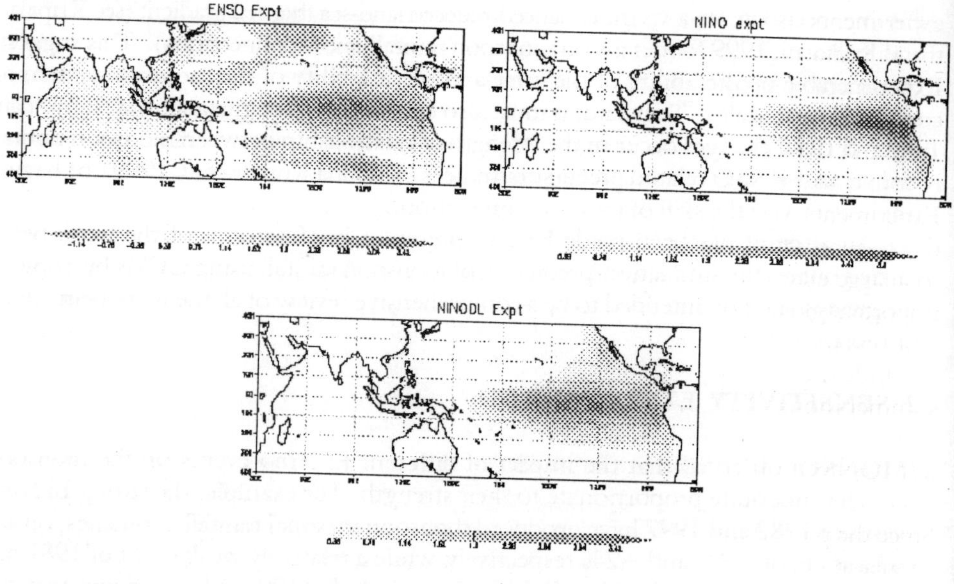

FIG. 2 SST patterns used in sensitivity experiments

characteristics of monsoon. Particularly, the ENSO related tele-connections in the model (ENSO-CTL) are found to be very realistic when compared to the observed features. As can be seen from the rainfall differences between these experiments in Figure 3, the response of the Indian monsoon is found to be quite different to the inter El Nino differences in the SST patterns in the Pacific Ocean. It appears from these

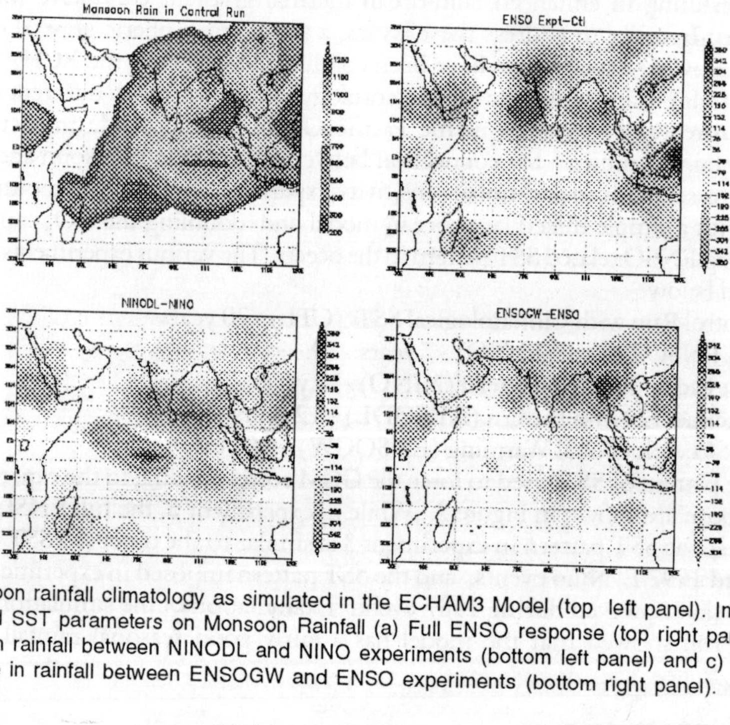

FIG. 3 Monsoon rainfall climatology as simulated in the ECHAM3 Model (top left panel). Impact of ENSO related SST parameters on Monsoon Rainfall (a) Full ENSO response (top right panel), (b) difference in rainfall between NINODL and NINO experiments (bottom left panel) and c) difference in rainfall between ENSOGW and ENSO experiments (bottom right panel).

experiments that though the magnitude of SST anomalies are relatively much smaller, the El Nino events with anomalies close to the dateline (similar to 1987, NINODL Experiment) produce much stronger impact on the Indian summer monsoon rainfall compared to those events with anomalies located more to the east (events similar to 1982 and 1997, NINO Experiment). Also, the impact of ENSO on the Indian monsoon is found to be much less in presence of strong continental warming (ENSOGW Experiment), suggesting that the enhanced land-sea gradient and an enhanced monsoon due to general warming could counter the negative impacts of El Nino on the monsoon to a large extent and thereby possibly weakening the connection between these two important global climate phenomena. Other sensitivity experiments are also conducted to assess the impact of El Nino related SST pattern and the recent increasing trend in the Indian ocean (climatological SSTs are imposed in other oceans) on the Indian summer monsoon. For brevity, these results are not discussed here.

3. MONSOON SIMULATION IN GCMS

Since the pioneering works of Manabe *et al.* (1974) and Hahn and Manabe (1975) on simulating Indian monsoon with atmospheric general circulation models (GCMs), the modelling community has made tremendous progress both in terms of improving the physical parameterizations as well as enhancing the model resolution. It has been shown in many studies that the state-of-the art GCMs can simulate the mean monsoon circulation reasonably well when forced with observed sea surface temperatures. However, simulation of monsoon rainfall and its interannual variability differ from one model to another as evidenced from the studies based on Atmospheric Model Intercomparison Project (AMIP) (Sperber and Palmer, 1994 and Gadgil and Sajani, 1998). This indicates that the simulation of monsoon circulation and associated rainfall are highly sensitive to resolution and physical parameterization schemes used in various models. Apart from this, the simulation of seasonal mean Indian monsoon rainfall is found to be sensitive to small changes in the initial conditions unlike some other regions in the tropics such as Sahel in Africa, north east Brazil etc. (Palmer and Anderson 1994). Though it is not intended here to present a comprehensive assessment of the status of monsoon simulation by GCMs, an effort is made to provide an idea of monsoon rainfall simulation and its sensitivity to the initial conditions using a 49 year (1949-1997) six member ensemble of MRI/JMA 98 GCM run with observed sea surface temperatures. This model with a horizontal resolution of 2.8° × 2.8° (see Yukimoto *et al.*, 2001 for more details on the model description) is run with different 1st January initial conditions taken from a 10 year climatological SST run. Ensemble mean seasonal climatology of winter (DJF) and monsoon (JJAS) seasonal rainfall as simulated in this model are shown in Figure 4 along with the observed rainfall(Xie and Arkin, 1996). As can be seen from the figure, the seasonal migration of tropical rain-belt from the winter hemisphere to the summer as well as the spatial structure of monsoon rainfall over the Indian sub-continent are well simulated by the model. Monsoon circulation, characterizing the low level westerly jet in the Somali region as well as the easterly jet at upper tropospheric levels, is also simulated very well in these runs (not shown). In order to examine the model's ability to simulate the interannual variability of monsoon rainfall over India, the simulated area averaged rainfall over the region 70E-95E,

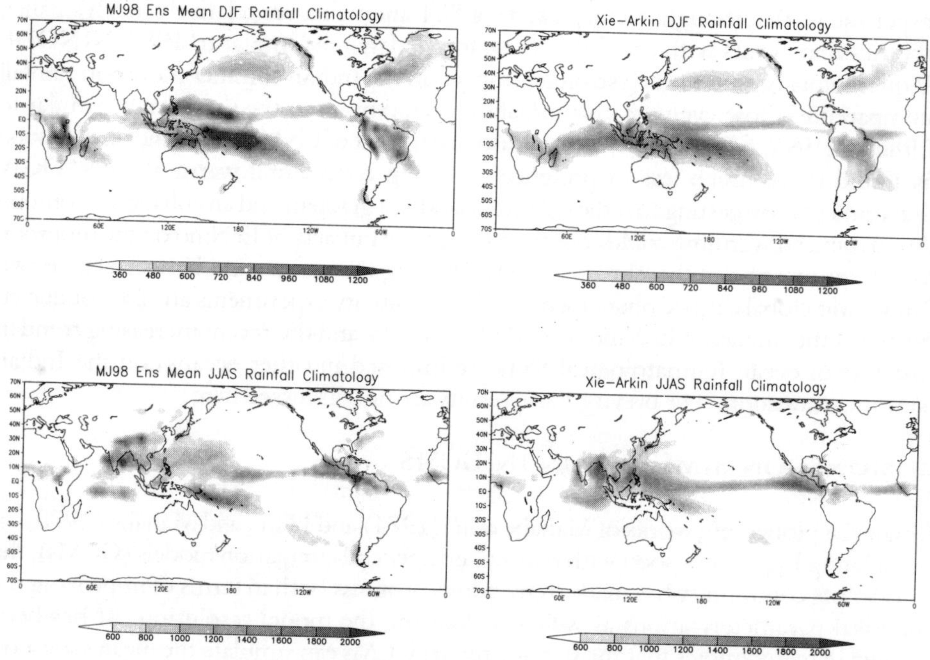

FIG. 4. Simulated ensemble mean rainfall (MRI/JMA98) and observed rainfall (Xie and Arkin) during winter (DJF) and monsoon season (JJAS).

8N-30N (land points only), for the entire six ensemble runs are shown along with the observed (Parthasarathy *et al.*, 1994) rainfall anomalies in Figure 5. The sensitivity of Indian monsoon to the initial conditions, particularly in this model, can be understood from the large spread that is present in most of the years.

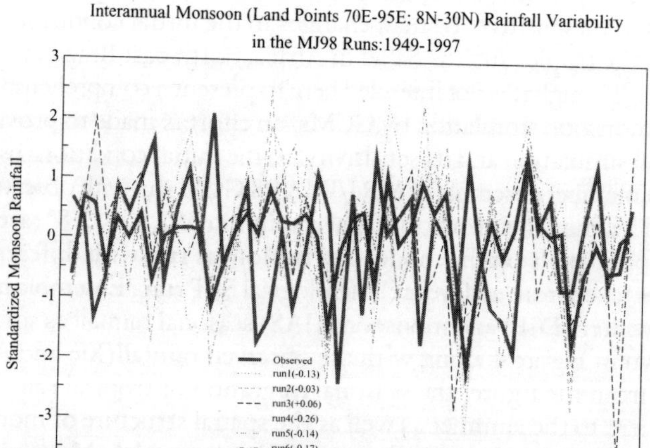

FIG. 5 Interannual monsoon (only land points 70-95E, 8-30N) rainfall variability in the MRI/JMA98 runs : 1949-97.

To study the impact of El Niño on Asian summer monsoon of 1997, an international project titled 'CLIVAR Asian/Australian monsoon Atmospheric General Circulation Model intercomparison" was carried out. Under the project, a set of 10 ensemble integrations for 2-years from September 1996 to August 1998, with observed boundary forcing of SST by 11 Atmospheric General Circulation Models (AGCM) have been carried out. We participated in the project for which Hadley Centre climate model (HadAM2b) has been used. The aim of this study was focused on Intercomparison of Asian summer monsoon of 1997 simulated by AGCMs. Observed seasonal mean precipitation over India in summer 1997 was normal (102%) in spite of strong El Niño. Most of the models simulated correct sign of precipitation anomaly over India when compared with observed although there is a difference in the magnitude. Spatial distribution of precipitation anomaly over 20°S – 40°N, 30°E-120°E region for summer 1997 simulated by 10 AGCMs is compared with observations. Strong negative anomaly over equatorial East Indian Ocean is noticed in observation, which is well simulated by all models except IAP (Institute of Atmospheric Physics ,China) and GFDL (Geophysical Fluid Dynamics Laboratory). All AGCMs showed skill in simulating Southern Oscillation Index (Mandke *et al.* 2001; Kang *et al.* 2002)

4. EXPERIMENTAL MONSOON FORECASTS WITH GCMS

Seasonal prediction of Indian Summer monsoon rainfall during June to September has been primarily carried out every year using statistical methods involving different precursors representing various facets of ocean-land-atmosphere system. Most of the precursors indicate the state and direction in which the ENSO is evolving at the beginning of the monsoon season, i.e., the end of May or early June. India Meteorological Department (IMD), provides a single number of rainfall percentage departure over India as a whole and classify it in three categories, Excess, Normal and Deficient. However, as mentioned in the introduction, Climate Modeling group at IITM, Pune, India has been attempting experimental ensemble seasonal forecasts for the monsoon rainfall since 1997 using two GCMs, the COLA (Center for Ocean-Land-Atmosphere Studies, Maryland, USA) GCM and the UK Meteorological Office (UKMO), London, model (Krishnan *et al.*, 2001; Mujumdar and Krishnan, 2001; Mandke et. al. 1999). A brief description of these two models are given in Table 1. An ensemble of 6 members were run using May mean SST anomalies persisted over the climatological SSTs of June to September. The COLA AGCM was integrated using the initial conditions (based on NCEP/NCAR reanalysis) at 00UTC of 26,27,28,29,30 and 31 May for the years 1997 to 2001. In case of UKMO model six initial conditions are taken from model dumps. Each member of the ensemble was integrated for the monsoon season. Some additional runs were also made in the similar manner but by using April SST and corresponding initial conditions for some years. The simulated and observed (Xie-Arkin rainfall) monsoon rainfall anomalies by these two models for the Indian region (over 5-30°N and 60-95°E) are shown in Figure6. As can be seen from the figure, the simulations made with the COLA model based on May persisted SSTs are generally in good agreement with the observed rainfall anomalies during this period. Apart from these, simulations were also made using observed SST during the monsoon season instead of using April/May persisted SSTs for all these years. It is noted that the

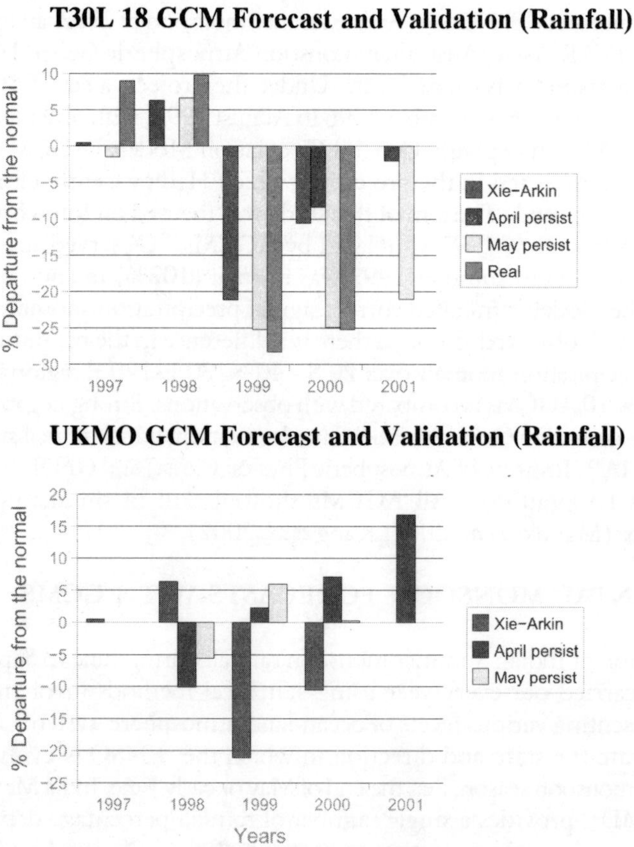

FIG. 6 Experimental seasonal monsoon rainfall forecast using COLA and UKMO GCMs during 1997-2001.

simulations based on the observed June-September SST boundary forcing produced more realistic monsoon rainfall and circulation features as compared to the May persisted SST experiments. The accurate SST information during the summer monsoon season, therefore, is necessary for improving the seasonal forecasting of the Indian summer monsoon rainfall. It would be worth mentioning here that these results are communicated to India Meteorological Department (IMD) to supplement their efforts of seasonal forecasts.

The Extended Range simulation of Indian summer monsoon using Extended Range Monsoon Prediction (ERMP) model is also being carried out by modeling group at Space Application Center ISRO, Ahamdabad, India. They have developed a methodology for the anomaly prediction of the average monthly rainfall during the monsoon season using ensemble integration of ERMP model (Pal et al 2001a, 2001b). ERMP model is derived from the NCMRWF-T80 (National Centre for Medium Range Weather Forecasting) model (from NCEP, USA). The boundary forcing used is weekly OI SST from NCEP instead of fixed boundary condition in the original T80 model. The extended range simulation has been carried out for June and July months with time lagged sequence of initial data 1-7 June taken from NCMRWF analysis. The June, July average of the atmospheric parameters are obtained using total average of the fields over 7 day ensemble runs through the period from start of the run to end of the month for June and for the whole month for July.

The model-simulated fields are compared with the analysis of NCMRWF, NCEP and IMD. The diagnostic analysis indicates that wind circulation patterns and total water vapor associated with progressive monsoon are reasonably simulated at different levels of the troposphere in all the ensemble run. The rainfall simulation is able to capture the broad monsoon features over Bay of Bengal and East Arabian Sea. However, simulated rainfall exhibits a common problem of rainfall locking, which refers to unrealistic concentration of precipitation over certain areas like warm land or over steep orographic features. Figure 7 shows the comparision of real time forecast of monthly average rainfall anomaly distribution over India with that of IMD observations

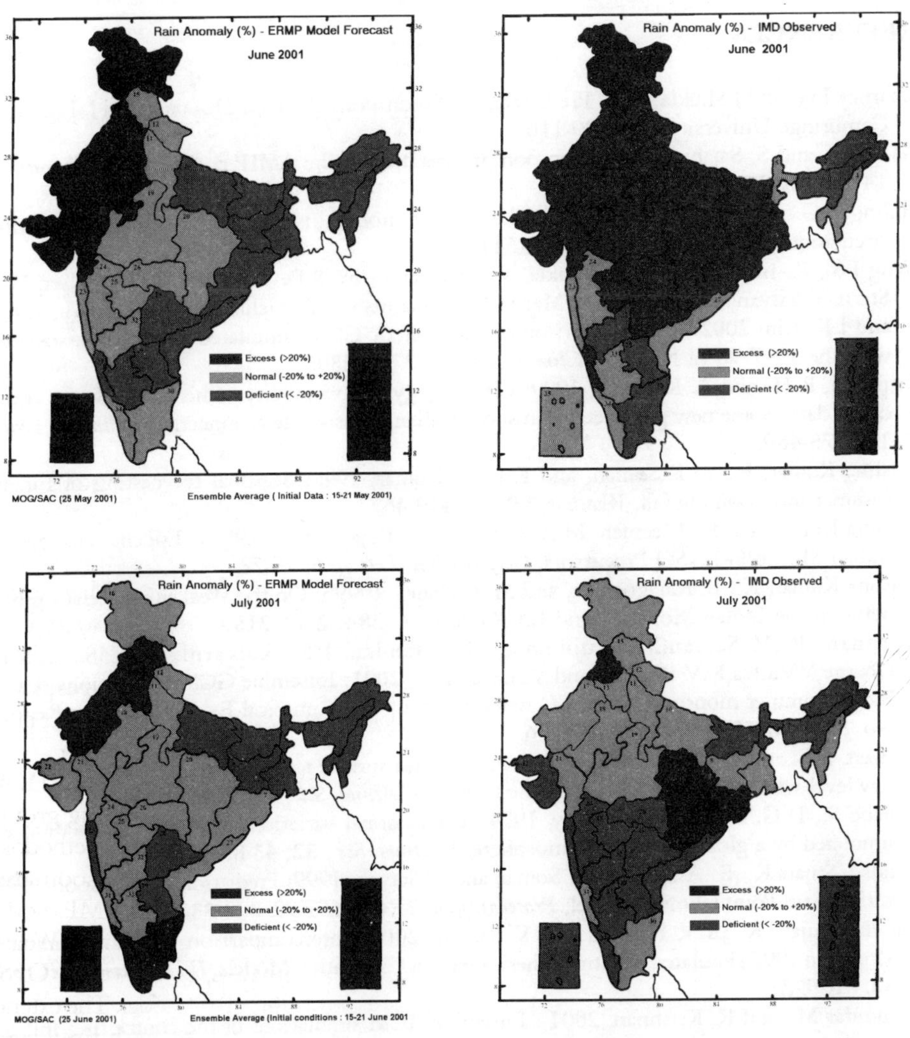

FIG. 7 Evaluation of Extended Range Prediction of monsoon rainfall for June and July of 2001.

for two months viz. June and July of 2001. The verification results show that in most cases, forecasts over 50% of subdivisions have been in agreement with the observations, in about 25% subdivisions forecasts have been one stage away and in 25% subdivisions they have been two stages away from observations. It is to be noted that the simulated rainfall over the zones with high variability has good resemblance with the observations. Further improvement of the extended range prediction crucially depends on the availability of accurate SST predictions on a near real time basis.

Acknowledgements. Part of the results presented in this note were done by one of the authors' Krishna Kumar in collaboration with Dr. Mojib Latif (MPI, Hamburg, Germany) and Dr. M. Sugi (MRI, Japan). The author (Krishna Kumar) wishes to acknowledge their help and support. Inputs on the extended range prediction from Drs. P.K. Pal and P.K. Thapliyal, SAC, Ahmedabad are thankfully acknowledged.

REFERENCES

Charney J.G. and J.Shukla 1981: Predictability of monsoons. *Monsoon Dynamics,* Ed J. Lighthill, Cambridge University Press ,99-110.

Gadgil S., and S. Sajani, 1998 : Monsoon precipitation in the AMIP runs, *Climate Dynamics*, 14, 659-689.

Hahn, D.G. and S. Manabe, 1975: The role of mountains in the south Asian monsoon circulation. *J. Atmos. Sci.*, **32**, 1515-1541.

Kang I.S., K. Jin, K.M. Lau, J. Shukla, V. Krishnamurthy, S.D. Schubert, D.E. Wailser, W.F. Stern, V. Satyan, A. Kitoh, G.A. Meehl, M. Kanamitsu, V.Ya. Galin, Akimasa, G. Wu, Y. Liu and J.K. Kim 2002: Intercomparison of atmospheric GCM simulated anomalies associated with the 1997-98 El Nino, *J. Climate*,**15**, 19, 2791–2805.

Kripalani, R.H. and A. Kulkarni, 1999: Climatology and variability of historical Soviet snow depth data: Some new perspectives in snow-Indian monsoon teleconnections, *Climate Dyn.*, **15**, 475-489.

Krishna Kumar, K., M.K.Soman, and K.Rupa Kumar, 1995: Seasonal forecasting of Indian summer monsoon rainfall, *Weather*, **50**, 12, 449-467.

Krishna Kumar, K., R. Kleeman, M.A. Cane and B. Rajagopalan, 1999a: Epochal changes in Indian Monsoon-ENSO Precursors, *Geophys. Res. Lett.*, **26**, 75-78.

Krishna Kumar, K., B. Rajagopalan, and M.A. Cane, 1999b: On the Weakening Relationship between the Indian Monsoon and ENSO, *Science*, **284**, 2156-2159

Krishnan, R.,V. Satyan,M. Mujumdar,S.K. Mandke, J.R. Kulkarni, A.K. Sahai,S.P. Gharge,V.Vaidya,K.V. Ramesh and Vinay Kumar, 2001: Ensemble GCM simulations of the Indian summer monsoon. CAS/JSC working group on Numerical Experimentation Report No. 31, *WMO/TD No. 1064*, 6.5-6.6.

Lindzen, R.S. and S. Nigam, 1987: On the role of sea surface temperature gradients in forcing low level winds and convergence in the tropics. *J. Atmos. Sci.*, **44**, 2418-2436.

Manabe S., D.G. Hahn, J. Holloway, 1974: The seasonal variation of tropical circulation as simulated by a global model of atmosphere. *J. Atmos. Sci.*, **32**, 43-83.

Mandke, Sujata K. ,K. Ashok, M. K. Soman and V. Satyan, 1999: Prediction of Monsoon 1998 with a Grid Point Climate Model, *Proceedings of TROPMET-99*, 328-333.

Mandke, Sujata K. ,K.V. Ramesh and V. Satyan, 2001: Intercomparison of Asian Summer Monsoon 1997 simulated by Atmospheric General Circulation Models, *IITM Research Report No. RR-092.*

Mujumdar M., and R. Krishnan, 2001 : Ensemble GCM simulations of the contrasting Indian summer monsoon of the 1987 and 1988, *IITM Research Report-89.*

Mujumdar, M. ,V. Vaidya, R. Krishnan, V.Satyan and A.R.Seshagiri, 2001: Assessment of Experimental Ensemble Seasonal Simulation of the Indian Summer Monsoon (1997-2000) using COLA GCM., *Proceedings of TROPMET-2001,* 275-281.

Pal P. K., P. K. Thapliyal and C M Kishtawal ,2001a : Extended range simulation of Indian summer monsoon using ERMP model, Part-A: Ensemble Analysis, *Scientific Report: SAC/ RESA/MOG/ASD/ERMP-SR/01/2001*, Space Applications Centre, Ahmedabad **380015**, India, 39pp.

Pal P. K., P. K. Thapliyal and C M Kishtawal, 2001b : Extended range simulation of Indian summer monsoon using ERMP model, Part-B: real time anomaly prediction, *Scientific Report : SAC/RESA/MOG/ASD/ERMP-SR/02/2001*, Space Applications Centre, Ahmedabad – 380015, India, 38pp.

Palmer T.N. and Anderson, D.L.T., 1994. Prospects for seasonal forecasting : A review. *Quart.J.Roy.Meteor. Soc.*, **120**, 755-794.

Pant, G.B. and Parthasarathy, B., 1981. Some aspects of an association between the southern oscillation and Indian summer monsoon. *Arch. Meteor. Geophys. Bioklimatol., Sr. B.*, **29**, 245-251.

Parthasarathy,B., Munot, A.A. and Kothawale, D.R., 1994. All-India monthly and seasonal rainfall series : 1871-1993. *Theor. Appl. Climatol.*, **49**, 217-224.

Rasmusson, E.M. and T.H. Carpenter, 1983. The relationship between eastern equatorial Pacific sea surface temperature and rainfall over India and Sri Lanka, *Mon. Wea. Rev.*, **111**, 517-528.

Shukla, J. 1987: Interannual variability of monsoons. *Monsoons*, Ed. J.S. Fein and P.L. Stephens, John Wiley and Sons, 399-464,.

Sikka, D.R., 1980: Some aspects of the large-scale fluctuations of summer monsoon rainfall over India in relation to fluctuations in the planetary and regional scale circulation parameters. *Proc. Ind. Acad. Sci. (Earth. Planet. Sci.)*, **89**, 179-195.

Sperber, K. and T.N. Palmer, 1994: Atmospheric Model Intercomparison Project: Monsoon Simulations. *Proc. Int. Conf. On Monsoon Variability and Prediction*. Trieste, WMO/TD-619, 601-608.

Yukimoto, S., M. Endoh, Y. Kitamura, A. Kitoh, T. Motoi and A. Noda, 2000: ENSO-like interdecadal variability in the Pacific Ocean as simulated in a coupled GCM. *J. Geophys. Res.*, **105**, 13945-13963.

Xie, P.A. and P. Arkin, 1996: Analysis of global monthly precipitation using gauge observations satellite estimates and numerical model predictions. *J. Climate*, **9**, 840-858.

Operational Models for Long-range Forecasts of Indian Summer Monsoon Rainfall

M. RAJEEVAN

India Meteorological Department
Pune-5. India
E-mail : *rajeevanm@hotmail.com*

ABSTRACT

The details of the operational models used by the India Meteorological Department for long range forecasts of Indian summer monsoon rainfall are discussed here. IMD uses parametric and power regression models based on 16 parameters. The model was updated in 2002, by replacing four predictors by new predictors. New predictors identified are Europe Pressure Gradient, South Indian Ocean SST, Arabian Sea SST. The average absolute error of the old 16 parameter model for the period 1991-99 was 7.2% while that of the updated model for the same period was 3.2%. Details of other experimental models used by IMD also are discussed.

Epochal changes in Indian monsoon predictability for the past 100 years have been examined using a statistical model. The periods of normal monsoon rainfall coincide with the period of weaker relationship between the predictors and Indian monsoon rainfall and also poorer model skill.

1. INTRODUCTION

The inter-annual variation of Indian Summer Monsoon Rainfall (ISMR) has many social and economic impacts. The long-range forecast of Indian summer monsoon rainfall plays a very crucial role in policy decisions in many vital sectors like agriculture, power, water resources etc. India was the first country to start operational long-range forecasts more than a century ago. The first operational forecast was issued on 4 June 1886 based on the observed relationship between Indian monsoon and Himalayan snow cover (Blanford, 1884). Walker (1923,1924) introduced the first objective technique, namely multiple regression. Since then, several forecast techniques were developed for long range forecasts of ISMR. During the early 1980s, models based on the Dynamic Stochastic technique were introduced (Thapliyal 1982). During the late 1980s a few models based on new techniques like Power Regression and Parametric were introduced (Gowariker *et al.* 1989,1991) for operational long-range forecasts of ISMR. During the recent years, a variety of models like Principal Component Regression (Rajeevan *et al.* 2000), Canonical Correlation Analysis (Rajeevan *et al.* 1999) Artificial Neural Network (Guhathakurta *et al.*1999, Rajeevan *et al.* 2000), Power Transfer

(Thapliyal 2001) were also developed. These models are also used for operational/ experimental long-range forecasts of ISMR.

In this review article, the details of these operational models used by IMD for the long-range forecasts of ISMR are discussed. The epochal variation of the predictability of ISMR is also discussed.

2. OPERATIONAL 16 PARAMETER MODELS

In 1989, India Meteorological Department (IMD) introduced the 16 Parameter Parametric and Power Regression models (Gowariker *et al.* 1989,1991) for long range forecasts of ISMR. This model consisted of 16 predictors, which include global and regional atmosphere-land-ocean parameters, covering the antecedent period from December to May. The operational forecast of IMD has two parts; a forecast for the category of monsoon (like normal/excess or deficient) and a forecast for the quantitative value of monsoon (June-September) rainfall over the country as a whole. The parametric model is used to prepare the monsoon categorical forecasts and the Power regression model is used for the quantitative monsoon rainfall forecasts, which has the model error of $\pm$ 4 %. The verification of the operational forecasts issued based on these models for the period 1989 to 1999 is given in Table1. The errors of the quantitative forecasts were higher than expected, particularly during the recent 4 years (1996-1999) though the largest error of 18 % occurred in 1994. The root mean square error

TABLE 1 Performance of the operational long range forecast model of IMD Period: 1989-2001

Year	Observed		Forecast	
	Category	ISMR % of LPA	Category	ISMR % of LPA
1989	Normal	101	Normal	102 (- 1)
1990	Normal	106	Normal	101 (5)
1991	Normal	91	Normal	94 (- 3)
1992	Normal	93	Normal	92 (1)
1993	Normal	100	Normal	103 (- 3)
1994	Normal	110	Normal	92 (18)
1995	Normal	100	Normal	97 (3)
1996	Normal	103	Normal	96 (7)
1997	Normal	102	Normal	92 (10)
1998	Normal	105	Normal	99 (6)
1999	Normal	96	Normal	108 (-12)
2000*	Normal	92	Normal	99 (-7)
2001*	Normal	91	Normal	98 (-7)

* Forecast from the updated model.

of the operational forecasts during this period (1989-1999) was 8.0 %, which is double the model forecast error of ± 4%. We have therefore critically examined the reasons for these large errors during the recent years and the results are discussed below. We have updated the 16- Parameter model by replacing four predictors by four new predictors. The details are given below.

Thapliyal (1997, 2001), Parthasarathy (1991), Hastenrath and Greischer (1993), Rajeevan (2001) have examined the secular variation of the rainfall-predictor relationships. All the studies have noted that the predictor-monsoon relationship varies with time, showing stronger relationship during some epochs and weaker relationships with some other epochs. Hastenrath and Greischer (1993) and Rajeevan (2001) also reported significant secular variations pose serious problems on the accuracy of the forecasts during the independent period. Recently, Kripalani and Kulkarni (1997) and Krishnakumar *et al.* (1999) have also addressed the weakening of the relationship between ENSO and ISMR. We have therefore, examined in detail the secular variation of all the 16 predictors of the model. It has been found that four predictors, 500 hPa ridge position in April, North India Minimum temperature (March), 10 hPa Zonal wind (January), and Darwin pressure (Spring)) have consistently lost the significant relationship with ISMR. Figure 1 shows the 21-year moving correlations of 500-hPa ridge position with ISMR. The horizontal line in Figure1 indicates the 95% confidence level for the 21- year period of data (0.42). It can be seen that the 500-hPa ridge was significantly correlated with ISMR till about 1982. However the correlations deteriorated drastically afterwards and even the sign of the correlation has been changed from positive to negative. In the next section, we discuss the newly identified predictors.

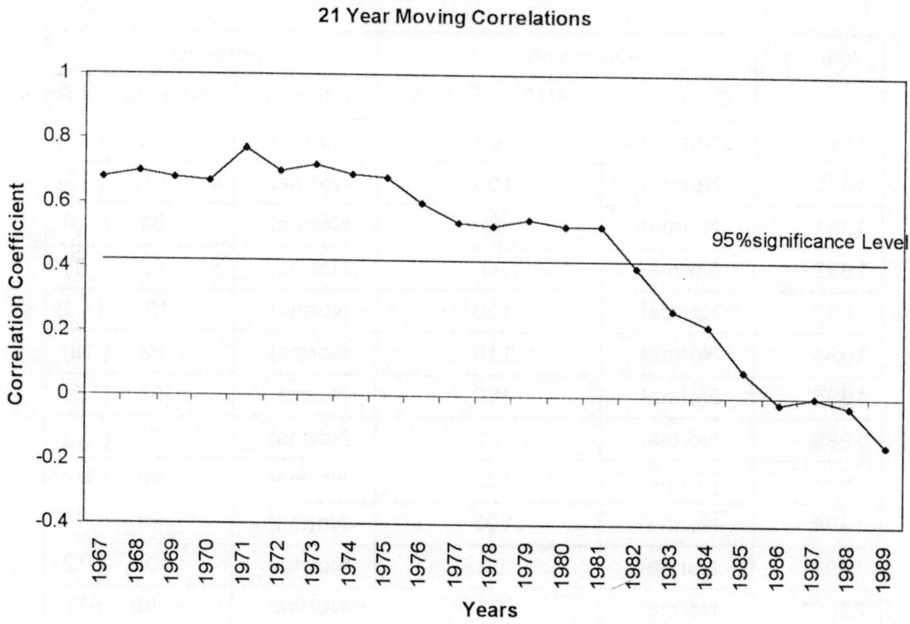

FIG.1. 21 Year moving correlations of 500 hPa Ridge. Data period: 1958-1999. The 95% significance level is shown as horizontal line.

2.1 Identification of New Predictors

Relationship of ISMR anomalies with equatorial Pacific Ocean SST anomalies is known for many years and has been extensively used in statistical models for long range forecasts of ISMR. Recently due to availability of more accurate and longer data sets of sea surface temperatures from the Indian ocean, the relationship of Indian ocean SST anomalies with Indian Summer monsoon rainfall has been examined in detail (Rajeevan *et al.* 2002, and Clark *et al.* 2000). These studies observed that the SST anomalies over Arabian Sea and South Indian Ocean are very crucial for the inter-annual variability of ISMR. Rajeevan *et al.* (2002) identified two areas where the SST anomalies during the winter and spring are positively correlated with ISMR anomalies. The SST anomalies over NW Arabian Sea (15° N-25° N, 50° E-100° E) averaged for January and February are positively and significantly correlated (C.C. = 0.44, significant at 99 % level, period 1958-99) with ISMR. Similarly, the SST anomalies over the south Indian ocean (15°S-30°S, 70°E-110° E) averaged for the months of February and March are also positively and significantly correlated (C.C=0.49, significant at 99 % level, Period 1958-99) with ISMR anomalies. Arabian Sea SST Index was negative during major deficient years like 1951, 1966, 1968, 1972, 1974, 1979, 1986 and 1987. Similarly this index was positive during the period 1970 and 1988. South Indian Ocean SST index was negative during major deficient years like 1951, 1965, 1966, 1968, 1974, 1979, 1982 and 1986. Thus these two new indices are promising as the predictors for long range forecasting of ISMR anomalies.

Rajeevan *et al.* (1998) have suggested that the zonal temperature gradient over western parts of Europe during January is significantly and physically related to Indian summer monsoon activity. It has been observed that during good monsoon years, the temperature gradient over NW Europe is directed towards equator. Hydrostatically, it means that the pressure gradient anomaly is directed towards the pole. During the deficient years, it has been observed that the poleward pressure gradient becomes weaker. This pressure gradient is ultimately related to anomalous mid-latitude activity including blocking activity etc during the spring season (Raman and Rao, 1981). Using the mean sea level pressure data (Basnett and Parker, 1997) an index was derived for the Europe Pressure gradient for the month of January. The pressure gradient is calculated as the difference of pressure anomalies averaged over two latitude bands, one averaged over the box, 60°N-70°N, 0° E-40°E and the other averaged over the box 30°N-45°N, 0°-40° E. The pressure index is positively and significantly correlated (0.34, significant at 99 % level for the period 1958-99) with ISMR anomalies. Below normal pressure gradient indicates below normal rainfall activity.

The 21-year moving correlations of the SST anomaly indices and the pressure gradient index with ISMR are shown in Figure 2. The 95% confidence level is shown as a horizontal dashed line. It can be seen that the south Indian Ocean SST and the pressure gradient is significantly correlated with ISMR throughout the period. But the Arabian Sea SST is significantly correlated with ISMR only after 1980s. During the recent years, this index is very significantly correlated with ISMR.

The fourth predictor used in the model is the Darwin Pressure tendency as defined by Shukla and Paolino (1983). This index is derived as the difference of mean sea level pressure of Darwin for April and January. This parameter has the correlation coefficient of – 0.55 (1958-99) which is significant at 99 % level. This parameter indicates that

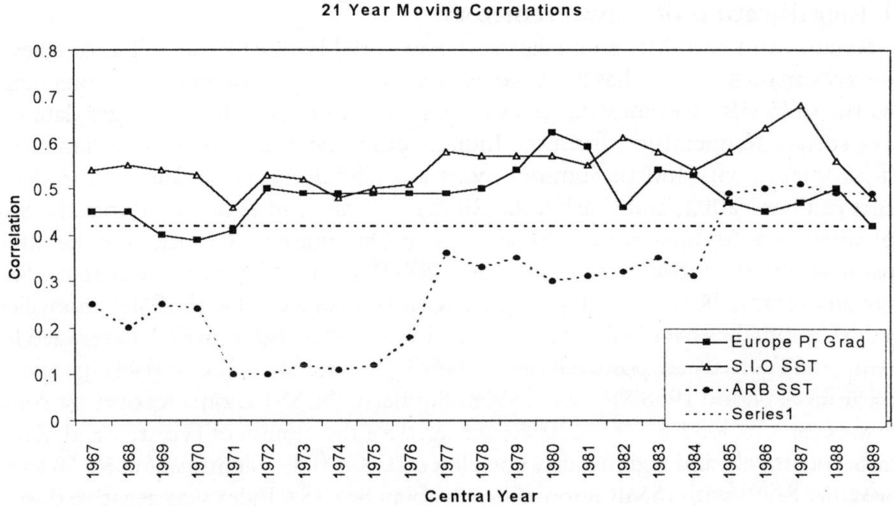

FIG.2 21 Year moving correlations of new predictors, Europe Pressure gradient (filled squares), South Indian Ocean SST (triangle) and Arabian Sea SST (filled circle). Period: 1958-1999.

when the Darwin pressure increases from January to April more than normal, than the ensuing ISMR is likely to be less than normal.

The revised list of 16 predictors are given in Table 2. Using these 16 atmosphere-land-ocean parameters, the parametric and power regression (PR) models were updated

TABLE 2 List of 16 Parameters used in the Updated 16 parameter Model

Parameter No.	Name of the Parameter
X_1	50 hPa East-West Ridge (January + February)
X_2 *	Darwin Pressure Tendency (April-January)
X_3 *	South Indian Ocean SST (February +March) (15 °-30 °S, 70 °–110 ° E)
X_4	East Coast Minimum Temperature (March)
X_5 *	Arabian Sea SST (January + February)
X_6	Central India Temperature (May)
X_7	N. H. Temperature (January +February)
X_8	N. H. Pressure (January to April)
X_9	Southern Oscillation Index (SOI) (March to May)
X_{10}	Indian Ocean Equatorial Pressure (January to May)
X_{11}	Himalayan Snow Cover (January to March)
X_{12}	Eurasian Snow Cover (December)
X_{13} *	Europe Pressure Gradient (January)
X_{14}	El Nino (Same Year)
X_{15}	El Nino (Previous Year)
X_{16}	Argentina Pressure (April)

* : New Predictors

by recalculating the model constants. Unlike the parametric model, which gives equal weightage to all 16 parameters, the PR model gives an appropriate weightage to each of the 16 predictors. The model tries to account for some kind of non-linear interactions among the predictors and the ISMR.

The comparison of the hindcast ISMR anomalies from the updated model with the operational forecasts of IMD for the period 1991-1999 is given in Table 3. From the Table, it can be seen that since 1994 the forecasts from the updated 16- parameter model were closer to the actual ISMR. With the updated 16-parameter model, comparatively much smaller errors were observed in 1994, 1997, 1998 and 1999. The average absolute error of the old 16-parameter model for the period 1991-99 was 7.2 whereas that of the updated model for the same period was 3.2 % only. This indicates large improvements in the performance of the updated model. This analysis showed that the updated 16-parameter model performs better than the old 16-parameter model, which was used by the IMD for the operational forecasts from 1989 to 1999. The forecasts from this updated model for the years 2000-2002 are given in Table1. The performance of IMD's operational long-range forecasts (actual and forecasts) for the period 1988-2001 is shown in Figure3.

TABLE 3 Comparison of hind casts from the updated 16-parameter model and Operational forecasts Period: 1991-1999. Forecast Errors are given in Brackets

Year	Actual% of LPA	Operational Forecast (Old 16 Parameter Model)	Hindcast with the updated 16 Parameter Model
1991	91	94 (-3)	93 (-2)
1992	93	92 (1)	95 (-2)
1993	100	103 (-3)	107 (-7)
1994	110	92 (18)	98 (-12)
1995	100	97 (3)	100 (0)
1996	103	96 (7)	105 (-2)
1997	102	92 (10)	101 (1)
1998	105	99 (6)	108 (- 3)
1999	96	108 (-12)	96 (0)
Absolute Mean Error (91-99)		7.0%	3.2%

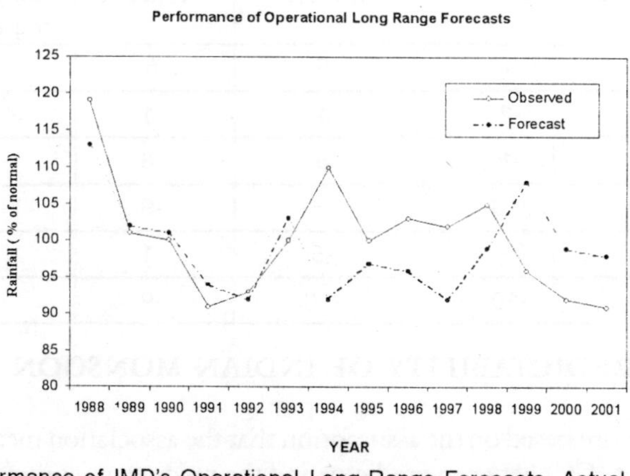

FIG.3. Performance of IMD's Operational Long-Range Forecasts. Actual (continous line) and Forecast (broken line). Period: 1988-2001

3. EXPERIMENTAL MODELS

In addition, models based on multiple regression, Principle Regression Analysis, Dynamic Stochastic Transfer, Artificial Neural Network and Power Transfer techniques also are used for preparing experimental long range forecasts of ISMR. The multiple regression model was developed using 6 predictors (50 hPa east-west ridge, Eurasian snow cover, East coast India temperature, Darwin Pressure Tendency, Argentina pressure (April) and South Indian Ocean SST). The model was developed with the data of the period, 1961-1995 and has the multiple correlation coefficient of 0.84. The Dynamic Stochastic Transfer Model (Thapliyal, 1982) was developed with East Coast India temperature as the input. The data for the period, 1961-2000 were used to develop the model constants. The Power Transfer model (Thapliyal, 2000) was developed with 10 predictors (50 hpa east-west ridge, East Coast India temperature, El Nino Previous year, Central India temperature, Southern Oscillation Index, N.H. pressure, Eurasian snow cover, N.H.Temperature, Argentina pressure, Darwin-Seychelles pressure difference). The principal component regression model (Rajeevan *et al.* 2000) and Artificial Neural Network model (Guhathakurta *et al.* 1999, Rajeevan *et al.* 2000) were developed with 8 predictors. The predictors are East Coast India temperature, Nino-3 tendency, NW Europe temperature, N.H.Pressure anomaly, May Minimum Temperature, South Indian Ocean SST, Arabian sea SST and 50 hpa east-west ridge. In the Principal component model, the first 5 significant principal components were used as the predictors in the linear multiple regression approach. The neural network model was trained with 3 hidden neurons. The conventional back propagation algorithm was used to train the model. The output was taken as the average of ensemble of 30 forecasts prepared with different initial weights.

Table 4 shows the performance of these experimental models for the independent years 1996-2001.

TABLE 4 Performance of Experimental Models: Forecast Errors in Percentage (1996-2001)

Year	Multiple Regression	DST	Power Transfer	Neural Network	Principal component regression
1996	7	9	5	6	2
1997	-1	-2	4	-1	-1
1998	-4	-1	5	-6	-11
1999	-12	-10	-5	-9	-12
2000	-1	-7	-5	-1	-7
2001	-10	-10	-10	-9	-12

4. CHANGING PREDICTABILITY OF INDIAN MONSOON

The statistical models are based on the assumption that the association measured by the correlation coefficient (CC) between predictor and the predictant, computed based on past data would persist in future also. However, secular variations between the predictors and ISMR have been noted. These variations have been linked to changes in the global and regional circulation patterns. These secular changes therefore pose serious challenge to long-range forecasting. Analysis of more than 20 known predictors has revealed that many of the predictors have lost the significant relationship with ISMR during the recent years (not shown).

Epochal changes in Indian monsoon predictability for the last 100 years have been examined using a statistical model developed with 100 years of data (1901-2000). Five predictors (NW India Minimum temperature in May, N.H.Temperature (January and February), Argentina Pressure (spring), Darwin Pressure Tendency and Nino 3 SST Index tendency (MAM-DJF)) for the period 1901-2000 have been used to develop a statistical model. These predictors represent the ENSO forcing, land surface conditions over Eurasia and the intensity of the heat low over NW India. The principal component analysis (PCA) of these 5 predictors was made and the resultant 3 significant principal components were further used to develop a multiple regression model. The results are shown in Figure 4, which shows the 11 year moving correlation between a) the first principal component and ISMR b) correlation coefficient between the actual and hindcast ISMR or skill of the model. The 11 year running mean of standard deviation anomaly (subtracted from long term mean of 10%) of ISMR also is shown.

The C.C. between the first P.C. and ISMR reflects the general relationship between the predictors and ISMR. This relationship was weak in 1930s and 1940s. During the recent years also this relationship was found to be weak. During the 1960s to 1990s this relationship was however very strong. Obviously, the skill of the model also shows similar type of variations. The model skill was positive during 1960s to 1990s. However, the model skill was found to be negative during 1930s and 1940s and also during the recent years. It is also interesting to note that the standard deviation of ISMR was found to be smaller (or ISMR remains within normal limits) during 1930s and 1940s

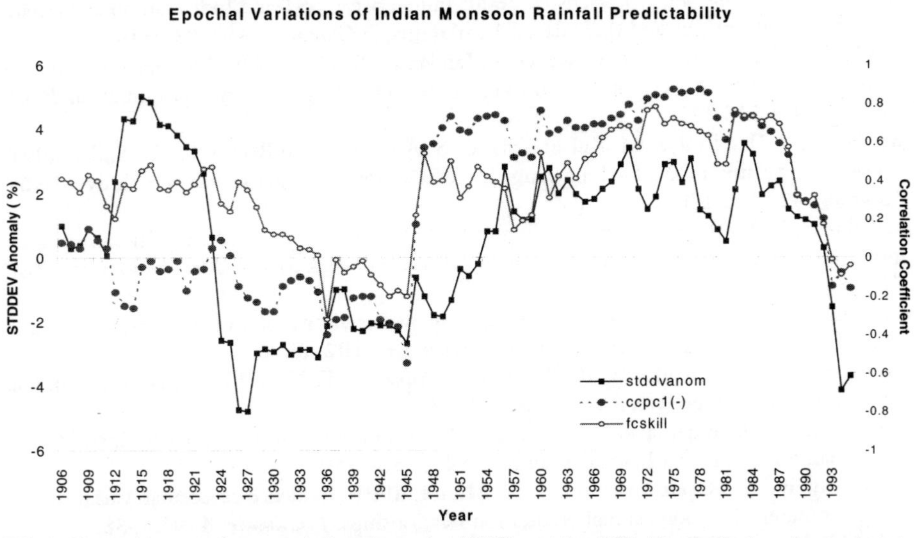

FIG.4. Eleven Year moving correlations between ISMR and the PC1 (Green line) and actual and forecast ISMR (red line). 11 year moving average of standard deviation anomaly is shown in blue line.

and again during the recent years. It is to be mentioned that ISMR was normal successively for last 12 years since 1989. Therefore the periods of normal monsoon rainfall coincide with the periods of weaker relationship between the predictors and ISMR and also poor model skill. This is curious because it is generally believed that statistical models do not show good skill when the inter-annual variation of monsoon is very large. However, here we have seen that good positive model skill was observed when the inter-annual variability of ISMR was also very large. This is because the predictor-ISMR relationship was also stronger during those periods when the inter-annual variability was very large. Thus stronger boundary forcing-ISMR coupling leads to large inter-annual variations of ISMR. When this coupling becomes weak, monsoon tends to remain normal as observed in the recent years. But during these periods empirical models based on these boundary-forcing parameters will show poor predictive skill.

Acknowledgements. I am thankful to Dr.R.R.Kelkar, DGM for nominating me to deliver the talk at the Indo-US workshop and for his kind encouragement and support. I am also thankful to Dr. R.K. Dube, ADGM (R) and Dr.V.Thapliyal, DDGM (WF) for encouragement and for providing facilities to carry out research work related to long range forecasting. I am also thankful to all staff members of LRF Section, for their sincere and dedicated support given during the course of this study.

REFERENCES

Basnett, T.A., and D.E. Parker, 1997: Development of the Global mean sea level pressure data set, GMSLP2, *Climatic Research Technical note No 79*, Hadley Centre, Meteorological Office, Bracknell, 16 pp.

Blanford, H.F, 1884, on the connection of the Himalayan snow with dry winds and seasons of droughts in India, *Proc. of the Royal Society of London*, **37**, 3-22.

Clark, C. O., J. E. Cole and P. J. Webster, 2000: Indian ocean SST and Indian summer rainfall: Prediction relationships and there decadal variability, *J. Climate*, **13**, 2503-2519.

Gowariker, V., V. Thapliyal, R. P Sarker, G. S.Mandal, D. R Sikka, 1989: Parametric and power regression models: New approach to long range forecasting of monsoon rainfall in India, *Mausam*, **40**, 115-122.

Gowariker,V., V. Thapliyal, S. M. Kulshrestha, G. S Mandal, N Sen Roy and D.R. Sikka: 1991: A power regression model for long range forecast of southwest monsoon rainfall over India, *Mausam*, **42**, 125-130.

Guhathakurta, P., M. Rajeevan, and V. Thapliyal,1999: Long range forecasting Indian summer monsoon rainfall by a Principal component Neural Network Model. *Met. and Atmos. Phys*, 71, 255-266.

Hastenrath, S., and L. Greischer, 1993: Changing predictability of Indian monsoon rainfall anomalies. *Proc. Ind. Acad. Sci, (Earth and Planet Sci)*, **102**, 35-47.

Kripalani, R.H. and A. Kulkarni, 1997: Climatic impact of El-Nino / La Nina on the Indian monsoon a new perspective, *Weather*, **52**, 39-43.

Krishnakumar, K., B. Rajagopalan, M. A. Cane, 1999: On the weakening relationship between the Indian monsoon and ENSO. *Science*, **284**, 2156-2159.

Parthasarathy, B., K. Rupakumar and A. A. Munot, 1991: Evidence of secular variations in Indian summer monsoon rainfall circulation relationships. *J. Climate*, **4**, 927-938.

Raman, C.R.V., and Y.P. Rao, 1981: Blocking highs over Asia and Droughts over India, *Nature*, **289**, 271-273.

Rajeevan, M., D. S. Pai, and V. Thapliyal, 1998: Spatial and temporal relationships between global land surface air temperature anomalies and Indian summer monsoon rainfall. *Met. and Atmos. Physics*, **66**, 157-171.

Rajeevan, M., V. Thapliyal, S.R. Patil, and U.S. De, 1999: Canonical Correlation Analysis (CCA) models for long-range forecasts of sub-divisional monsoon rainfall over India. *Mausam*, **50**,145-152.

Rajeevan, M., P. Guhathakurta, and V. Thapliyal, 2000: New models for long range forecasts of summer monsoon rainfall over northwest and Peninsular India. *Met. and Atmos. Physics*, **73**,211-225.

Rajeevan, M., 2001: Prediction of Indian summer monsoon: Status, problems and prospects, *Curr. Science*, **11**,1451-1457.

Rajeevan, M., D.S. Pai and V. Thapliyal, 2002: Predictive relationships between Indian Ocean Sea surface temperatures and Indian summer monsoon rainfall. *Mausam*, **53**, 337-348.

Shukla, J. and D.A. Paolino, 1983: The southern oscillation and long range forecasting of the summer monsoon rainfall over India. *Mon. Wea. Rev.*, **111**, 1830-1837.

Thapliyal, V., 1982: Stochastic Dynamic model for long range forecasting of summer monsoon rainfall in peninsular India. *Mausam*, **33**, 399-404.

Thapliyal, V., 1990: Long range prediction of summer monsoon rainfall over India: Evolution and development of new models. *Mausam*, **41**, 334-346.

Thapliyal, V., 1997: Preliminary and final long range forecast for seasonal monsoon rainfall over India. *J. Arid Environ.*, **36**, 385-403.

Thalpiyal, V., 2001: Long-range forecast of summer monsoon rainfall over India: Evolution and development of new power transfer model. *Proc. Ind. National. Sci. Acad.*, **67 A**, 343-359.

Walker, G. T., 1923: Correlation in seasonal variation of weather. *Memories of India. Met. Dept.*, **24**,75-131.

Walker, G. T., 1924: Correlation in seasonal variation of weather, IX: A further study of world weather. *Memories of India Met. Dept.*, **24**, 275-332.

Monsoon's Feedback on ENSO Evolution[1]

SUMANT NIGAM[2]

Department of Meteorology and ESSIC

University of Maryland, College Park, Maryland, USA

E mail: nigam@atmos.umd.edu

ABSTRACT

Asian summer-monsoon heating anomalies are parameterized in terms of concurrent El Nino SST anomalies and used as additional forcing in the Cane-Zebiak Pacific ocean-atmosphere anomaly model in order to isolate the Asian summer monsoon's feedback on ENSO evolution. Analysis of model integrations having interactive Asian-sector heating anomalies reveals that monsoon's feedback on the Pacific surface winds leads to increased ENSO occurrence — an extra ENSO event every 20 years or so. Examination of the ENSO distribution with respect to the peak SST-anomaly in the eastern equatorial Pacific shows increased El Niño occurrence in the 2.2B3.6K range along with a modest reduction in the 0.6B1.2K range, i.e., a population-shift due to the strengthening of weak El Niños in the interactive monsoon integration. The monsoon feedback also results in a wider period-distribution of ENSO variability, but with El Niños peak-phase remaining seasonally locked with the Northern winter months. These findings provide modeling support for the positive nature of the monsoon-ENSO feedback suggested by previous empirical and diagnostic modeling studies. The feedback is generated primarily by the diabatic heating changes over both land and ocean in the Asian Tropics.

The variability of the Asian monsoon is however not all related to ENSO: Other modes of SST variability in the Pacific, Atlantic, and Indian ocean-basins can also be influential on the Asian hydro climate (rainfall, stream flow, drought indices). The influence can be manifest through decadal variations of the monsoon-ENSO linkage, which ought to be reevaluated using indices that reflect monsoon rainfall variations over both land and ocean.

1. INTRODUCTION

The El Niño / La Niña phenomena — opposite phases of the Southern Oscillation — have been the focus of intense analysis, coupled modeling, and prediction research because of their widespread impact on atmospheric circulation and global precipitation

---

[1]The modeling analysis was performed by Chul Chung, as part of his doctoral dissertation at the University of Maryland. Dr. Chung is now at the Scripps Institution of Oceanography, La Jolla, CA.

[2]Professor of Meteorology and Earth System Science, Room 3403, Computer and Space Sciences Bldg., University of Maryland, College Park, MD 20742-2425. Email: *nigam@atmos.umd.edu*

during both winter and summer seasons (Bjerknes 1969; Horel and Wallace 1981; Philander 1985; Ropelewski and Halpert 1987; Nigam *et al.* 1999). The extratropical circulation anomalies in the Pacific / North American sector during Northern winter (Horel and Wallace 1981; DeWeaver and Nigam 2002), and the Asian summer-monsoon rainfall anomalies (Rasmusson and Carpenter 1983) are leading examples of El Niño Southern Oscillation's (ENSO) impact on midlatitude and tropical climates.

The phrase 'Asian summer monsoon' inclusively represents the seasonal circulation and rainfall over the Asian continent and the tropical Indian ocean basin. Inclusion of land-masses and adjoining seas in the monsoon-domain follows from the canonical definition of the monsoon which involves seasonal land-sea contrasts in sensible and, ultimately, deep latent heating (e.g., Ramage 1971; Webster and Yang 1992)[3].

The link between Asian summer monsoon and ENSO — the two planetary-scale climate systems — has been examined using both observations and dynamical models. Rasmusson and Carpenter (1983) showed that extreme summertime droughts in India tend to occur during the moderate-to-strong El Niño years, and later, Bhalme *et al.* (1990) showed the eastern equatorial Pacific SST to be inversely related to the Indian summer-monsoon rainfall. However, the monsoon-ENSO linkage, itself, exhibits variability on decadal time scales, being notably weak in the recent two decades (Kumar *et al.* 1999). The impact of Asian monsoon circulation and precipitation anomalies on subsequent ENSO evolution has also been investigated: Empirical analysis (Meehl 1987; Yasunari 1991; Webster and Yang 1992) and diagnostic modeling experiments (Nigam 1994) indicate that Asian summer monsoon rainfall anomalies impact Pacific trade winds, and potentially, ENSO evolution. Lag-correlation analysis does shows the Indian summer-monsoon rainfall variations to be more strongly (and negatively) correlated with the central/eastern tropical Pacific SST anomalies in the post summer-monsoon season than during any other period (Verma 1992; Lau and Yang 1996), indicating the possibility of positive feedback between Asian summer-monsoon variations and ENSO evolution. The active role of the Indian monsoon in Southern Oscillation's evolution was first recognized by Sir Gilbert Walker (1923, 1924), who found the Southern Oscillation index during June–August to be more strongly correlated with the following winter's index (+0.8) rather than with the preceding winter's one (-0.2).

Historical analyses of monsoon–ENSO linkage are based on rainfall over the Indian subcontinent, for reasons of data availability. (The Asian monsoon region is however much larger, as noted above, and the sensitivity of the linkage to domain definitions needs to be investigated). Table 1 summarizes the findings of various linkage analyses, by binning them into contemporaneous, and lead/lag categories. It is interesting to note that the ENSO-leading correlations are rather modest, indicating the limited potential of ENSO-based predictions for the all-India averaged rainfall, at least. The contemporaneous correlations are stronger, having been close to -0.6 over an extended period; but have recently fallen in magnitude to 0.3, as noted earlier. Interestingly, the phenomena are correlated most strongly in the post-summer monsoon season: The Indian summer rainfall correlation with the autumn/fall ENSO indices is in the -0.6 to -0.8 range, indicating, the considerable influence of monsoon variability on ENSO evolution. ENSO is however not the only major phenomenon to be influenced by monsoon variability.

[3]Ramage (1971), for example, defines the monsoonal region to include the oceanic and land-surface regions between 20°W–170°E and 20°S–40°N (see figure 1a in Webster and Yang 1992).

TABLE 1 : Summary of Monsoon-ENSO Correlations

Correlation Type	Correlation Strength	Studies
Simultaneous	-0.30 to -0.60	Sikka 1980 Pant and Parthasarathy 1981 Rasmusson and Carpenter 1983 Shukla and Paolino 1985 Ropelewski and Halpert 1987 Verma 1992 Webster and Yang 1992 Kumar, Rajagopalan, and Cane 1999 Krishnamurthy and Goswami 2000
ENSO leading (by a season, generally)	-0.05 to -0.35	Walker 1923, 1924 Khandekar 1979, 1982 Webster and Yang 1992
Monsoon leading (by a season, generally)	-0.60 to -0.80	Walker 1923, 1924 Angell 1981 Meehl 1987 Yasunari 1991 Verma 1992 Webster and Yang 1992 Lau and Yang 1996

The Mediterranean climate (and its variability) was recently shown to be also linked to the Asian summer monsoon (Rodwell and Hoskins 1996). The linkage is through the Rossby wave response and associated vertical motion that are generated by the gigantic monsoon heat source in the Asian subtropics. In fact, there is a suggestion that European floods in the summer of 2002 were, in part, a consequence of this year's anemic Indian monsoon; the largescale subsidence to the northwest of the core monsoon region, was apparently weaker than in normal/abundant monsoon years.

Thus, while the search for monsoon predictors continues, the summer monsoon, itself, remains quite influential for remote climates, both contemporaneously (as over the Mediterranean) and in later seasons, when the influence is transmitted through the oceans (as for ENSO). This property of the Asian/Indian summer monsoon was known to early investigators. Almost, half a century ago, Sir Charles Normand succinctly summarized the role of the Indian summer monsoon in Southern Oscillation's evolution in his presidential address to the Royal Meteorological Society in 1953; the address was titled 'Monsoon Seasonal Forecasting':

"Unfortunately for India, the southern oscillation in June-August — at the height of the monsoon — has many significant correlations with later events and relatively few with earlier events...The Indian monsoon therefore stands out as an active, not a passive feature in world weather, more efficient as a broadcasting tool than as an event to be forecast."

2. MONSOON'S FEEDBACK ON ENSO

Asian summer monsoon variability, particularly when ENSO influenced, warrants inclusion in theories of ENSO evolution, and in ENSO prediction models, to reflect the monsoon's feedback; 'feedback' implies a two-way interaction that is generally spaced in time. The pertinent question that motivates this section can be posed as follows:

If monsoon's feedback on ENSO was known since the 1920s, why has Asian Monsoon variability not been included in models of ENSO evolution and prediction?

The answer must lie in our incomplete understanding of the feedback mechanisms; Table 2 summarizes the current status.

TABLE 2: Monsoon-ENSO Feedback Mechanisms

Influence Direction	Proposed Mechanisms	Investigators	Understanding
Monsoon ⇒ ENSO	From impact on surface winds in the western tropical Pacific	Yamagata and Matsumoto 1989 Wainer and Webster 1996 An 1996 Chung and Nigam 1999 Kirtman and Shukla 2000	**Mature**: Conceptually, easy to understand and model
ENSO ⇒ Monsoon	Spring/summer ENSO heating anomalies regulate the low-level monsoon westerlies and moisture influx	Nigam 1994	**Incomplete**: Conceptually, more difficult to understand and model: modeling depends on representation of relatively subtle interactions
	From meridional ITCZ displace-ments in the western Pacific	Ju and Slingo 1995	
	From interaction of the Walker and regional Hadley circulations	Goswami 1998	

Mechanisms involved in shaping ENSO's influence on the summer monsoon are not well (or fully) understood, and their representation in models is dependent upon the models' ability to mimic relatively subtle interactions of the physical components of the climate system. On the other hand, the monsoon's influence on ENSO is transmitted through the impact on surface winds in the western tropical Pacific – a somewhat more straightforward dynamical response of monsoon heating variations in the Asian sector (land and ocean). Creative modeling strategies are thus required for modeling the monsoon's feedback on ENSO. A number of issues must however be dealt with before a modeling analysis strategy can be devised. For example,

- Is the summer monsoon influenced only by the concurrent ENSO SST (heating) anomalies, or are the preceding period anomalies also important, potentially, through their impact on the Asian land surface and Indian oceanic circulation, both of which have at least a season-long memory? Since producing realistic ENSO variability is, itself, challenging for most coupled GCMs, it is unlikely that these models can be trusted to reasonably represent ENSO's concurrent or delayed influence on the Asian summer monsoon.
- Asian summer monsoon variability is however not all related to ENSO: A significant portion, apparently, results from internal dynamics (Palmer *et al.* 1992), and from interactions with the underlying land surface properties, such as soil moisture (Meehl 1994) and the Eurasian snow cover (Dickson 1984; Vernekar *et al.* 1995). The ENSO-unrelated monsoon variability can also impact ENSO evolution, contributing, perhaps, to the aperiodicity of the oscillation (Wainer and Webster 1996). Atmospheric GCMs however cannot be counted upon to produce realistic internal variability, since many of them have difficulty in producing even the climatological rainfall pattern over the monsoon region. The northward advance of monsoon rainfall over the Indian subcontinent is not captured in many GCMs; the result is a lop-sided land-ocean rainfall distribution.
- Mechanisms producing ENSO's influence on the summer monsoon are not well understood, as noted in Table 2: Diagnostic modeling of circulation during contrasting rainfall years (1987, 1988) shows the ENSO-related spring-to-summer heating anomalies in the Pacific to modulate the low-level monsoon westerlies, and thus, the large-scale moisture flux into Indochina, while the flux convergence (or rainfall) is controlled in this analysis by the anomalous stationary waves, arising from the interaction of zonal-mean zonal flow anomalies and Himalayan-Tibetan orography (Nigam 1994). Ju and Slingo's (1995) analysis of GCM simulated monsoon variability corroborates the importance of the spring season ENSO heating anomalies, notably, those arising from latitudinal displacements of ITCZ in the western Pacific. The general relevance of these mechanisms has not been ascertained, and the extent of their representation in GCM simulations is largely unknown.

The impact of anomalous Asian summer monsoon on ENSO evolution has been examined by Yamagata and Masumoto (1989), Wainer and Webster (1996), and An (1996). Except for the first study, these analyses do not address the feedback issue since the monsoon heating anomalies are prescribed, and therefore unresponsive to the state of the Pacific SSTs. The structure of monsoon heating anomalies is also highly idealized in these investigations. On the other hand, monsoon's feedback on ENSO evolution is implicitly included in Barnett *et al.*'s (1993) hybrid coupled model, but isolating the feedback would be difficult because the Pacific surface wind stress anomalies are parameterized using Pacific SST anomalies.

3. MODELING STRATEGY

The Cane-Zebiak Pacific ocean-atmosphere anomaly model (the CZ model; Zebiak and Cane 1987) — a widely used intermediate-level model for ENSO simulation and prediction research — is chosen as the investigative tool as none of the proposed monsoon-ENSO linkage mechanisms posit an active role for the Indian Ocean SST anomalies. The CZ model was also preferred since it can generate quasi-realistic ENSO variability. The modulation of ENSO characteristics (amplitude, frequency, episode duration etc.) from inclusion of the Asian summer monsoon feedback in the CZ model will be determined.

3.1 Parameterization of Monsoon Heating Anomalies

The difficulties in dynamically modeling ENSO's impact on the monsoon are circumvented by parameterizing the Asian sector heating anomalies in terms of the concurrent Pacific basin SST anomalies. The monsoon heating anomalies that arise from ENSO's influence are the target, given our interest in the feedback. Influence of precursor land-surface and Indian ocean states may however also be implicitly present in the contemporaneous linkage because of canonical aspects of ENSO evolution.

The ENSO-covariant Asian monsoon heating anomalies are extracted from a Rotated Principal Component Analysis (RPCA) of combined interannual variability of the tropical Pacific SST, and the 1000mb zonal and meridional winds and 400mb diabatic heating in the near-global tropics during the 1979-1997 summer months of June, July and August. The winds and diagnosed heating are obtained principally from the ECMWF reanalysis, i.e., data for 15 of the 19 summers comes from the reanalysis fields. The RPCA technique extracts recurrent modes of combined variability by simultaneously analyzing the structure of auto-covariance and crosscovariance matrices, and its efficacy in extracting the truly-coupled modes increases with the number of variables in the combination (Nigam and Shen 1993). The individual variables are put on par by dividing each of them by the square-root of the sum of the temporal variance over that variable's spatial grid. The leading 10 loading vectors are orthogonally rotated using the VARIMAX criterion (e.g., Horel 1981), which destroys the spatial orthogonality of the vectors but leaves their temporal orthogonality intact.

The mid-tropospheric heating anomalies in the Asian sector, included additionally in the Gill type CZ atmospheric model (Gill,1980), are determined at each model time-step from the projection of model's SST anomaly on SST-components of the leading modes of interannual variability.

The projection is calculated using the "least squares" method, applied over the SST sub-space of the chosen rotated modes of combined variability. The leading mode of summertime variability is displayed in Figure1: the SST anomaly structure and the principal component (i.e., time series) indicate the obvious linkage of this pattern with ENSO variability. The 1982/83, 1987, and the phenomenal 1997 El Niño are all captured, including the anomalous evolution of the 1982/83 event; the weak SST-cooling during 1981 and 1994-96 summers, a somewhat stronger cooling during 1984-85, and the 1988 La Niña are also evident in the time series. The 400mb heating anomalies depict the southward retreat of the ITCZ over the central/ eastern Pacific sector, and an impressive reduction of deep heating (convective precipitation) over the eastern Indian ocean and the northern tropics of the western Pacific during summer of the El Niño year. There is also some evidence of heating (precipitation) redistribution over Indo-China, with reduced deep heating over northern India, and strengthened heating over the central and northeastern China; note that the heating reduction over northern India is in qualitative accord with Rasmusson and Carpenter's (1983) analysis.

A substantial attenuation of heating is also evident over Central America, along with hints of a weakened Mexican monsoon.

The analysis of summertime interannual variability reveals that ENSO variability is not fully captured by the leading mode (Figure 1). The second-leading mode (not shown) is also linked to interannual SST variability in the central equatorial Pacific.

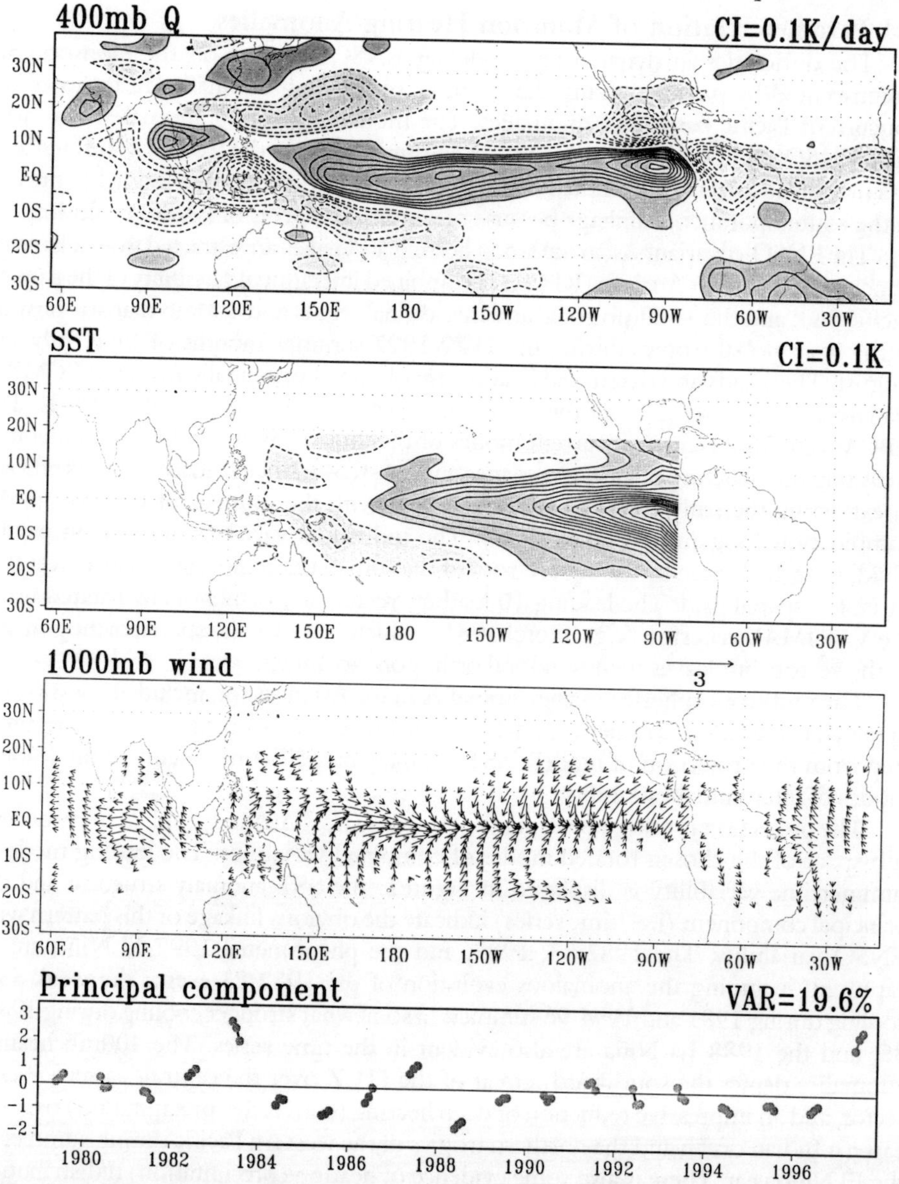

FIG. 1. The leading mode of combined interannual variability of tropical Pacific SST, and the 1000mb zonal and meridional winds, and 400mb diabatic heating (Q_{400}) in the near-global tropics during the 1979-97 summer months of June, July and August: a)Q_{400} , b) SST, c) V_{1000}, and d) the common principal component. The contour interval and shading threshold is 0.1K/day for heating and 0.1K for SST. This leading rotated mode is associated with ENSO variability, and explains about 20% of the combined domain averaged variance.

It explains 10% of the combined variance, and represents variability resulting from the departure of individual El Niño / La Niña evolutions from their canonical pattern, and, possibly, some aspects of decadal time-scale variability; the shortness of the data record is a concern, though. Our heating parameterization is thus based on the *two* leading modes of combined variability, which are both ENSO related. The ENSO related Asian sector heating anomalies were inserted at each model time-step [$\Delta t = (1/36)$ year] as additional forcing in the Gill-type atmospheric model component of the CZ model, during the summer months of June, July and August.

4. MONSOON'S FEEDBACK ON CZ MODEL'S ENSO

The CZ coupled model computes SST and atmospheric heating anomalies over the tropical Pacific (135°E - 85°W and 19°S -19°N), while surface winds are obtained from the global Gilltype atmospheric model driven by the tropical Pacific heating anomalies. The model's heating domain is expanded in both the zonal and meridional directions (56°E - 11°W and 31°S - 39°N) to include the Asian, and potentially, the American monsoon regions. The original and expanded heating domains are shown in Figure 2, which also shows the Asian monsoon region (shaded) where the empirically parameterized heating anomalies are specified during June, July and August months. Over the tropical Pacific, however, the 'expanded domain CZ model' generates heating anomalies all through the year, exactly as in the original model; thus if the parameterized Asian sector heating were set to zero, the modified model would reduce to the original CZ model.

Monsoon's feedback on the CZ model's ENSO characteristics is evaluated from comparison of extended model integrations. The 95,000 year long "control" (original CZ model) and "monsoon" (CZ model with interactive Asian-sector heating anomalies) runs archive the monthly SST, zonal wind, upper-layer thickness (or thermocline depth), and deep heating in the equatorial Pacific and Asian sectors. The Japan Meteorological

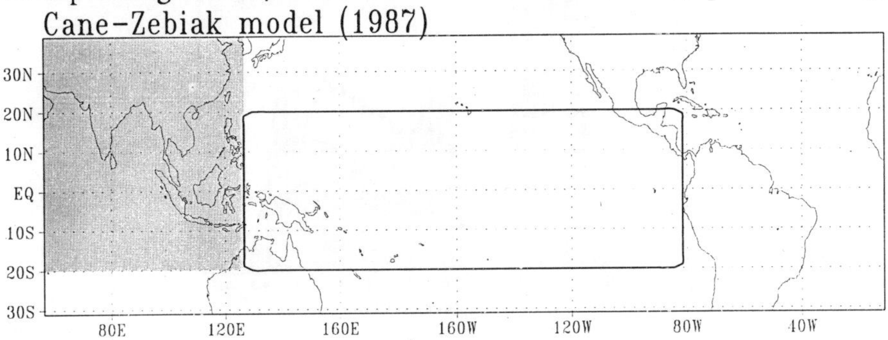

FIG. 2. The original and expanded heating domains in the CZ coupled model. The unshaded rectangle is the original model domain in which SST and atmospheric heating anomalies are computed over the Pacific basin, while the shaded area to the left represents the region in which Asian summer-monsoon heating anomalies are additionally specified in the "monsoon" runs.

Agency's SST index, which is the SST anomaly averaged in the 4°S-4°N and 150°W-90°W equatorial box, is used to monitor ENSO variability in model integrations.[4] The standard deviations of the JMA SST index are 1.359K and 1.465K in the control and monsoon runs, respectively; the increased standard deviation is indicative of the positive feedback between Asian summer monsoon and ENSO. The modulation of ENSO characteristics are analyzed for warm and cold events separately, and these are identified using the 5-month running average of the JMA SST index (hereafter, simply the JMA index); a warm event is defined to occur in the CZ model integrations when the JMA index exceeds 0.5K for at least 6 consecutive months, while the cold event occurrence is based on this index being less than -0.4K for at least 6 successive months.

4.1 Impact on CZ Model's ENSO Frequency

The distribution of warm and cold events w.r.t. the peak JMA index is examined in Figure 3, which shows the distributions for both the control and monsoon runs using a 0.2K index interval. In the control run, there are 38,205 ENSO events, with the warm events distributed over a much wider amplitude range than the cold events; the

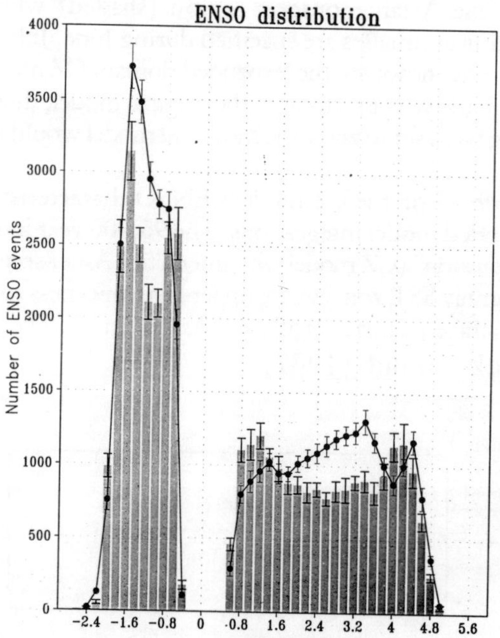

FIG. 3. Distribution of the number of ENSO events w.r.t. their peak JMA index in the 95,000-year control (bar graph) and monsoon (solid dots connected by line) runs; there are 38,205 ENSO events in the control run, and 42,655 events in the monsoon run. Each error bar represents the standard error of the corresponding interval-mean at the 95%confidence level; see Chung and Nigam (1999) for additional details on error estimation.

[4]The JMA and NINO3 SST indices differ only in the extent of latitudinal averaging C

model's bias towards stronger warm events is also evident from the sizeable population of warm events in the 3.0-5.0K index range. Interestingly, both the warm and cold event distributions exhibit bi-modality in the control case, but this feature is considerably attenuated in the 'monsoon' distribution, which is displayed using solid dots (connected by a line).

The 95,000-year monsoon run has 42,655 ENSO events, which represents, approximately, a 12% increase in ENSO occurrence over the control case. The increase in ENSO occurrence is statistically significant because the standard error of the number of ENSO events in our 95,000-year long control-case sample is estimated to be less than 400.[5] A comparison of the two distributions shows that ENSO occurrence is enhanced in the monsoon run primarily in the 2.2-3.6K index range in case of warm events, and the -1.0K to -1.6K range in case of cold events. In both ranges, the enhancements are statistically significant as the error bars representing ±2.101 SD (the standard deviation, SD, was inferred from the variance of ENSO occurrence in 19 non-overlapping 5,000-year long sub-samples, and the factor 2.101 denotes the 95% confidence level in *t*-statistics at 18 degrees of freedom) are well separated. The concurrent reduction in the number of warm events in the 0.6-1.2K range indicates that the above noted population increase results, in part, from a population-shift, due to strengthening of El Niños in the monsoon run. Interestingly, the inclusion of ENSO related Asian sector heating in the CZ model results in a diminution of the bi-modality seen in the control case distributions.

In summary, there are more ENSO events in the monsoon run, with the average frequency increasing from 0.40/year in the control case to 0.45/year in the monsoon run, i.e., an extra ENSO event every 20 years, or so. The ENSO events will also likely be strong, e.g., with peak JMA indices between 2.2-3.6K in case of warm events, and between -1.0K to -1.6K for cold events. A comparison of the two SST power spectrums indicates reduced power in the 45-50 month period range in the monsoon run. The power reduction in the center of the 2-7 year ENSO period range is however compensated by modest power increases in the wings, particularly in the 30-45 month band. The broadening of the power spectrum is however not due to a breakdown in the seasonal phase locking feature of ENSO evolution in the monsoon run.

4.2 Impact on CZ model's ENSO evolution

An assessment of the monsoon's feedback on ENSO evolution poses somewhat of a dilemma, given the significant changes in the ENSO strength-distribution between the control and monsoon runs. If composites based on the peak JMA index are compared, the potential impact on SST amplitudes may not be clearly revealed; on the other hand, if compositing is based on the percentile range of the peak JMA index, the impact on evolution may be obscured as ENSOs with different peak amplitudes can have somewhat different evolutions.

We begin by displaying the composited evolution of the warm events whose peak JMA indices are within the 23rd and 27th percentiles; note that this range corresponds to a peak JMA index range of 1.41-1.57K in the control distribution and 1.71-1.89K in the monsoon case, and has 781 and 861 members, respectively. Figure 4 displays the composited equatorial (4°S-4°N) SST evolutions over a 3-year period beginning in January of the year preceding the one in which the JMA index peaks. The evolution

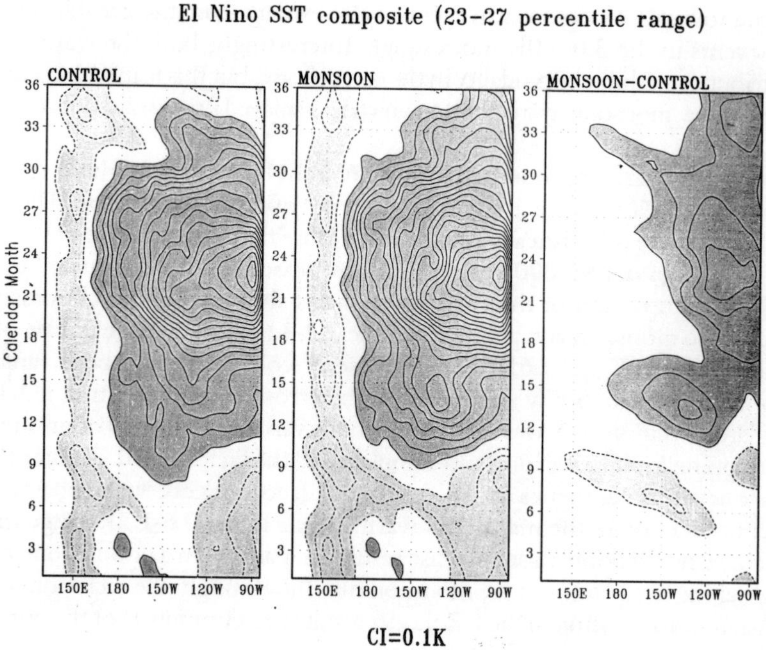

FIG. 4. El Niño SST evolution in the 4°S-4°N equatorial band, based on compositing in the 23-27 percentile range of the peak JMA index: a) control run b) monsoon run, and c) monsoon-control. The composite is based on 781 events (out of 19,571 warm events) with peak JMA SST index in the 1.41-1.57K range in the control case, and 861 events (out of 21,501 warm events) with JMA index in the 1.71-1.89K range in the monsoon case. Note that peak El Niño SST amplitudes are attained close to month-24, i.e., at the end of year-2; the month markings are at the middle of the months. The contour interval and shading threshold is 0.1K in all three panels, and the zero-contour is omitted as before.

difference, shown in the right panel, shows warmer eastern Pacific SSTs in the monsoon run, certainly in the post month-18 period (i.e., beyond June of the El Niño year); during El Niño's mature phase in December (month-24), the peak SSTs are warmer in the monsoon composite by ~0.4K (or ~25%), consistent with the above noted correspondence between the percentile range and peak JMA indices in the two runs. The composited El Niño in the monsoon run is not only stronger but also somewhat longer lived as evident from the presence of warm SST anomalies well into the 'post mature-phase' summer season in the control case (left panel), which generates another round of positive feedback from the ENSO related Asian sector heating anomalies.

The warmer El Niño SSTs in the 23rd -27th percentile range of the monsoon run's distribution suggest only that inclusion of the ENSO related monsoon heating enhances the relative population at higher peak JMA index values. Does this imply a general strengthening of El Niño events in the monsoon run? Not necessarily, unless there is evidence of both a relative population increase at higher index values and a decrease in relative population at lower index values, i.e., of a population shift. Although the distributions shown in Figure 4 are unnormalized, they are strongly indicative of a population-shift in the monsoon run, and hence of strengthening of "weak" El Niños from the feedback of ENSO related Asian summer monsoon heating.

5. CONCLUDING REMARKS

- Asian summer monsoon's feedback on ENSO evolution needs to be corroborated using more complex (and skillful) coupled ocean-atmosphere models. If confirmed, it needs to be included in theories and models of ENSO variability.
- Feedbacks of the Australian monsoon and the year-round American monsoons on ENSO evolution also need to be ascertained.
- Asian summer monsoon's linkage with lower frequency interannual variability remains largely unexplored. Does decadal variability in the Pacific and Atlantic basins, and the southern Indian ocean exert significant influence on the Asian hydroclimate (rainfall, streamflow, drought indices)? And, if so, what would be its feedback? An analysis pre-requisite here would be the availability of coupled ocean-atmosphere-land GCMs that provide quasi-realistic simulations of hydroclimate variability.
- The monsoon-ENSO relationship has been shown to vary on decadal time scales, with the linkage being notably weak in recent decades. The analysis undertaken here for parameterization of the ENSO-related, Asian sector, summer heating anomalies indicates that ENSO influences, primarily, the deep convection (rainfall) over the tropical Indian ocean — in accord with the findings of Troup (1965) which showed the tenuous nature of the subcontinental precipitation based monsoon-ENSO linkage; Indian subcontinent being close to the node in the summertime surface pressure distribution associated with the Southern Oscillation. The monsoon-ENSO linkage, and its variability, thus needs to be reevaluated using indices that reflect monsoon rainfall variability over both land and ocean.

Acknowledgements. The author would like to thank Dr. S.V. Singh (Head, NCMRWF, New Delhi, India) for the invitation to attend the Indo-US Workshop, and for his hospitality. Organization of a workshop on weather and climate by an operational weather forecasting center (such as NCMRWF) is a reflection of the lessening distinction between weather forecasting and short-term climate variability and prediction. It is to NCMRWF's credit that it has taken the lead in recognizing the advantages of this viewpoint, particularly, in context of monsoon modeling and prediction.

The author would also like to thank the American Meteorological Society for permission to reproduce figures 1-4 from a *Journal of Climate* paper, titled, *Asian Summer Monsoon–ENSO Feedback on the Cane-Zebaik Model ENSO* (September 1999, Volume 12, pages 2787-2807, authors: Chul Chung and Sumant Nigam).

REFERENCES

An, S., 1996: A critical examination of the current theories on the ENSO dynamics using an intermediate coupled ocean-atmosphere model. *Ph.D. dissertation*, 171pp. Seoul National University.

Barnett, T. P., M. Latif, N. E. Graham, M. Flügel, S. Pazan, and W. White, 1993: ENSO and ENSO-related predictability. Part I: Prediction of equatorial Pacific sea surface temperature with a hybrid coupled ocean-atmosphere model. *J. Climate*, 6, 1545-1566.

Bhalme, H. N., A. B. Sikder, and S. K. Jadhav, 1990: Coupling between the El Niño and planetary-scale waves and their linkage with the Indian monsoon rainfall. *Meteo. Atmos. Phys.*, **44**, 293-305.

Bjerknes, J., 1969: Atmospheric teleconnections from the equatorial pacific. *Mon. Wea. Rev.,* **97**, 163-172.

Chung, C., and S. Nigam, 1999: Asian summer monsoon—ENSO feedback on the Cane-Zebiak Model,ENSO. *J. Climate*, **12**, 2787-2807.

DeWeaver, E., and S. Nigam, 2002: Linearity in ENSO's atmospheric response. *J. Climate*, **15,** 2446-2461.

Dickson, R. R., 1984: Eurasian snow cover versus Indian monsoon rainfall—An extension of the Hahn-Shukla results. *J. Climate Appl. Meteor.*, **23**, 171-173.

Gill, A. E., 1980: Some simple solutions for heat-induced tropical circulations. *Quart. J. Roy. Meteor. Soc.*, **106**, 447-462.

Goswami, B. N., 1998: Interannual variation of Indian summer monsoon in a GCM: External conditions versus internal feedbacks. *J. Climate*, **11**, 501-522.

Horel, J. D., and J. M. Wallace, 1981: Planetary-scale atmospheric phenomena associated with the Southern Oscillation. *Mon. Wea. Rev.*, **109**, 813-829.

_____, 1981: A rotated principal component analysis of the interannual variability of the Northern Hemisphere 500 mb height field. *Mon. Wea. Rev.*, **109**, 2080-2092.

Ju, J., and J. Slingo, 1995: The Asian summer monsoon and ENSO. *Quart. J. Roy. Meteor. Soc.*, **121**, 1133-1168.

Kirtman, B. P., and J. Shukla, 2000: Influence of the Indian Summer Monsoon on ENSO. *Quart. J. Roy. Meteor. Soc.,* **126**, 562, 213-240.

Kumar, K. K., B. Rajagopalan, M.A. Cane, 1999: On the weakening relationship between the Indian Monsoon and ENSO. *Science*, **284**, 2156-2159.

Lau, K.-M., and S. Yang, 1996: The Asian monsoon and predictability of the tropical ocean-atmosphere system. *Quart. J. Roy. Meteor. Soc.*, **122**, 945-957.

Meehl, G. A., 1987: The annual cycle and interannual variability in the tropical Pacific and Indian Ocean regions. *Mon. Wea. Rev*., **115**, 27-50.

_____, 1994: Coupled land-ocean-atmosphere processes and South Asian monsoon variability. *Science*, **265**, 263-267.

Nigam, S., and H.-S. Shen, 1993: Structure of oceanic and atmospheric low-frequency variability over the tropical Pacific and Indian oceans. Part-I: COADS Observations. *J. Climatol.*, **6**, 657-676.

_____, 1994: On the dynamical basis for the Asian summer monsoon rainfall-El Niño relationship. *J. Climate*, 7, 1750-1771.

_____, M. Barlow, and E.H. Berbery, 1999: Pacific decadal SST variability: Impact on U.S. drought and streamflow. Eos, Transactions, *American Geophysical Union*, **Vol. 80**, 51 (Dec. 21), 621-625.

Normand, C., 1953: Monsoon seasonal forecasting. *Quart. J. Roy. Meteor. Soc.*, **79**, 463-473.

Palmer, T. N., C. Brankovic, P. Viterbo and M. J. Miller, 1992: Modeling interannual variations of summer monsoons. *J. Climate*, **5**, 399-417.

Philander, S. G. H., 1985: El Niño and La Niña. *J. Atmos. Sci.*, **42**, 2652-2662.

Ramage, C. S., 1971: Monsoon Meteorology. Vol. 15. *International Geophysics Series*, Academic Press.

Rasmusson, E. M., and T. H. Carpenter, 1983: The relationship between eastern equatorial Pacific sea surface temperatures and summer monsoon rainfall over India and Sri Lanka. *Mon. Wea. Rev.*, **111**, 517-528.

Rodwell, M. J., and B. J. Hoskins, 1996: Monsoons and the dynamics of deserts. *Quart. J. Roy. Meteor. Soc.*, **101**, 193-202.

Ropelewski, C. F., and M. S. Halpert, 1987: Global and regional-scale precipitation associated with El Niño/ Southern Oscillation. *Mon. Wea. Rev.*, **115**, 1606-1626.

Sikka, D. R., 1980: Some aspects of the large-scale fluctuations of summer monsoon rainfall over India in relation to fluctuations in the planetary and regional scale circulations. *Proc. Ind. Acad. Sci. (Earth & Planet Sci.)*, **89**, 179-195.

Troup, A. J., 1965: The Southern Oscillation. *Quart. J. Roy. Meteor. Soc.*, **91**, 490-506.

Verma, R. K., 1992: ENSO-Monsoon linkages as evidenced from Pacific SST correlations with monsoon precipitation. TOGA Notes, 6, 1-3.

Vernekar, A. D., J. Zhou, and J. Shukla, 1995: The effect of Eurasian snow cover on the Indian monsoon. *J. Climate,* **8**, 248-266.

Wainer, I., and P. J. Webster, 1996: Monsoon/El NiñoBSouthern Oscillation relationships in a simple coupled ocean-atmosphere model. *J. Geophys. Res.*, **101**, C11, 25,599-25,614.

Walker, G.T., 1923: Correlation in seasonal variations of weather. III: A preliminary study of world weather. *Mem. India Meteorol. Dept.*, **24**, 75-131.

_____, 1924: Correlation in seasonal variations of weather. IV: A further study of world weather. *Mem. India Meteorol. Dept.*, **24**, 275-332.

Webster, P. J., and S. Yang, 1992: Monsoon and ENSO: Selectively interactive systems. *Quart. J. Roy. Meteor. Soc.*, **118**, 877-900.

Yamagata. T., and Y. Masumoto, 1989: A simple ocean-atmosphere coupled model for the origin of a warm El Niño Southern Oscillation event. Phil. Trans. R. Soc. London, A329, 225-236.

Yasunari, T., 1991: The Monsoon YearCA new concept of the Climate Year in the tropics. *Bull. Amer. Meteor. Soc.*, **72**, 1331-1338.

Zebiak, S. E., and M. A. Cane, 1987: A model El Niño-Southern Oscillation. *Mon. Wea. Rev.,* **115**, 2262-2278.

High Resolution Numerical Modeling on Tropical Cyclone Structure and Track

CHI-SANN LIOU

Naval Research Laboratory Monterey, CA 93943-5502

E-mail: liou@nrlmry.navy.mil

ABSTRACT

The Navy operational mesoscale model, Coupled Ocean/Atmosphere Mesoscale Prediction System (COAMPS™), is enhanced for high-resolution tropical cyclone forecast. A tropical cyclone initialization system, which is patterned after the operational tropical cyclone initialization system designed for the Navy Operational Global Prediction System (NOGAPS), has been developed for the higher resolution COAMPS. In this initialization system, observed tropical cyclone position and structure are included in the COAMPS analysis through 41 synthetic observations. The synthetic observations are constructed by adding a tropical cyclone circulation onto a large-scale environmental flow. The tropical cyclone circulation at each of 4 quadrants is computed from a tropical cyclone structure model that uses observed maximum wind speed and 35 and 50-knot wind radii. The environmental flow is obtained from NOGAPS analysis with tropical cyclone circulation filtered out through a spectrum truncation. The degree of filtering depends upon the averaged 35-knot wind radius of the cyclone. Major improvements made to the new system for COAMPS are (a) increasing horizontal and vertical resolution of the synthetic observations, (b) including asymmetric structure of tropical cyclones in the synthetic observations, (c) improving gradient-balanced height calculation by considering environmental mean wind, and (d) determining the environmental flow of a tropical cyclone by considering the cyclone size.

Case studies have been conducted to test and evaluate the new tropical cyclone initialization system for COAMPS. The synthetic observations generated by the new system show stronger inner circulations and weaker outer circulations as compared with those generated by the operational method. The wind analysis using the new set of synthetic observations displays much stronger and tighter tropical cyclone circulation than that using the operational synthetic observations for both 15 km and 5 km resolution grids. The 48 h wind forecast initialized with the new set of synthetic observations also reveals stronger and tighter tropical cyclone circulation as compared with the forecast initialized with the operational set. The comparison of the tropical cyclone structure with observations indicates that the new initialization system significantly improves the tropical cyclone structure both in the COAMPS analysis and forecast. The error statistics of 29 COAMPS track forecasts shows that the new tropical cyclone initialization system made moderate improvement to the 48 h track forecasts. The improvement is more significant in the initial period and is reduced somewhat in the later periods.

1. INTRODUCTION

Tropical cyclone track forecast errors have decreased in a slow, but steady fashion over the past several decades in both the Atlantic and Pacific basins. This increase in skill, especially in the past 10 years, is primarily due to improvements made to numerical modeling systems over the same period. The improvements include increased model resolution, better data assimilation, more completed model physics, and application of increased satellite products and data to tropical cyclone track prediction problems. To further improve the operational tropical cyclone forecast, we have conducted research on ensemble tropical cyclone track forecast and high-resolution tropical cyclone modeling. The goals of the high-resolution modeling effort are to reduce the number of large track error forecasts and to provide skillful tropical cyclone structure and intensity forecasts. This paper focuses on improvements made to tropical cyclone structure and track forecast through the development of a high-resolution tropical cyclone initialization system applied for the high-resolution numerical modeling.

2. HIGH RESOLUTION MODEL

The high-resolution mesoscale model selected for the high-resolution tropical cyclone forecast is the Coupled Ocean/Atmosphere Mesoscale Prediction System (COAMPS). COAMPS is developed at Naval Research Laboratory (Hodur, 1997) and has been operational for more than four years. COAMPS includes an atmospheric data assimilation system comprised of data quality control, analysis, and initialization; a nonhydrostatic atmospheric model component; and an ocean model (under development). The data analysis is accomplished by a multivariate optimum interpolation method (Barker, 1992) that is being replaced by a 3-dimensional variational adjustment method. The atmospheric forecast model is a fully compressible nonhydrostatic model on terrain following height coordinates. It uses multiple-level nested grids to achieve high-resolution modeling at selected areas. The fully compressible governing equations are integrated by split-explicit and semi-implicit time schemes together with the Davis type lateral boundary treatment. The model contains parameterizations for surface fluxes, subgrid scale mixing, cumulus parameterization, radiation, and explicit moist physics. In operational applications, COAMPS is running twice a day in several areas of the world with 2- or 3-level nested grids of 81/27/9 km resolutions. COAMPS has demonstrated good skills in operational mesoscale weather forecast over many areas of the world, and in case studies for various atmospheric phenomena including mountain waves, land-sea breezes, terrain-induced circulations, mesoscale convective systems, coastal rainbands, frontal systems, and tropical cyclones.

3. SYNTHETIC OBSERVATIONS

Since there is no enough observations to properly describe tropical cyclone circulation, numerical model forecast of tropical cyclones relies on special methods to initialize tropical cyclone circulation. In COAMPS, the initial tropical cyclone position, structure and intensity are incorporated into initial conditions by analyzing synthetic observations that are generated according to the latest observed position, structure, and intensity of tropical cyclones. To accommodate the higher resolution of COAMPS, a tropical cyclone initialization system that is patterned after the operational tropical cyclone initialization system designed for the Navy Operational Global Prediction System (NOGAPS) has been developed for the higher resolution COAMPS. The major improvements made to the new system for COAMPS include (a) increasing horizontal and vertical resolution for the synthetic observations, (b) including asymmetric tropical cyclone circulation in the synthetic observations, (c) improving gradient-balanced calculation for height by considering environmental mean wind, and (d) determining environmental flow of a tropical cyclone by considering its size. In the new system, the structure of each tropical cyclone is represented by 41 rather than 13 higher resolution synthetic observations. The synthetic observations include the information of wind speed and wind direction at 7 pressure levels from 1000 mb to 400 mb, and height and temperature at 5 levels from 1000 mb to 700 mb. They are constructed by adding tropical cyclone circulation onto a large-scale environmental flow. The tropical cyclone circulation is computed from a modified Rankine vortex tropical cyclone structure model,

$$V_{tc} = V_{max}\left(\frac{r}{R}\right), \quad r < R$$

$$= V_{max}\left(\frac{R}{r}\right)^{\alpha} \eta\phi, \quad r > R$$

where V_{max} is the maximum wind speed, R is the radius of the maximum wind, η is an assigned vertical weighting function, and ϕ is an assigned radius distance weighting function. With observed maximum wind speed, 35- and 50-knot wind radii from warning messages issued by the Joint Typhoon Warning Center (JTWC), the parameters of V_{max}, R and α can be determined with which the tropical cyclone circulation can be computed. An empirical value for α is assumed if one of the radii is missing from the message, and an empirical value for R is further assumed if both radii are missing. The cyclone circulation is computed at 8 octants so that asymmetric structure will be included in the tropical cyclone circulation, if the warning messages indicate so. The environmental flow is computed from NOGAPS analyses with tropical cyclone circulation filtered out by spectrum truncation. The degree of filtering depends upon averaged 35-knot wind radius of each tropical cyclone. We use triangular 20-wave truncation for cyclones with averaged 35-knot radius less than 222 km, and use 15-wave truncation for larger cyclones with the 35-knot radius greater than 222 km. After the wind has been determined at the 41 points, the height is computed by gradient wind balance, The temperature is then computed from the height by hydrostatic balance.

4. EVALUATION

Case studies have been conducted to test and evaluate the new tropical cyclone initialization system for COAMPS. We have selected for detailed evaluation two tropical cyclones in 2000, Saomai (wp22) and Bopha (wp24), which appeared together on 8 September. The cyclone wp22 was a large and strong storm with a 350 km radius for

35-knot wind and a maximum wind speed of 90 knots, while the cyclone wp24 was a small and weak storm with a 130 km radius for 35-knot wind and a maximum wind speed of 50 knots. Synthetic observations generated by the operational and new tropical cyclone initialization systems are compared for the cyclones wp22 (Figure1) and wp24 (Figure 2). In both cases, synthetic observations from the new system show stronger inner circulation and weaker outer circulation as compared with the operational system. The wind and height are also better fitted to gradient wind balance in the new system. Triply nested COAMPS with 45/15/5 km resolution is used to analyze these synthetic observations together with other observations, and then to make 48 h forecast. The wind analysis using the new set of synthetic observations displays much stronger and tighter inner tropical cyclone circulation than that using the operational set for both resolutions of 15 km (Figure 3) and 5 km (Figure 4). The 48 h wind forecast initialized with the new set of synthetic observations also shows stronger and tighter tropical cyclone circulation (Figure 5). The comparison with observations reveals that the new initialization system significantly improves the tropical cyclone structure both in COAMPS analysis and forecast. The tropical cyclone tracks of the wp22 (Figure 6) and wp24 (Figure 7) are also better predicted by COAMPS with the new initialization system. The better-initialized tighter and stronger inner circulation from the new initialization system is responsible for a better simulation on the binary interaction (Fujiwara effect) between the two cyclones that leads to better the COAMPS 48 h track forecasts.

We have conducted more than 50 COAMPS 48 h forecasts to evaluate the track forecast with the new initialization system. Track forecast improvements are found to be more significant for cases that the tropical cyclone interacts with other weather systems such as the subtropical high or other cyclones. The error statistics of 29 COAMPS track forecasts (Figure 8) shows that the new tropical cyclone initialization system made moderate improvement to the 29 track forecasts. The improvement is more significant in the initial period and then reduced somewhat in the later time

5. SUMMARY

We have developed a high-resolution initialization system for high-resolution tropical cyclone analysis and forecast in COAMPS. The high resolution indeed improves the tropical cyclone structure analysis and forecast. For cases that the track is sensitive to the size and strength of the initial circulation, the high-resolution initialization system also makes significant improvement in the track forecast. We now concentrate our efforts in improving the forecast model for the high-resolution tropical cyclone forecast. We are developing methods to move inner meshes following selected tropical cyclones so that their circulations are computed by the highest model resolution. We are evaluating different cumulus and microphysics parameterization schemes to select best packages for simulating tropical cyclone structure and intensity. We are also developing methods to better assimilate satellite observations for tropical cyclone initialization.

Acknowledgements. This work is supported by the Office of Naval Research through program PE-0602435N and Space and Naval Warfare System Command through program PE-0603207E.

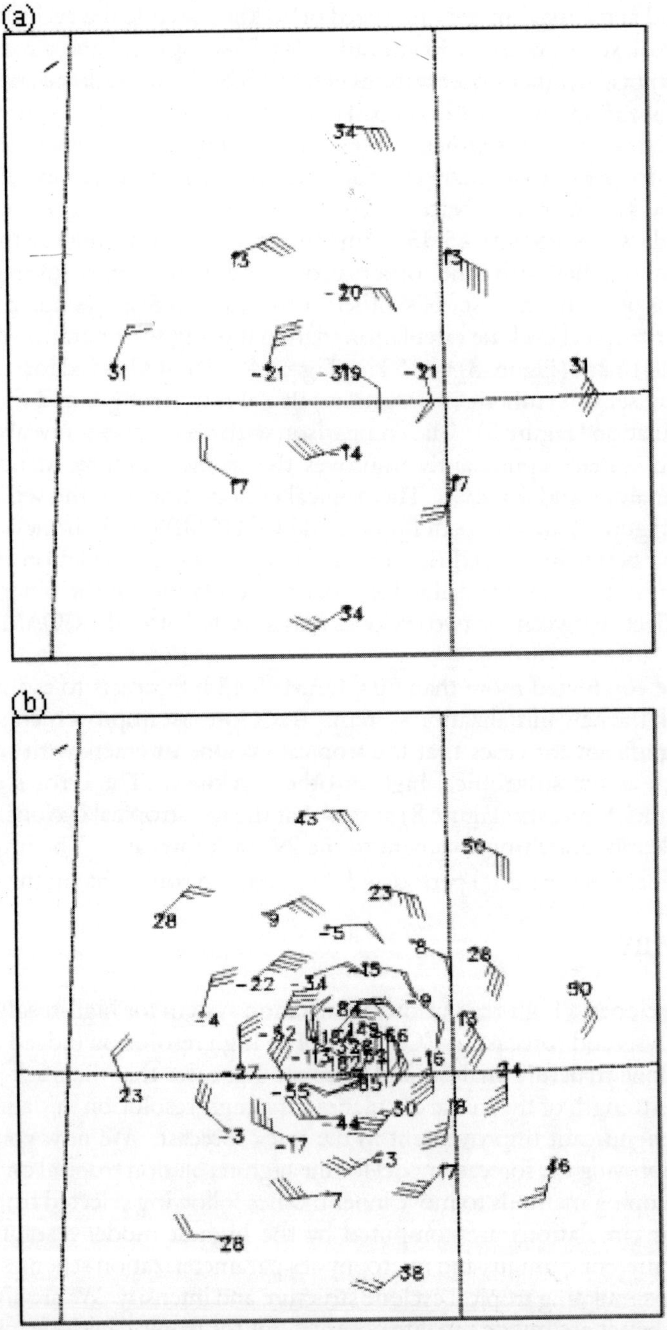

FIG. 1. Comparison of synthetic observations created within 6-degree distance c cyclone wp22 by the (a) operational and (b) new initialization system. Th observed maximum wind speed is 90 kt and 35-kt radii are 180, 180, 150 ar 150 nautical miles in 4 quadrants.

(a)

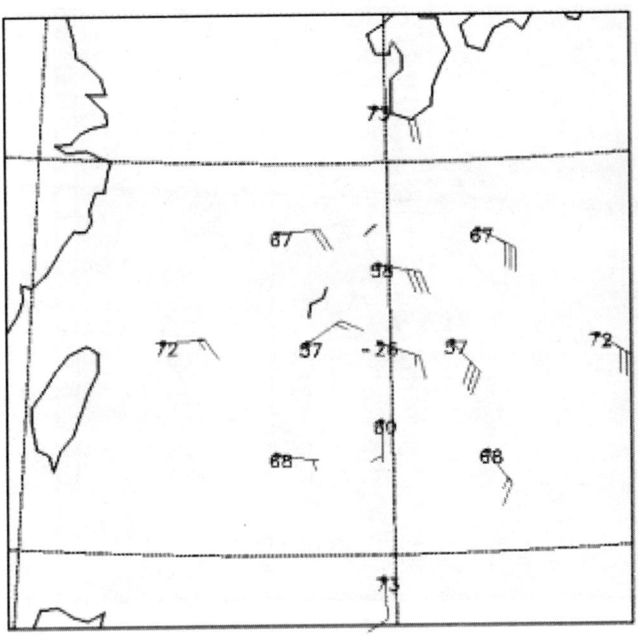

(b)

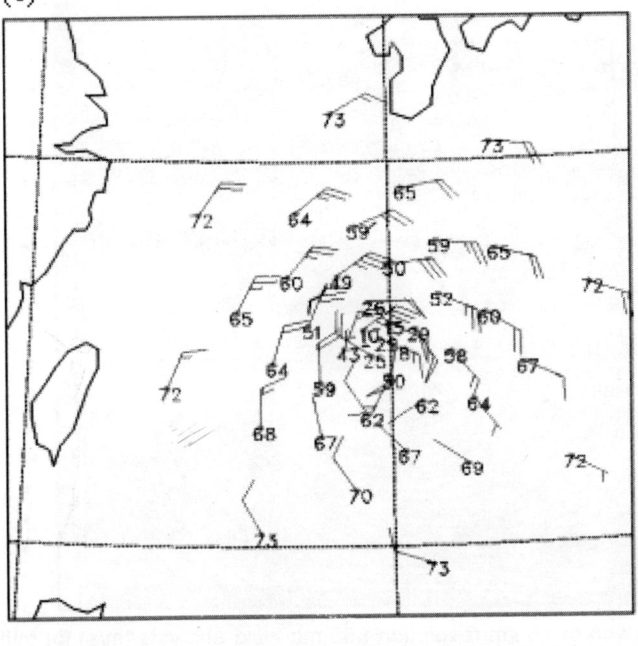

FIG. 2. Same as Fig. 1, except for cyclone wp24 and the observed maximum wind speed is 50 kt and 35-kt radii are 65, 65, 35 and 35 nautical miles in 4 quadrants.

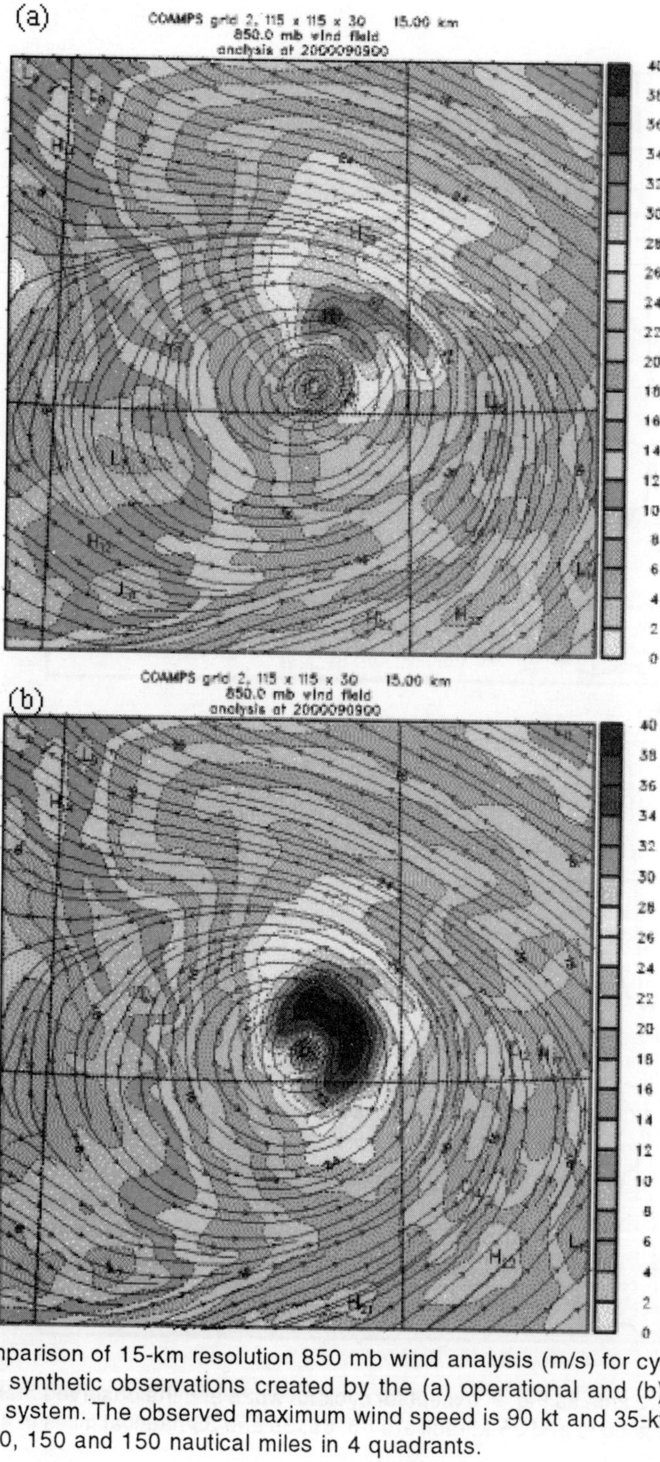

FIG. 3. Comparison of 15-km resolution 850 mb wind analysis (m/s) for cyclone wp22 using synthetic observations created by the (a) operational and (b) new initialization system. The observed maximum wind speed is 90 kt and 35-kt radii are 180, 180, 150 and 150 nautical miles in 4 quadrants.

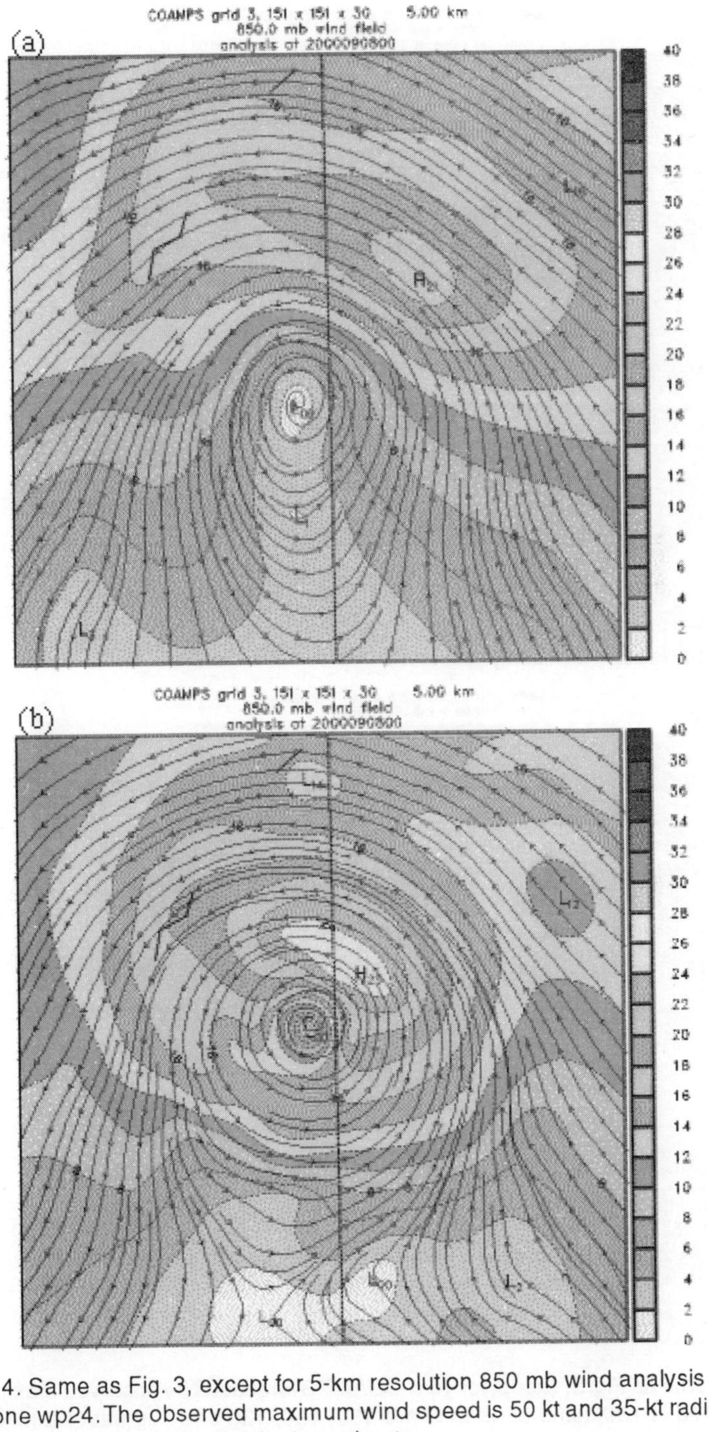

FIG. 4. Same as Fig. 3, except for 5-km resolution 850 mb wind analysis (m/s) of cyclone wp24. The observed maximum wind speed is 50 kt and 35-kt radii are 65, 65, 35 and 35 nautical miles in 4 quadrants.

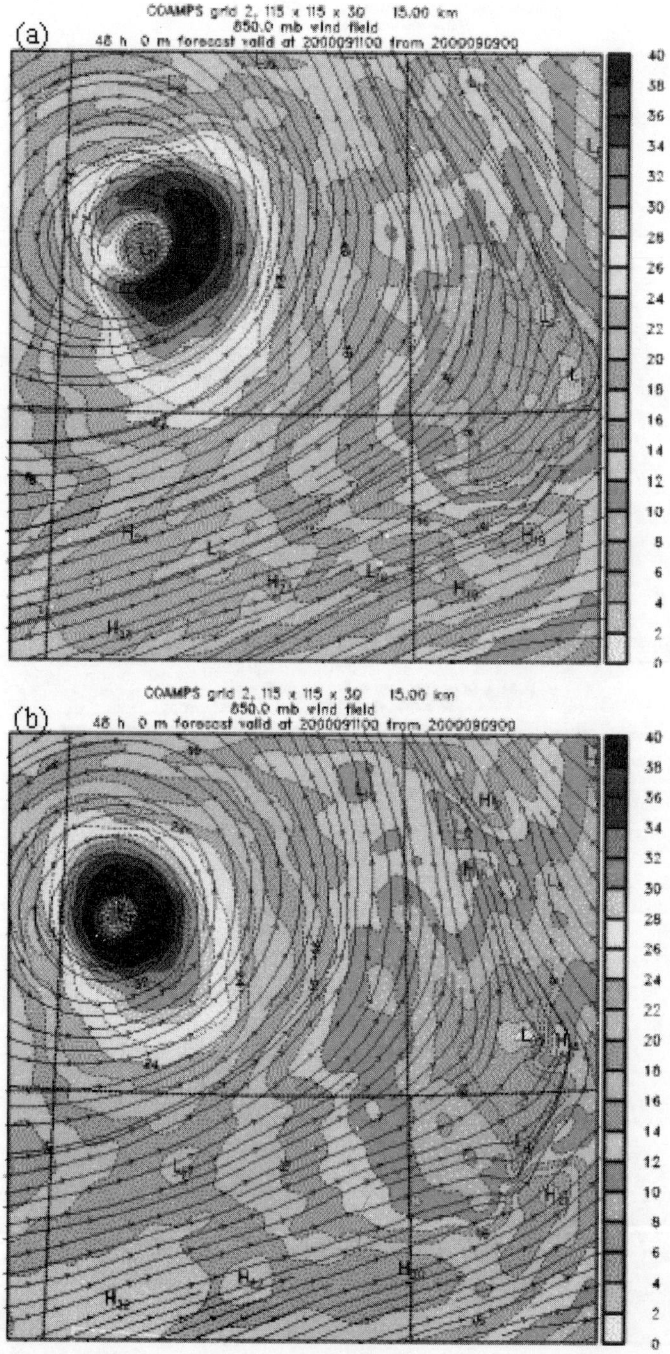

FIG. 5. Same as Fig. 3, except for 48-h COAMPS forecast. The observed maximum wind speed is 130kt and 35-kt radii are 180, 180, 130 and 130 nautical miles in 4 quadrants.

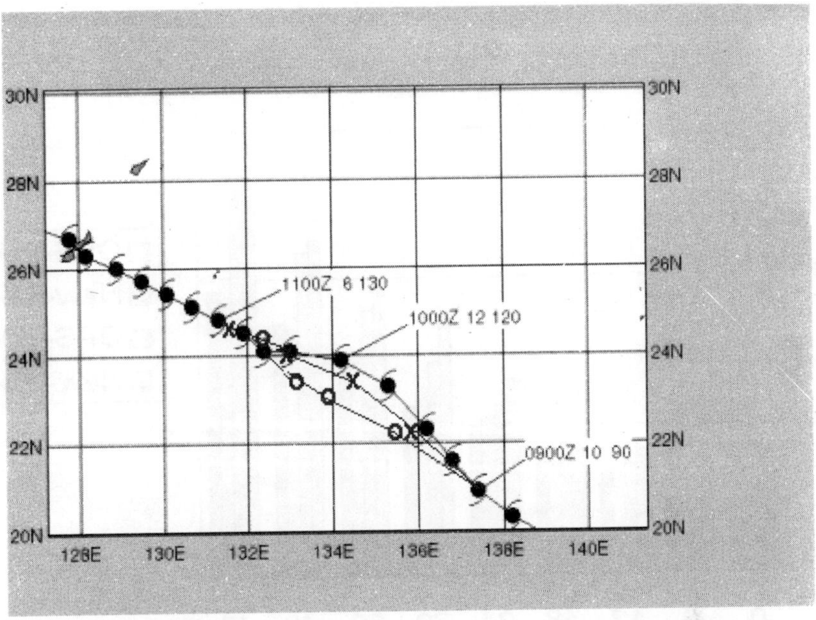

FIG. 6. COAMPS 48 h track forecast of cyclone wp22 starting from 0000 UTC 9 September 2000 initialized by synthetic observations from operational (circled) and new system (crossed).

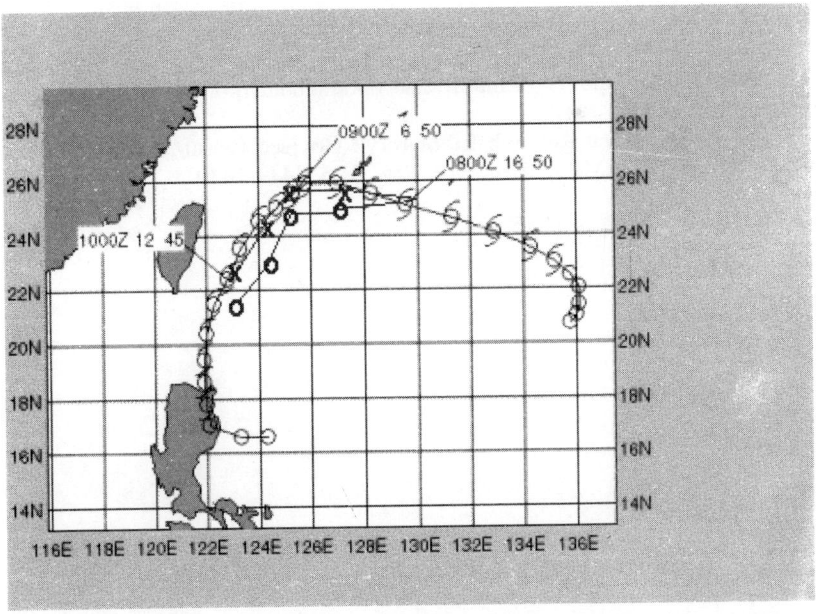

FIG. 7. Same as Fig. 6, except for cyclone wp 22 starting from 0000 UTC 8 September 2000.

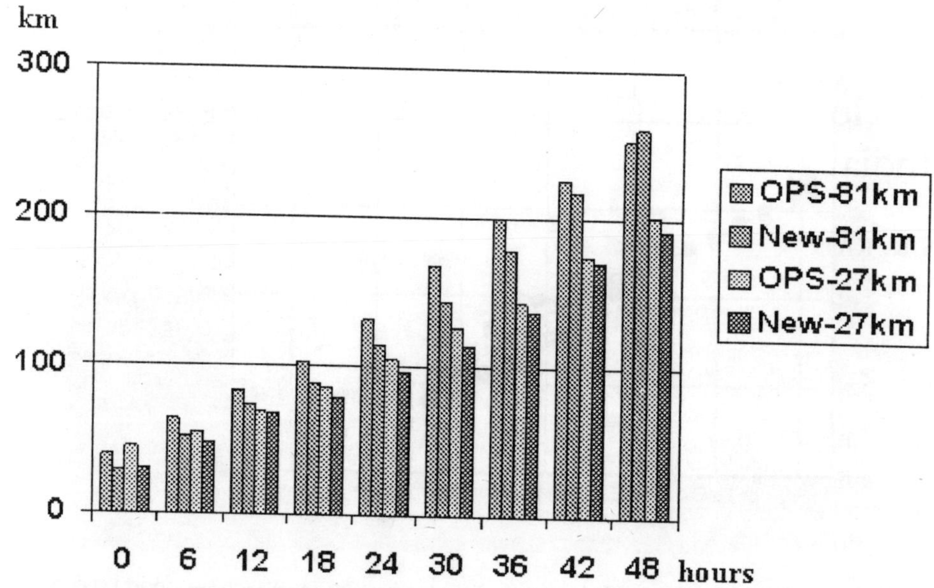

FIG. 8: Track error statistics of 29 COAMPS forecasts for 2-level nested grids of 81/27 km with tropical cyclones initialized by operational and new synthetic observations.

REFERENCES

Barker, E., 1992: Design of the Navy's multivariate optimum interpolation analysis system. *Wea. Forecasting*, 7, 220-231.

Hodur, R., 1997: The Naval Research Laboratory's Coupled Ocean/Atmosphere Mesoscale Prediction System (COAMPS). *Mon. Wea. Rev.*, **125**, 1414-1430

Cyclone Related Disaster Management Efforts in India

S.R. KALSI

India Meteorological Department, Mausam Bhavan, New Delhi

ABSTRACT

Tropical cyclone (TC) is one of the natural hazards that is often accompanied with damaging winds, torrential rains leading to floods and occasionally storm surge that inundates the low-lying areas and becomes the worst killer of the man-kind. TCs acquire dimensions of huge disasters when they threaten the inhabitants without any prior warning. India Meteorological Department (IMD) has made significant contribution to provide advance warnings in regard to the TCs. This paper reviews the efforts made by IMD for improving TC analysis and forecasting system. We take holistic view of disaster management and list out the institutional arrangements established in this context for effective follow up actions to minimize loss of life and property.

1. INTRODUCTION

About 7 percent of the total global crop of Tropical Cyclones (TCs) are spawned in the north Indian Ocean. They threaten the security and safety of life and property in the Indian Sub-Continent by frequently devastating densely populated coastal areas. The recent experience in Orissa and Gujarat clearly demonstrates the vulnerability of our people. There is a continuing need to improve the disaster prevention and preparedness strategy. IMD has built up the capacity for providing forecasts of TCs. It has strengthened the observing network, telecommunication system for expeditious collection of data and improved its forecasting system. It has established and improved linkage with the disaster management agencies so that effective follow up actions are taken to reduce the loss of life and property. An attempt is made to review the efforts made in regard to the TC monitoring and forecasting in Section 2 on analysis and prediction. Emphasis is made on the data resources available to TC forecasters in IMD. Damaging aspects of heavy rains, strong winds and storm surges are included in Section 3. The warning systems and other issues of disaster management including the institutional arrangements are discussed in Section 4. Section 5 is devoted to future scenario and includes some of the recommendations of High Pwered Committee (HPC). Their implementation may bridge the gaps in the process of management of TC induced disasters. Finally, the areas of possible co-operation figure in Section 6.

2. ANALYSIS AND PREDICTION

TC is characterized by two predominant scales of atmospheric motion: large scale (1000 Km) and TC scale (100 Km). We need to have various observational systems to resolve the scales associated with the TC vortex, its surrounding bands and the large scale environmental flow in which the TC is embedded. For an appropriate appreciation of the TC prediction problem, we require the knowledge of availability and distribution of the data resources.

2.1 Tools

TCs spend most of their life cycle out in the deep but data sparse oceanic regions. They have been analyzed over the years with data mainly from ships of opportunity. But this data can not result in an accurate positioning of TC center. At the same time, there has been considerable decrease in the ship observations from the Bay of Bengal. Ships supplement the standard land based measurements. IMD operates 550 surface observatories. Drifting buoys have been deployed elsewhere from aircraft in advance of a hurricane and successfully reported surface pressure, wind speed at 1-meter level, and ocean temperature at sea levels. Department of Ocean Development, Govt. of India has also started a very important national data buoy programme. They have since deployed a set of 12 moored buoys in the northern Indian Ocean to provide meteorological and oceanographic data. This data network is going to be augmented in the next couple of years.

IMD operates network of 65 pilot balloon observatories and 35 RS stations to meet requirements of defining the vertical structure of the TCs and also of the surrounding environment. A mesosphere, stratosphere, troposphere (MST) radar has also been installed at Thirupatti in India. Another profiler developed by Sameer has been deployed at IMD Pune. Another important source of upper level data is the aircraft's reports (AIREPS). Increasing number of commercial jet aircraft are equipped with the Aircraft Meteorological Data Relay (AMDAR) systems. These data are being made available for forecasting and also being used by the IMD Limited area Analysis and Forecasting System (LAFS).

We have installed a group of 10 Cyclone Detection Radars (CDRs) over the Indian coastline (Figure 1). Indian radars are providing useful estimates of storm centres upto a range 300 – 400 Km. Raghavan (1997) gave an exhaustive review of radar observations of TCs in Indian seas. It has been used in operational TC tracking & forecasting and also to understand the structure of TCs. It has also enabled us to have an estimate of the recent 6-12 hour motion of the storm which is a key predictor in short term track prediction. IMD has installed Doppler Weather Radar (DWR) facility at Chennai. Two more DWRs for Kolkata and Sriharikota are being installed and will be commissioned soon.

TC forecasters everywhere use Dvorak (1984) technique to estimate storm location and intensity. It has been found to provide realistic estimates even for TCs in the Bay of Bengal as well as Arabian Sea. Dvorak technique is also used for forecasting intensity of TCs. INSAT data has also been used for TC intensity analysis and forecasting (Kalsi, 2002). Figure 2 shows the highly organized cloud pattern of the Super TC that struck Orissa on 29 October 1999 near Paradip. A sharp, dark and clear eye is seen in the

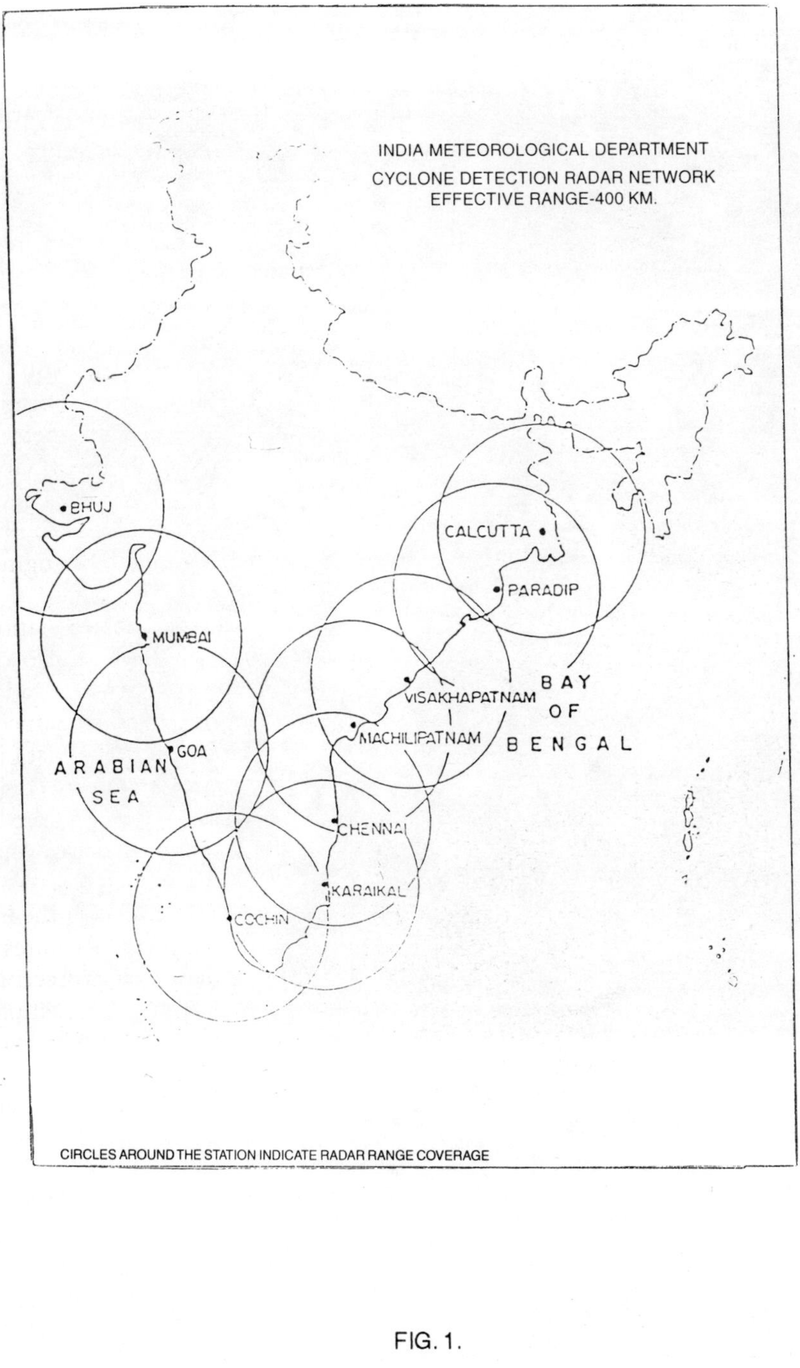
INDIA METEOROLOGICAL DEPARTMENT
CYCLONE DETECTION RADAR NETWORK
EFFECTIVE RANGE-400 KM.
BHUJ
CALCUTTA
PARADIP
MUMBAI
VISAKHAPATNAM
BAY
OF
BENGAL
MACHILIPATNAM
GOA
ARABIAN
SEA
CHENNAI
KARAIKAL
COCHIN
CIRCLES AROUND THE STATION INDICATE RADAR RANGE COVERAGE

FIG. 1.

BAY_991025 29-OCT-99 03:30Z VISIBLE INSAT1D VIS STRETCH
IMDPS IMD NEW DELHI
E80
E85
E90
E95
E100
E105
N25
N20
N15
N10
N5

Fig. 2

satellite imagery surrounded by the deep overcast. Complete analysis of this system was made almost on the basis of the satellite imagery only. IMD has been producing producing Cloud Motion Vectors (CMVs) at 0000,0600 & 1200 UTC but unfortunately low level CMVs are not available for TCs. Efforts are also on to build a geo-stationary satellite purely for met purposes, which will have capabilities both for sounding and imaging missions by 2003 time frame. Infrared data from geo-stationary satellite is being used to estimate convective precipitation using Arkin (1983) technique.

Data from METEOSAT-V deployed along 63° E longitude are also available for the Indian Ocean, but very important observations from satellites such as Defense Meteorological Programme Satellite (DMSP) and Earth Research Satellite (ERS) are still not available on real time basis to TC forecasters in India. Joshi *et al.* (1999) has reviewed the utility of the data from meteorological satellites including TRMM and ERS for the analysis and forecasting of TCs. Kalsi *et al.*(2002) have used SSM/I. TRMM, ATOVS and Quick look data to study three intense TCs that developed over the north Indian Ocean during 1999.

2.2 Techniques

Though IMD commenced Numerical Weather Prediction (NWP) techniques nearly 3 decades back, these have still not reached the desired levels of prediction. There is still reliance on the conventional empirical analysis and forecasting methods. Data from various sources cited above from India and neighborhood is collected in the TC forecast offices.It is subjectively analysed in conjuction with satellite and radar images to perform diagnostic analysis. Intensity analysis is carried out using Dvorak(1984) technique which gives signals for forecasting further development. Lack of adequate number of reliable insight regarding structure of wind field encountered in TCs. Forecasts and warnings based on the empirical analysis delineate most vulnerable areas. Impacts likely to be made by the severe weather are included in the warning bulletins.

2.3 Motion of Tropical Cyclones

TC track prediction techniques can be broadly booked under two categories:- i) subjective & ii) deterministic. Synoptic, Satellite, Radar Method etc. come under subjective techniques. Statistical and dynamical methods are broadly categorised as Objective or deterministic techniques.

2.3.1 Synoptic Method

The most widespread synoptic pattern recognition technique involves the relative locations of the cyclone with the sub tropical ridge and upper level trough in the mid-latitude westerlies. Generally, though not always, a strong sub-tropical ridge located poleward of the cyclone will result in continued movement westward. Similarly presence of a westerly trough in the upper level west of the TC is an excellent indicator of re-curvature. We continue to depend on this conventional method of forecasting movement of TCs

2.3.2 Dynamical Methods

A number of techniques are in vogue the world over. Only the operational models in use in IMD are given below. LAFS is run operationally by IMD. Prasad (1997) reviewed the numerical techniques for numerical forecasting of TCs. The scheme basically generates radial distribution of surface pressure within the vortex from an empirical formula proposed by Holland(1980). Basic inputs for generating the surface pressure are the parameters like central pressure of the storm, its environmental pressure, radius of maximum wind, current position, movement and intensity of the storm, which are inferred from the surface synoptic charts and satellite imagery. Recently a Quasi-Lagrangian Model (QLM) for cyclone track prediction in the Indian seas has been implemented in IMD. The model is an adapted version of the hurricane prediction model of the National Center for Environmental Predictions (NCEP - erstwhile National Meteorological Center) Washington. The model provides track forecasts out to 36 hours at present. The initial analysis and lateral boundary conditions are generated from the outputs of T-80 model operationally run by National Centre for Medium Range Weather Prediction (NCMRWF) New Delhi. The TC track forecast with this model have been found to be more reliable. However, the track forecast for 24 hours is still over 100 km as per the Table enclosed.

TABLE Track Prediction Errors

Year	Period	24 Hrs		36 Hrs	
		Mean Position Error (Km)	Angular Deviation between observed and predicted track vectors (RMSE in Deg.)	Mean Position error (Km)	Angular Deviation between observed and predicted track vectors (RMSE in Deg.)
1998	20-22Nov. 6-10 Jun.	146.2 140.1	17.4 32.4	242.6 205.2	12.7 27.9
1990	16-20May 15-18 Oct. 26-30 Oct.	97.4 151.4 107.7	12.1 13.7 19.2	133.0 426.5 185.7	12.2 20.7 23.0
2000	15-19 Oct. 27-30 Nov.	105.0 114.8	25.8 05.7	237.1 161.5	23.8 15.5
2001	21-28 May	90.01	15.2	160.0	20.1
	24-27 Sep. 14-17 Oct.	10.01 20.0	10.1 12.3	180.0 210.0	15.1 22.0
	MEAN	118.3	16.4	214.1	19.3

Observed track vector: Initial (at T0) to observed (at T0 + 24h or T0 + 36h) positions
Predicted track vector: Initial (at T0) to predicted (at T0 + 24h or T0 + 36h) positions

3. TROPICAL CYCLONE DISASTERS

Though some of the TCs may have some positive effects in bringing the most needed rains over drought stricken areas, the TC-caused disasters are always the first in terms of death toll and economic losses. These disasters are mainly caused by the torrential rains, strong winds and whooping storm surges that devastate the low-lying coastal areas.

3.1 Heavy Rains

Heavy rains occur in the core area and surrounding bands. They also occur in remote areas quite away from the TCs. If the land-falling TC stagnates or moves slowly, successive outbreaks of heavy rains over the same area may cause flooding. Rainfall amount of 30 cm accumulated over 24 hours is not a rare feature. Rainfall is generally very heavy and spread over a large area thus leading to excessive amount of water that leads to flooding. The heavy rains waterlog the ground and cause softening of the earth due to soaking. This contributes to weakening of the tank embankments , the leaning over of the utility poles or collapse of the pole type structures.

3.2 Strong Winds

TCs are among the most severe storm disasters because of the strength of the surface winds around the center. Strongest winds are found in the eyewall regions of the TCs. Wind speeds greater than 120 Kmph are characteristics of intense TCs. All physical structures are vulnerable to the extreme pressures exerted by winds and thus collapse or get damaged. The typical damages to buildings are due to failure of roofing systems, windows and doors, uprooting of trees, blowing away of the thatched huts etc.

3.3 Storm Surges

One of the peculiar characteristics of TCs having a very high damage potential is storm surge which is generally misnamed as tidal wave in India. Storm surges are an abnormal rise of seawater caused by a number of factors associated with cyclones. The destruction caused by the surge is tremendous. Houses are the worst affected. A devastating cyclone struck Orissa coast of India near Paradip on 29 October 1999. It was termed as Super TC with an estimated wind speed of over 250 kmph. It paralyzed the entire State apparatus killing more than 9887 people. A huge surge of about 5 to 6 m above the astronomical tide was experienced along with very heavy rains in which cumulative total over the course of the storm period exceeded 60 cm at some places. This storm surge penetrated inside upto a distance of about 15 Km from the coast.

4. OTHER ISSUES OF DISASTER MANAGEMENT

The economic losses and deaths can be minimized if the monitoring and forecasting capability for the TCs is improved upon. Another key link is warning procedures and strategies, which are necessary to make optimum use of the meteorological skill to combat disasters. We have already talked about the monitoring and forecasting capability of IMD. A holistic view of other aspects of disaster management is given below.

4.1 The Four-stage Warning System

A two stage warning system for tropical cyclones was in vogue in IMD, in which warnings were issued to state government officials in two stages: the first stage warning known as "Cyclone Alert" issued at least 48 hours in advance of the expected commencement of adverse weather over the coastal areas and the second stage warning known as "Cyclone Warning" issued at least 24 hours in advance. These warnings continue to be issued giving the latest position of the cyclone, its intensity (Maximum sustained surface wind speed) and likely time and point of landfall together with storm surge height and type of damage expected.

In addition, with effect from the beginning of the cyclone season of 1999 (April-May) a special bulletin known as "Pre-cyclone watch" has been introduced. This bulletin contains early warning about the development of a cyclonic disturbance in the north Indian Ocean, its likely development into a tropical cyclone and the coastal belt likely to experience adverse weather is indicated. This early warning bulletin is issued by the Director General of Meteorology and is addressed to the Cabinet Secretary and other Senior Officers of the Government of India including the Chief Secretaries of concerned maritime states.

Lastly to cover the devastating impacts of the cyclones on the inland areas, fourth stage of warning on the 'post-landfall scenario' has also been introduced. This commences about twelve hours before landfall and continues till such time the cyclone-force winds are expected in the interior areas. At this stage, the district collectors of all interior districts besides the coastal areas likely to be affected are included in the bulletin. After the weakening of the cyclone into a depression stage, a final message on 'de-warning' is issued to all concerned.

4.2 Cyclone Bulletins and Warnings

In addition to 4-Stage Cyclone Warning bulletins the following bulletins and warnings are issued by the Cyclone Warning Centres for their respective areas of responsibility:

- Weather bulletins for Merchant Navy Ships
- Broadcast of forecast / warning bulletins through INMARSAT, EGC Bulletin for Indian Navy;
- Port warnings;
- Fisheries warnings;
- Four stage warnings for Central and State Government Officials;
- Bulletins for broadcast through AIRs for general public;
- Warning for registered recipients of the warnings;
- Bulletins for Press;
- Warnings for Aviation;

4.3 Communication

This implies more than simply dissemination. Communication is complete only after the information is received and understood. An effective communication system is vital component of any effective warning system. Local officials and the public must first hear or see hazard warning information before they are in any position to take avoidance action.

Department has established a number of communication channels for the collection and exchange of meteorological data, bulletins within the country and on the Global Telecommunication System (GTS) of the World Meteorological Organization for exchange of data with other countries.

IMD has launched its own website (www.imd.ernet.in) where a large number of weather related products including severe weather warnings are hosted for access by various agencies and public. Cyclone warnings are communicated to Govt. officials and other parties by high priority Telegrams, T/P's, Telex and Telephones. Police Wireless are also used as and when possible. The general public, the coastal residents and fishermen are warned through State Govt. officials and broadcast of warnings through AIRs and telecast programme in national and regional hook up. The cyclone warning are issued to State Govt. officials in two stages.

In addition to existing mode of dissemination of cyclone warnings to various State government, P&T officials, port officials etc. through high priority telegrams, telephones and Telex by I.M.D. a new scheme, Cyclone Warning Dissemination System (CWDS) using INSAT has been implemented.

During time of need, IMD makes use of Police Wireless, coastal Radio of DOT, CAD AFTN links, and Railway microwave links. It is also broadcasting two bulletins for safety at sea using GMDSS system operating through INMARSAT.

4.4 Response

Institutional arrangements must exist between the Meteorological Organization and the crisis Managers. Such an arrangement already exists between IMD and Disaster Management Agencies both at the Centre and the States. Crisis Management Groups have been constituted at the Central and State levels in which representatives from IMD provide the most critical information for suitable follow-up actions. There are three Area Cyclone Warning Centres (ACWCs) established at Kolkata, Chennai and Mumbai which cater not only to the requirements of cyclone warning in the respective States in which they are located but also exercise technical control on the activities of the Cyclone Warning Centres (CWCs) established at Bhubaneswar, Visakhapatnam and Ahmedabad. The cyclone warning work is co-ordinated and supervised by office of Deputy Director General for Weather Forecasting, Pune. A Cyclone Warning Division has been set up at IMD HQ New Delhi to advise the Govt.at the apex levels on matters related to cyclones. The crucial information from the concerned meteorological office flows down to the Control Rooms set up in this regard both at the centre and the States where Cyclone Distress Mitigation Committees (CDMCs) have also been set up.

The lead in the disaster management process is usually taken by the meteorological community which galvanizes the decision makers/crisis managers to take appropriate actions regarding the preparedness at different levels of warning. There are many lessons to be learnt from recent calamity in Orissa State of India. Efforts have to be continued not only for improvements in forecasting and warnings but also to enhance the preparedness and awareness of the people so as to:

- Enhance risk management and social and economic stability
- Reduce loss of life and property
- Reduce damage to infrastructure.

5. FUTURE SCENARIO

IMD has always tried to update the technology required to enable the cyclone warning service to reach state-of-the-art level. IMD has ongoing Projects for future development in the field of Space Meteorology, Telecommunication and Instrumentation.

Government of India has constituted a High Powered Committee (HPC) to review the existing norms of the Disaster Management Scheme. It is coming up with several suitable suggestions for improvements in effective Disaster management practices at the Centre and the State levels. One of the ideas being floated is to set up Special Task Force equipped with state of the art technology and nationwide computerised data base. Community has to be increasingly involved in the process of disaster management and to create greater awareness in the community, voluntary organisation such as NGO's have to step in. A partnership with media has to be built up for wider coverage. IMD has since submitted a report on water and climate related hazards to the HPC which is coming up with several recommendations. Some of them are as follows:

- It is imperative to modernize the computerized weather forecasting system by introducing very fine resolution numerical models for tropical cyclone predictions. This may require ocean-atmosphere coupled models with suitable parameterization schemes to comprehend intense precipitation, strong winds and storm surges well in advance. This would essential;ly need upgrading of the computing facility in IMD.
- Deployment and networking of adequate number of Doppler Radars and augmentation of the buoys in the north Indian Ocean would facilitate improvement in analysis and prediction of cyclones.
- Appropriate information in user defined and implementable format and based on scientific inputs/adequate data is required for decision making.
- Mechanisms for coordination between different agencies should be formulated in light of the Cyclone Review Committee recommendations.
- A robust and sound database of the area/region giving the landuse/landcover, population, socio-economic data infrastructure (like road, rail network, hospital etc.), geography, contours etc. should be maintained at the national, state and local levels. It should have data on previous disasters. An analysis (through GIS) of the likely impacts may be carried out for different intensities of the disaster events.
- Provision of satellite phones and back up power supply systems may be ensured.
- The response essentially lies with the state/local administration asnd a specific plan of action of evacuation and relief should be prepared with specific agencies responsible for each action. The role of each of the agency (food, water, shelter, transport, communication etc.) needs to be spelt out clearly.
- For the dissemination of the warning messages, communication systems which have very broad public reach such as radio, television, community warning (Cyclone warning dissemination systrm- CWDS) should be used. Use of local warning systems such as siurens, loudspeakers, door to door visits which are community based approaches, should be maximized. The dissemination of warnings to the fisherman in the sea can be through satellite direct broadcast (such as Worldspace).
- Existing road, aviation and other linkages may be reviewed for easy mobilization of resources in the disaster area. Provision of adequate number of cyclone shelters has to be made in the cyclone prone coastal areas.

6. AREA OF CO-OPERATION

The cyclones continue to threaten the security and safety of the coastal population. From the hydro meteorological point of view the followings are the proposed areas for collaboration:

- Utilization of satellite data for determination of cyclone intensity (maximum sustained wind), outer core wind strength, structure of cyclone and rainfall intensity
- Specification of synthetic vortex for use in the numerical model assimilation of the synthetic vortex in the initial analysis using GFDL MMM model, assimilation of non conventional data such as SSM/I, Scatterometer data etc.
- Simulation studies of cyclone track prediction and landfall processes
- Development of a dynamical tropical cyclone storm surge interface model
- Heavy rainfall forecasting
- Development of networking for an efficient cyclone warning system

Acknowledgements. Author is grateful to the Director General of Meteorology for providing him the opportunity of participating in the Workshop on Weather and Climate Modeling organized under Indo-US S & T Forum.

REFERENCES

Arkin, P.A., 1983: A diagnostic precipitation index from IR satellite imagery. *Trop. Ocean-Atoms. Newslett.*,17, 5-7.

Dvorak, V.F., 1984: Tropical Cyclone Intensity Analysis using satellite data. *NOAA, Tech. Rep., NESDIS*, **11**, 44pp.

Holland, G.J. 1980: An analytical model of the wind and pressure profiles in hurricanes, *Mon. Wea. Rev.*, **108**, 1212-1218.

Joshi, P.C., B.M. Rao, and C.M Kishtwal, 1999: Prospects of meteorological satellites applications: India, Meteorology beyond 2000. *Proc. of TROPMET-99*,154-164

Kalsi, S.R., 2002: Use of satellite imagery for tropical cyclone intensity analysis and forecasting. *Met Monograph, Cyclone Warning Division*. No, 1/2002 , 100 pp. [Available from 0/o Director General of Meteorology , Lodhi Road, New Delhi-110003]

Kalsi, S.R., Raymond Zehr, Mark DeMaria and John Knaff, 2002: An Indo-US Joint study of Tropical Cyclones of north Indian Ocean. *Study Report available with India Meteorological Department and NOAA/NESDIS.*

Prasad, K., 1997: Prediction of Tropical Cyclones by numerical models- A Review. *Mausam*,**48**, 225-238

Raghavan, S., 1997: Radar Observations of Tropical Cyclones over the Indian Seas. *Mausam*, .**48**,2, 169-188

Location Specific Forecast: Interpretation of Outputs from Global and Meso-scale NWP Models

ASHOK KUMAR, PARVINDER MAINI, L.S. RATHORE & S.V. SINGH

National Center for Medium Range Weather Forecasting
Department of Science and Technology
Mausam Bhawan, Lodi Road, New Delhi-110003, India
E-mail: ashok@ncmrwf.gov.in

ABSTRACT

Location specific weather forecast is prepared by using the output from the global circulation model (GCM). The procedure for preparation of forecast is a man-machine mix approach. Direct model output (DMO) forecast, statistical interpretation (SI) forecast and forecast information for different synoptic systems provided by the graphical output from the GCM are the basic components used for it. Data feed back from different stations and information from satellite imageries was also used while preparing the forecast. Statistical Interpretation(SI) forecast is obtained by using Perfect Prog Method(PPM) in which analysis of upper air circulation fields from GCM around the location of interest is used. Direct Model Output(DMO) forecast is obtained from the prediction of surface weather elements from the GCM. Finally an evaluation of the forecast skill for monsoon 2001 and some of the future plans are presented.

1. INTRODUCTION

A deterministic location specific forecast of surface weather parameters using the output from the Global and Meso-scale NWP models can be obtained as Statistical Interpretation(SI) forecast, Direct model output(DMO) forecast, Man-Machine mix approach or automated forecast preparation by an expert system using the information from SI &DMO forecast and other forecast information of synoptic systems.

Location specific forecast for surface weather parameters are highly dependent upon local orographic conditions. In order to have deterministic location specific forecast for these parameters, it is very essential to develop statistical interpretation(SI) forecast models by using advanced mathematical and statistical techniques. As such models have the inbuilt accounting capability for the interaction between such local orographic conditions and upper air circulation. SI forecast models are developed by using Model Output Statistics(MOS) and Prefect Prog Method(PPM) models. Station specific and region specific equations can be developed. If NWP model forecast error is large or model is changed, then forecast needs to be made bias free by using past one or two seasons data.

Local weather forecast for different surface weather elements can also be obtained directly from NWP models by using the forecast at the nearby grid points. Station specific forecast can be obtained from the nearest grid or by interpolating from the nearby grids. Region specific forecast can also be obtained for different agro-climatic regions(agro-climatic zones) by averaging over the model grid points in the region. Such forecast also needs to be made bias free by using the past one or two seasons data.

Designing the operational local weather forecast system either by man-machine mixed approach or by developing the expert system by using the above type of deterministic location specific forecasts for surface weather elements is the final important step for location specific weather forecast.

2. OPERATIONAL LOCAL WEATHER FORECAST SYSTEM AT NCMRWF

First deterministic local weather forecast bulletin was issued on 23rd march 1991 from NCMRF. The forecast issued was prepared by a Man-Machine mix approach, using the information basically from Direct Model Output (DMO),forecast information for different synoptic systems provided by the graphical output from the R-40 Global circulation model. Data feed back from different stations and information from satellite imageries was also used while preparing the forecast. The forecast was issued for next three days and for average cloud amount, total rainfall, average wind speed, predominant wind direction, maximum and minimum temperature trends all pertaining to next 24 hours starting from 8.30 AM for each day.

The work on development of statistical interpretation (SI) forecast system (PPM Technique) for maximum/minimum temperature and rainfall was started in 1990 by a group of scientists at NCMRWF. Analysis data obtained from ECMWF was used for developing the models. SI forecast is being obtained along with the operational run of the NWP model (R-40, & later T-80 since 1993) and is used while preparing final operational location specific forecast by the group of forecasters since 1992 (Figure 1.) (Kumar et.al.,2000)

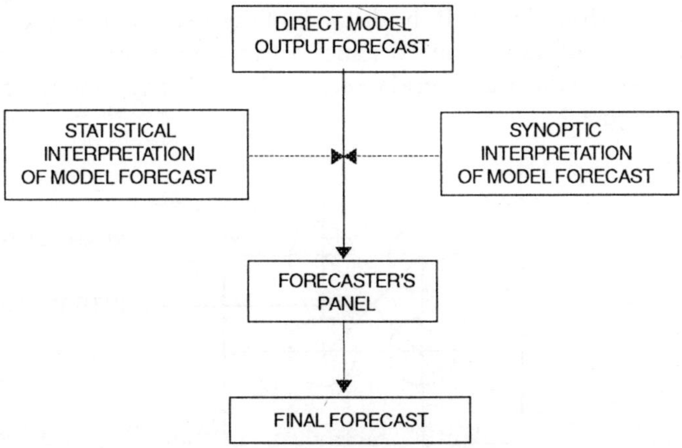

FIG. 1. An integrated scheme for preparation of local weather forecast (man-machine-mix approach)

At present the forecast is being issued for all the weather parameters mentioned above for next four days along with total rainfall during the next week for 81 stations on every Tuesday and for 47 stations on every Friday. The forecast table used for preparation of the operational local weather forecast is shown in Table 2.

3. DMO FORECAST

The forecast values for different surface and upper air weather elements can be obtained at the 256 X 128 gaussian grids of T 80 model. The longitudinal distance between any two grid points is 1.40625 degrees while the latitudinal distance varies from 1.38847 to 1.40021. Forecast values are available at each time step which is 15 minutes. Hence for a particular day 96 forecast values are obtained and model is run for 7days starting from 0000UTC initial conditions.

From 256X128 gaussian grids of T-80 model, an Indian window of size (25X25) covering India is considered starting from 38.52 degree latitude and 66.09 degree longitude to 4.90 degree latitude and 99.84 degree longitude. The model output is then obtained at each time step at these 625 grid points for the following six surface weather elements.

— Surface Pressure (mb)
— Rainfall Rate (mm/s)
— Zonal Wind Component (at 10ft or 3.048 mt) (m/s)
— Meridional Wind Component (at 10ft or 3.048 mt) (m/s)
— Temperature (at 4.5ft or 1.3716 mt) (°c)
— Specific Humidity (g/g)

3.1 Nearest Grid and Interpolated Forecast Values

As the forecasts are obtained at gaussian grids and not at a particular location, hence the simplest way to get forecast at a specific location is to use the interpolated value from the four grid points surrounding it. But if the location is very near to a grid point, then the forecast at that grid point can also be taken as the forecast for the location. In order to decide as to which forecast among the two should be given more weightage for a location, it is necessary to know the distance of the location from the four grid points surrounding it. If the distance of the location from the nearest grid is less than one-fourth of the diagonal distance between any two grid points, then more importance is given to the nearest grid forecast values otherwise the interpolated value is considered, Figure 2.

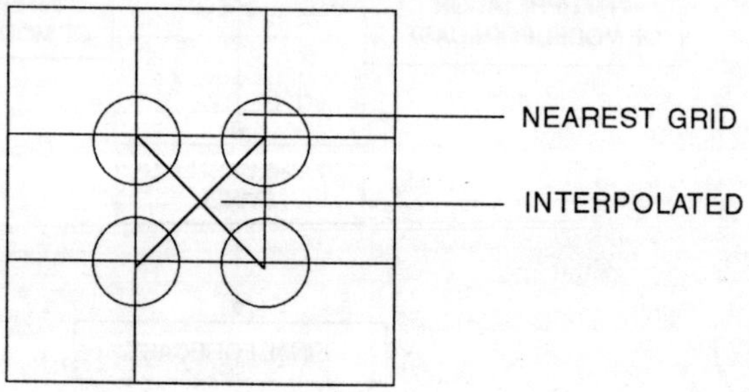

FIG. 2. Area considered around a grid point for deciding the relative importance of nearest grid and the interpolated DMO forecast values for a particular location

TABLE 2. Forecast table used for preparation of operational local weather forecast

NATIONAL CENTRE FOR MEDIUM RANGE WEATHER FORECASTING
STATION:LUDHIANA
DATE:14-12-2001 TIME: 0300 UTC

COORDINATES: 30.92 N 75.85 E
ALTITUDE: 247. meters
BASED UPON 0000 UTC ANALYSIS FOR:13-12-2001

To Dr. S. S. HUNDAL
NODAL OFFICER,AGRO ADVISORY SERVICE UNIT
LUDHIANA ,PUNJAB

Sr. No.	WEATER	DIRECT MODEL OUTPUT (INTERPOLATED) ALT: 654M					DIRECT MODEL.OUTPUT (NEAREST GRID POINT)* 31.52N 75.94E ALT: 1067.M					STATISTICAL INTERPRETATION				FINAL			
		13-12 24hr	14-12 48hr	15-12 72hr	16-12 96hr	17-12 120hr	13-12 24hr	14-12 48hr	15-12 72hr	16-12 96hr	17-12 120hr	13-12 24hr	14-12 48hr	15-12 72hr	16-12 96hr	14-12 24hr	15-12 48hr	16-12 72hr	17-12 96hr
1	MSL PRESSURE hp	1023. 1023.	1023.	1022.	1022.		1023.	1022.	1022.	1022.	1024.								
2	CL COVER okta E M	4 0	1 4	8 3	7 1	8 0	6 0	1 7	8 6	8 1	8 0					6	6	7	4
3	PROB. OF PRECIP.PR ECIPITATIONmmPPM eqn PRECIP.	0.0	0.0	2.1	0.4	0.1	0.0	0.0	3.9	0.5	0.1	0.32 0.0 0.0	0.60 0.1 0.0	0.68 0.2 0.2	0.54 0.1 0.0	5	5	0	0
4	WIND SPEED kmph	5	6	11	9	5	6	8	13	9	4					4	5	3	3
5	WIND DIRECT.deg	25	66	103	103	25	65	74	112	107	62					90	110	110	110
6	MAX. TEMP. deg cel	26.0	25.8 -0.2	21.9 -3.9	24.7 2.8	24.7 0.0	25.4	24.5 -0.9	20.2 -4.3	23.5 3.3	23.7 0.2	22.1	21.6 -0.5	21.0 -0.6	22.2 1.2	0	-3	+2	0
7	MIN. TEMP. deg cel	12.5	13.7 1.2	16.6 2.9	17.1 0.5	13.2 -3.9	12.7	13.9 1.2	16.5 2.6	16.4 -0.1	12.4 -4.0	7.7	10.2 2.5	11.5 1.3	12.0 0.5	+1	+2	0	-3
8	R.H.MAXIMUM(%)	100	100	59	59	89	84	100	66	66	94								
9	R.H. MINIMUM(%)	0	0	39	28	0	0	0	45	30	0								
10	WIND DIR. FREQ. 0—45 45—90 90—135 135—180 180—225 225—270 270—315 315—360	 45.8 16.7 0.0 0.0 0.0 0.0 28.1 9.4	 6.3 70.8 0.0 0.0 0.0 0.0 12.5 10.4	 0.0 37.5 62.5 0.0 0.0 0.0 0.0 0.0	 0.0 35.4 51.0 13.5 0.0 0.0 0.0 0.0	 64.6 21.9 0.0 0.0 0.0 0.0 0.0 13.5	 15.6 68.7 0.0 0.0 0.0 0.0 5.2 10.4	 0.0 62.5 27.1 8.3 2.1 0.0 0.0 0.0	 0.0 3.1 96.9 0.0 0.0 0.0 0.0 0.0	 0.0 29.2 43.8 20.8 6.3 0.0 0.0 0.0	 25.0 38.5 16.7 0.0 1.0 2.1 12.5 4.2	WEEKLY CUMULATIVE RAINFALL FORECAST FOR NEXT WEEK:- MODEL OUTPUT: 2.6 mm, FINAL: 10mm							

3.2 Forecast

DMO forecast values of both types i.e. nearest grid and interpolated, for each location of interest are obtained. Four day forecast for the following parameters is obtained by using forecast values at each time step of 15 minutes.

Average Mean Sea Level Pressure
Cloud Amount (Morning and Evening)
Rainfall (24 hours accumulated)
Maximum Temperature
Minimum Temperature
Average Wind Speed
Predominant Wind Direction
Maximum Relative Humidity
Minimum Relative Humidity

Here, the validity of the forecast values for a particular day is for the subsequent 24 hours starting from 0300UTC(0830 Local time) of that day. As at NCMRWF the T 80 model is run only for 7 days based on 0000UTC analysis, hence 24,48,72,96 and 120 hours forecasts are obtained. For maximum and minimum temperatures the trends from 24 to 48,48 to 72 ,72 to 96 and 96 to 120 are also obtained. In order to have the correct assessment of the predominant wind direction, the wind direction for a particular day is divided into eight classes and frequency of each class is obtained. The forecast values are shown in the first two columns of the forecasting table (Table 2) prepared.

4. STATISTICAL INTERPRETATION (SI) FORECAST

4.1 Statistical Interpretation Approach

Surface weather elements like rainfall, maximum temperature and minimum temperature are highly dependent on local topographic and environmental conditions. As in NWP models and particularly in GCM it is very difficult to include each and every aspect of these local conditions at all the locations, hence direct model output (DMO) forecast obtained from NWP models may have problems. But upper air circulation at a specific location is not so dependent on local conditions and can be obtained from the analysis or forecast of a GCM easily. Moreover uncertainties involved in the relationships between the circulation and surface weather can be formally expressed by a suitable statistical relation for a particular location. Hence, a statistical relation developed between upper air circulation around the location of interest and observed values of the surface weather element at the location, by using carefully chosen predictors and suitable statistical techniques, will be accounting for the effect of these local conditions. This indicates that Statistical Interpretation forecast so obtained will have better skill as compared to DMO forecast. Basically two methods are used for Statistical Interpretation forecast. These are Perfect Prog Method(PPM), (Klienet al.,1959) and Model Output Statistics(MOS),(Glahn and Lowry,1972).

In PPM Approach, a statistical relation is derived that relates large sample of observed surface elements(predictands) to concurrent observed surface and upper air reports i.e. analysis(predictors).In order to get a forecast for the appropriate valid time, values of the predictors obtained from NWP models is substituted in the relation developed. This approach assumes that the model forecasts are "perfect". Hence this approach is having a disadvantage that it does not account for systematic biases and errors of the model. This problem can be solved by using the unbiased model forecast, which can be obtained just on the basis of model analysis and forecast data for last one or two months. A major advantage of this method is that stable forecasting relations can be derived from a long period of record. Its forecast improves as NWP model forecast is improved.

In MOS approach, a statistical relation is developed using observed surface weather elements(predictands) and NWP model forecasts for upper air circulation variables (predictors).As such separate MOS relation is developed for a particular day forecast. To make operational forecasts, MOS relations are usually applied to the same dynamical model that provided the developmental sample. It has an advantage that MOS relations account for some of the biases and systematic errors found in the NWP model. A disadvantage of this technique is that Model Output data for a sufficiently longer period is required so as to derive a stable relation. Hence it can not be applied immediately when a new NWP model is made operational. Also if the NWP model undergoes a major change the MOS relations will have to be developed again.

4.2 Data

In order to cover all type of variability for a season, at least five to six years of data is required. As sufficient data is not available from NWP model operational at NCMRWF, hence PPM models are developed during monsoon and winter season by using six years (1985-1990) of analysis data obtained from the ECMWF. The data used is at 2.5x2.5 latitude-longitude grid. The period of the monsoon season is taken as June-August or June-September depending upon the location of the station and for winter season it is taken as December-February.

Predictands chosen for development of PPM models are the observed values of Quantitative Precipitation (QP),Probability of Precipitation(POP),Maximum Temperature and Minimum Temperature. Due to high variability in the distribution of rainfall values model is developed by taking cube root of quantitative precipitation. PoP values are obtained from the observed rainfall values as 1 for a rain case and 0 for a no rain case.

Forty seven meteorological parameters are chosen as possible set of predictors (Table 3),these includes basic analysis fields such as geopotential height, temperature, u-component and v-component of wind, vertical velocity and relative humidity at 1000,850,700 and 500 hPa. Besides these fields, many other derived weather parameters obtained by using the known meteorological relations are also considered. These derived parameters include vorticity, advection of vorticity, temperature gradient, advection of temperature, advection of temperature, advection of temperature gradient, thickness, mean sea level pressure(MSLP),1000-500 hPa precipitable water(PPW),saturation deficit(SD) and rate of change of moist static energy(RMSE).

Rainfall reported at 0300UTC on a particular day is the accumulated rain in the past 24 hours. Similarly maximum and minimum temperatures recorded at 0300UTC on a particular day pertains to past 24 hours. On the other hand the predictor values are available only at 0000UTC and 1200UTC. Keeping the above facts in view, the

TABLE 3. Meteorological parameters considered as possible predictors

Parameter	Level(hPa)
Relative humidity	1000,850,700,500
Temperature	1000,850,700,500
Zonal wind component	1000,850,700,500
Meridional wind component	1000,850,700,500
Vertical velocity	1000,850,700,500
Geopotential	1000,850,700,500
Saturation deficit	1000-500
Precipitable water	1000-500
Mean sea level pressure	——
Temperature gradient	850-700, 700-500
Advection of temperature gradient	850-700, 700-500
Advection of temperature	1000,850,700,500
Vorticity	1000,850,700,500
Advection of vorticity	1000,850,700,500
Thickness	850-500
Horizontal water vapour flux divergence	1000-500
Mean relative humidity	1000-500
Rate of change of moist static energy	1000-500

reference times at which the predictors are to be considered for different predictands is as given below.

-0000 UTC of the same day for minimum temperature
-1200 UTC of the previous day for maximum temperature
-0000 UTC of the same day, 1200 UTC of the previous day
and average of the two for rainfall(QP and POP)

A "day" in this case is the calendar date on which 24-hour rainfall is reported as shown in Figure 3 .This means that there are 47 predictors each for maximum and minimum temperatures and 141(47 × 3) predictors for Rainfall.

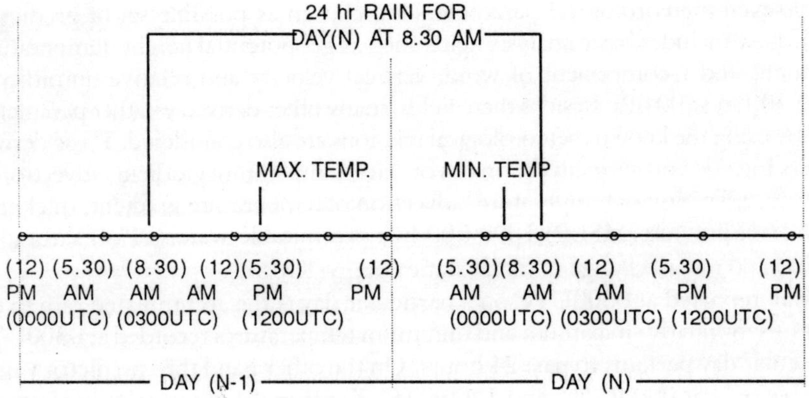

FIG. 3. Reference time for predictors

4.3 Forecast Equations

Development of forecast equations involves two major steps. The first step is the interpolation of predictor fields to the station location and the second step is the development of multiple regression equations through a screening procedure. Since the predictors are available at a regular($2.5^\circ \times 2.5^\circ$) grid and the stations are located irregularly, the predictor field ought to be interpolated to the station location.

4.3.1 Predictor's Value at the Station Location

Value of a particular predictor at a station is obtained by using the values of that predictor at the nine grids surrounding the station, Figure 4,(Woodcock,1984).This is achieved by employing canonical correlation(Rousseau, 1982).In this technique the first canonical variate, which is the best linear combination of the values of a predictor at the nine grid points that has maximum correlation with the predictand, is taken as the value of a particular predictor at the station. These canonical variates are found for each of the predictors to give a new set of potential predictors.

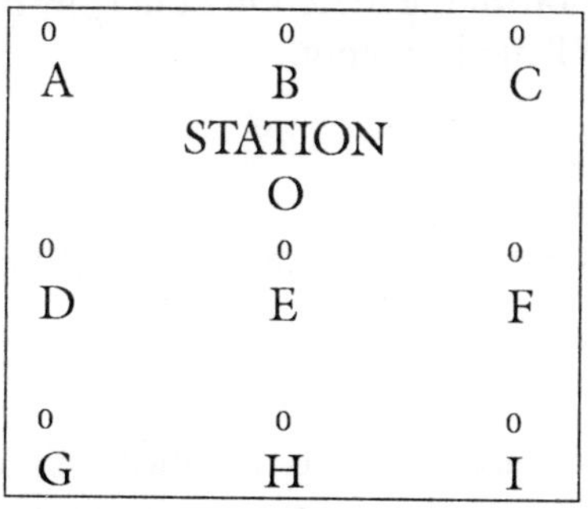

FIG. 4. The grids considered around a station for canonical variates

4.3.2 Predictors Selected

Predictors that explain most of the variance are selected from the set of all potential predictors by using a stepwise selection procedure. In this procedure, selection of predictors is terminated if the new candidate predictor contributes less than a critical value, to the percentage of variance explained by the predictors already selected(Tapp et al.,1986).In order to have a significant percentage of variance explained by the predictors selected and to have less noise in the predictions, this critical value is taken as 1.0% for maximum/minimum temperature and probability of precipitation and 0.5 % for quantitative precipitation(Kumar and Maini,1993,1999).

Generally two to three variables are selected for prediction of maximum temperature in which 1000 500 hPa saturation deficit and 850-hPa temperatures are most important predictors. In case of minimum temperature three to seven variables are selected and 850 hpa temperature and 850 500 hpa thickness are generally selected. Five to ten variables are selected for PoP and five to fourteen variables are selected for QP. Mean relative humidity and 850 hpa meridional wind component play an important role in case of both PoP and QP. 850 hPa vorticity is also an important variable in the prediction of QP.

These selected predictors are then used for developing the PPM models. Simple linear regression equations of the form(1) are obtained relating one predictand to the set of selected predictors.

$$Y = a_0 + a_1X_1 + a_2X_2 + ...a_nX_n \tag{1}$$

Where a_i's are the multiple regression coefficients and X_i's are the values of the predictors at the station. Here Y provides the probability of precipitation for a given set of predictors.

4.4 SI Forecast

For obtaining SI forecast from these PPM models, the post processed output at the same resolution i.e.(2.5 × 2.5),from the NWP model operational at NCMRWF is used. Earlier R 40 model with resolution (2.5 × 2.5) was operational at the centre (Kumar and Maini ,1996). Later in 1993 T 80 model with higher resolution was made operational. Hence post processed output of T 80 model is used here.

SI forecast for 24,48,72 and 96 hrs, is obtained for maximum and minimum temperature, QP and POP by substituting the X_i's in Eqn(1) by the respective forecast values from the operational NWP model output.

Hence

$$Y_{24} = a_0 + \sum_{i=1} a_i X_i \tag{2}$$

where n is the number of selected predictors.Similarly Y_{48}, Y_{72} and Y_{96} are also obtained.

5. SKILL OF FORECAST

The skill of the operational forecast had always been verified at the end of each monsoon and winter season. The results had been found as quite encouraging over the years as is indicated by the ratio scores for rainfall over last ten years during monsoon season (Table 3) and last six years during winter season (Table 4). The skill of SI forecast and that of DMO forecast had also been verified over the years. The results had been found to be considerably good and quite encouraging for all the forecast hours. SI forecast had been found to be an improvement over DMO and having skill almost as good as final operational local weather forecast. The results for monsoon 2001 are shown in Tables 5,6,7 and 8 respectively.

TABLE 4 Ratio scores (%) of Yes/No Precipitation Forecast for past six winter seasons (December, January & February)

STATION	RATIO SCORE (%)					
	1994-95	1995-96	1996-97	1998-99	1999-2000	2000-2001
DELHI	93	90	94	88	87	92
FAIZABAD	-	95	92	96	-	94
HISAR	90	97	89	91	89	94
LUDHIANA	77	79	92	91	85	90
PANTNAGAR	77	82	97	-	76	85
SRINAGAR	-	72	87	79	-	62

TABLE 5 Ratio Score (%) of Yes/No rainfall forecast for monsoon seasons (June, July, August and September) of 1992-2001

STATION	RATIO SCORE(%)									
	1992	1993	1994	1995	1996	1997	1998	1999	2000	2001
DELHI	60	69	76	73	67	59	67	79	62	61
HISAR	70	70	63	70	58	59	49	75	68	66
JABALPUR	64	72	75	81	69	64	57	69	65	74
RAIPUR	62	69	76	71	59	59	60	63	60	66
TRICHUR	79	72	81	82	75	63	81	75	77	81
UDAIPUR	74	73	75	79	61	66	61	67	62	62

TABLE 6 Skill Score for rainfall (June, July, August and September) of 2001

STATION	DMO		SI		FINAL	
	RATIO	HK	RATIO	HK	RATIO	HK
DELHI	56	0.22	65	0.32	61	0.26
JABALPUR	68	0.17	61	0.21	74	0.42
RAIPUR	63	0.13	70	0.31	66	0.21
RANCHI	60	0.04	66	0.30	67	0.21
SOLAN	60	0.27	64	0.26	64	0.33
UDAIPUR	50	0.26	68	0.34	62	0.14

TABLE 7 Skill Score for minimum Temperature (June, July, August and September) of 2001

STATION	DMO		SI		FINAL	
	CORR.	RMSE	CORR.	RMSE	CORR.	RMSE
DELHI	0.15	1.96	0.25	2.08	0.20	1.84
JABALPUR	0.58	1.42	0.57	1.41	0.59	1.38
RAIPUR	0.38	1.41	0.41	1.18	0.39	1.36
RANCHI	0.70	1.08	0.70	0.99	0.72	1.08
SOLAN	0.76	1.96	0.66	2.63	0.81	1.72
UDAIPUR	0.72	1.72	0.69	1.55	0.73	1.70

TABLE 8 Skill Score for maximum Temperature (June, July ,August and September) of 2001

STATION	DMO		SI		FINAL	
	CORR.	RMSE	CORR.	RMSE	CORR.	RMSE
DELHI	0.50	2.48	0.45	2.19	0.39	2.42
JABALPUR	0.69	2.31	0.68	2.21	0.70	2.24
RAIPUR	0.78	1.98	0.71	2.34	0.76	2.12
RANCHI	0.55	1.81	0.56	1.87	0.63	1.71
SOLAN	0.24	2.61	0.42	2.00	0.20	2.36
UDAIPUR	0.78	2.07	0.83	1.76	0.76	1.99

6. FUTURE PLANS AND RESEARCH WORK BEING UNDERTAKEN

SI forecast has the desirable skill and exhibits the potential of being used for automation of location specific forecast, if improved by using better analysis data and advanced mathematical-statistical techniques. Development of new SI forecast system using the T-80 model analysis data and neural network, logistic regression and discriminant analysis is being undertaken.

Automation of operational location specific weather forecast using bias free forecasts(Kalman filter) directly obtained from operational NWP model, seasonal mean errors and operational SI forecast is being attempted.

7. FIELDS OF RESEARCH WORK TO BE UNDERTAKEN

Collaboration is required (possibly with MDL and FSL,NOAA,USA) to gain the knowledge on the techniques used for developing the procedures and operational systems for:

a) SI forecast: mathematical-statistical techniques used for MOS/PPM, modified MOS and Ensemble MOS
b) Bias free forecast: Kalman filter or Seasonal Mean estimates.
c) Expert system for location specific forecast.
d) Preparation of automated local weather forecast bulletin.

Acknowledgements. Authors express their special thanks to other scientists of application division for their active participation in the forecaster's panel and contribution to the final operational forecast.

REFERENCES

Glahn H.R. and Lowry D.A.,1972, "The Use of Model Output Statistics (MOS) in Objective Weather Forecasting", *J. Appl. Meteor.*,11,1203 1211.

Klien W.H.,Lewis B.M.,and Enger I.,1959, "Objective Prediction of Five Day Mean Temperature During Winter" *J. Meteorol.*,16, 672 682.

Kumar Ashok and Parvinder Maini,1993: An experimental medium range local weather forecast based upon global circulation model at NCMRWF.TROPMET 1993, Indian Met.Society, New Delhi, pp. 48 55.

Kumar Ashok, Parvinder Maini and S. V. Singh 1999: An operational model for forecasting probability of precipitation and YES/NO forecast. *Wea. Forecasting*, **14**, 38-48.

Kumar Ashok, ParvinderMaini, L.S. Rathore and S.V.Singh, 2000: An operational medium range local weather forecasting system developed in India. *Int. J. Climatol.*, **20**, 73-87.

Kumar Ashok and Parvinder Maini,1996: Statistical Interpretation of General Circulation Model: A Prospect for Automation of Medium Range Local Weather Forecast in India.*Mausam*,47, 227 234.

Rousseau D.,1982 : Work on the statistical adaptation for local forecasts in France, Statistical interpretation of NWP products. Seminar/workshop, ECMWF,pp. 347 364.

Tapp R.G., F. Woodcock and G.A. Mills,1986 :Application of MOS to precipitation prediction in Australia", *Mon.Wea.Rev.*,114, 50 61.

Woodcock F.,1984 :Australian experimental Model Output Statistics forecasts of daily Maximum and Minimum Temperature", *Mon.Wea.Rev.*, 112, 2112-2121.